Plain bearing design handbook

Plain bearing design handbook

R. J. Welsh
WhEx, CEng, FIMechE, FIMarE, MNECInst

Butterworth-Heinemann
Linacre House, Jordan Hill, Oxford OX2 8DP
225 Wildwood Avenue, Woburn, MA 01801-2041
A division of Reed Educational and Professional Publishing Ltd

 A member of the Reed Elsevier plc group

OXFORD   AUCKLAND   BOSTON
JOHANNESBURG   MELBOURNE   NEW DELHI

First published 1983
Reprinted 1999

British Library Cataloguing in Publication Data
Welsh, R. J.
    Plain bearing design handbook
    1 Bearings (Machinery)
    I Title
    621.8'22   TJ1061

Library of Congress Cataloguing in Publication Data
Welsh, R. J. (Robert James), 1902–
    Plain bearing design handbook
    Includes index
    1 Plain bearings (Machinery) – Design and
    Construction   I Title
    TJ1063.W44   1983   621.8'22   83–7719

ISBN 0 408 01186 6

Printed and bound by
Antony Rowe Ltd, Chippenham, Wiltshire

Preface

The production of this book arose from a realisation of three facts about plain bearing design. These were:

(a) That existing books on the subject tended to deal with one or other particular type of bearing and lacked detached objective guidance about which type(s) to adopt for any given duty.

(b) That the design calculations for most practical applications of plain bearings can seldom be based on accurately known data but, instead, have to take account of many factors not under the control of either the bearing designer or the bearing manufacturer. These 'unknown factors' include, for example, the extent to which journal and bearing manufacturing tolerances may be cumulative or compensatory in any given case, the accuracy of assembly (particularly after any dismantling for machine maintenance), the care exercised in the machine usage, the quantity, quality and cleanliness of the lubricant employed, and the degree of protection afforded from the elements, etc.

(c) That, because of the above, the success of plain bearing design depends largely on what P. L. Jones once described as 'the accurate estimation of a large number of immeasurables'. Indeed, it falls neatly into that class of design work that Lord Hilton called 'the art of knowing what you can get away with'.

This book aims to assist designers of all types of machinery by dealing with the more important immeasurables affecting bearing performance, explaining their nature in relatively simple terms and giving quantitative guidance on what one can usually 'get away with' in specified circumstances. To preserve complete objectivity, Appendix 1 gives a summary of the merits and demerits of ball and roller bearings in relation to plain bearings.

For the benefit of students, or others encountering plain bearing design for the first time, a complete chapter is devoted to the theory of film-lubricated bearings – thus enabling readers to appreciate where theory and practice in these matters concur, overlap, or contradict.

Although all the views expressed are those of the author, he would nevertheless like to acknowledge the extent to which his understanding of plain bearing problems has been developed from discussions with his friends in the engineering industry, such as Phil Holligan, Alfred Hill, Jim Narracott, Roger Joy, Peter Whitton and Bob White, to name but a few. Some have kindly helped by reading the manuscript and offering advice.

Contents

Round figure conversions

(SI-Technical metric-Imperial)

Energy

J	kp.m	ft.lb	kJ	kcal	Btu	MJ	kW hr
1	0.102	0.74	1	0.239	0.95	1	0.278
9.81	1	7.24	4.187	1	3.97	3.6	1
1.36	0.138	1	1.055	0.252	1		

Force

N	kp (kgf)	lbf
1	0.108	0.225
9.81	1	2.2
4.45	0.454	1

Length

m	ft	mm	inch	μm	micron	micro-inch
1	3.28	1	0.0394	1	1	40
0.305	1	25.4	1	0.025	0.025	1

Mass

kg	lb
1	2.2
0.454	1

Oil flow

dm^3/s	litre/min	gpm
1	60	13.2
0.0758	4.55	1

Power

kW	CV	hp
1	1.36	1.34
0.735	1	0.986
0.746	1.01	1

Pressure or stress

MPa MN/m^2 N/mm^2	kp/cm^2 (kgf/cm^2) (atm)		lbf/in^2 psi
1	10.2	10	145
0.0981	1	0.981	14.22
0.1	1.02	1	14.5
0.0069	0.07	0.069	1

Note: bar is used for fluid pressure
MPa is used for stress and bearing pressures
For practical purposes –
  1 daN/cm^2 = 1 kp/cm^2
         = 1 bar = 1 atmos.

pV factors

kW/m^2	kp/cm^2.cm/s	lbf/in^2 × ft/min
1	1.01972	28.55
0.981	1	28
0.035	0.036	1

1 kW/m^2 = 1 kPa × m/s

Torque

Nm	kp.m	lbf.ft
1	0.102	0.74
9.81	1	7.24
1.36	0.138	1

Velocity

m/s	ft/min
1	197
0.0051	1

Viscosity

dynamic		kinematic	
mPa s	cP	mm^2/s	cSt
1	1	1	1

Chapter 1

Introduction

Purpose

The purpose of this book is explained in the Preface.

Layout

Although all aspects of bearing design are to some extent interrelated, the book has been divided into 16 chapters, each devoted to one aspect of the design.

Shaft versus journal

In line with standard practice, the term 'shaft' is used only when referring to a complete item, such as a crankshaft or camshaft; the term 'journal' is used only when referring to that part (of the shaft) forming the rubbing surface of the bearing.

Range of sizes and speeds

In general, the book confines its specific recommendations to cases where the journal diameter is between 10 and 200 mm ($\frac{1}{2}$–8 in), and the journal speed is between about 60 and 6000 rev/min, but broad guidance is given on applications lying outside these limits.

Index and cross-references

For ease of reference, all sections (with two exceptions) and paragraphs are numbered as follows: chapter number followed by section (or paragraph) number. To prevent confusion, section numbers are in roman numerals, paragraph numbers in arabic numerals. For example, 7(ii) signifies the second section in chapter 7; 13.03 the third paragraph in chapter 13. The index thus directs readers to the actual paragraph they need, and not merely to the appropriate page. The pages are also numbered on this system.

Illustrations

The illustrations and tables are numbered on the same principle, so that figure 6.10 is the tenth illustration in chapter 6.

Units

SI units are used throughout, but equivalent values for both 'technical metric' and Imperial are given everywhere in parentheses after the SI values. Each of these three sets of values is expressed in the forms normally employed by engineers so that, for example, an Imperial pressure is abbreviated as psi and not as the more academic form of lbf/in^2, although the latter is the preferred British Standard and ISO abbreviation.

Chapter 2

Recommended design procedure

2.01   It is assumed that, before trying to design an actual bearing for a real commercial application, the designer will have acquired enough knowledge of bearing design to qualify him for this task. If this is not so, then it is recommended that before getting down to any details he should first read through this handbook so that he may obtain a broad appreciation of the problems with which he will have to contend. An hour or two spent in this way is likely to be well repaid by the time and money saved in producing a satisfactory design.

2.02   For the designer who has already acquired an appropriate background of bearing design knowledge, it is recommended that his first step in any bearing design exercise should be to copy the check list shown in figure 2.1, leaving ample space against each heading to make a record of the 'decision' reached – whether it be a dimension, a material specification, the name of a proprietary supplier, or a descriptive phrase. Many items on this check list will not be under the control of the bearing designer but will form part of the specification to which he has to work.

2.03   If the load on the bearing is to be subject to any significant variation in its magnitude or direction of application, or if the shaft has to operate at more than one rotational speed, then the second step should be to decide the 'equivalent steady load and speed' to employ for design calculation purposes. This may be done by reference to Chapters 4(ii), 4(iii) and/or 4(iv), whichever may be appropriate. It should be noted that the equivalent steady load and speed will not be the same for all types of bearing, and that there will be different values for film-lubricated, porous and dry-rubbing bearings. All three sets of values should be computed unless the type of bearing has already been firmly laid down in the original specification.

2.04   Unless the designer has been given no choice, his next step ought to be to decide on the general type of bearing to employ – i.e. hydrodynamically-lubricated, boundary-lubricated, porous-impregnated, dry-rubbing, or perhaps even a hydrostatic bearing or some other device strictly outside the scope of this book. Guidance on making this choice will be found in chapter 5.

2.05   If the decision is to use a film-lubricated bearing, reference to paragraph 3.25 will enable several more decisions to be taken and recorded on the check list.

2.06   Whichever type of bearing has been selected, there will still be many blanks in the decision column of the check list, and these should now be examined – not in the order in which they appear in the check list, but

3

Figure 2.1   Check list of variables involved in bearing design

Variable (Chapter number in parentheses)	Decision	Variable (Chapter number in parentheses)	Decision
Total load on bearing (4)		Journal diameter (6(i))	
Amount of end thrust (11)		Width of bearing (6(ii))	
Speed of revolution (4(ii))		Journal material (7(i))	
Type of rotation: continuous, reversing or oscillating (4(ii), 12)		Type and grade of bearing material (7(ii))	
Uniformity of load: (constant or variable) (4(iii))		Surface finish of journal (7(iii))	
Direction of load: constant or variable (4(iv))		Accuracy of journal alignment (6(iv))	
Frequency of starts and stops (14)		Manufacturing facilities available (standards of tolerances, etc.) (6(v))	
Duration of idle periods (14)		Choice between bush and split bearing (5, 10)	
Load when stationary (14)		Choice between thick-walled and thin-walled shells (8)	
Load when starting (14)		Surface finish of bearing (7(iii))	
Acceptable frequency of bearing replacement (14)		Journal clearance in bearing (6(iii))	
Type of lubrication: splash, pressure-fed or other (13(iii))		Method of securing bearing in housing (8)	
Grade of lubricant (if any) (13(i))		Thermal expansion coefficient of housing material (4(v))	
Supply pressure and lubricant flow volume (13(i))		Number, shape and depth of oil grooves (13(ii))	
Method of filtering lubricant (13(v))		Prevention of oil loss and/or dirt ingress (13(iv))	
Method of cooling lubricant (13(vi))		Anticipated standard of maintenance (14)	
Period between oil changes (14)		Environmental temperature (4(v))	
Risks of corrosion from environment (7)			

preferably in the order in which they are treated in chapter 5 onwards. The appropriate chapters in which to look for comments or guidance are indicated on the sample check list Fig 2.1, but reference to the index may also save time on occasion. Where guidance is weak, use common sense.

2.07   It will, of course, not be possible to deal with all items in strict sequence, since many aspects of design are interrelated and cannot be judged in isolation. This will become apparent as the design proceeds but, being a problem common to nearly all design work, it should present no special difficulty to an experienced designer.

2.08   At any stage in a design, it may become apparent that trouble is arising from some unsuitable aspect of the specification, a change to which could improve the final result. Under these circumstances, it may be necessary, or at least highly desirable, to make out a case for submission to some higher authority in the design hierarchy. The possibility of taking such action should always be borne in mind if the design appears to be running into such problems as impracticable loadings, manufacturing tolerances, or thermal expansion effects.

2.09   The final step in any bearing design exercise ought to be a critical review of the whole check list, not merely to ensure that nothing has been overlooked, but also to guard against any of the 'decisions' being contradictory or mutually incompatible.

Chapter 3

Theory of film-lubricated plain bearings

3.01   To understand the underlying principles of plain bearings, it is useful to start at the beginning, and to examine the conditions obtaining before the journal commences to revolve. Figure 3.1 depicts a bearing that was originally flooded with oil but has now stood for some hours while carrying a load, W. The clearance has been exaggerated for the sake of clarity.

3.02   The oil will have been almost entirely squeezed out from the portion of the bearing directly under the load, and there will be direct metal-to-metal contact over an area around the point A. The exact circumferential extent of this dry area will depend principally on the magnitude of the load in relation to the elasticity of the shaft and journal materials, much as the area of contact between a car tyre and the road depends on the weight of the car and the pressure of air in the tyre.

3.03   Although most of any loose oil will have drained away, a certain amount will adhere by surface tension to both the shaft and the bearing, and for some distance on each side of the point A these two films will have coalesced into a common pocket of oil, as indicated between the points B and C. In a real bearing, the clearance will be so small that this complete filling of the clearance space will probably extend right round the circumference.

3.04   If now the journal is slowly revolved in a clockwise direction, it will have a high coefficient of friction on the dry area where it is in contact, and will thus tend to roll up the side of the bearing, as shown in Figure 3.2, until it slips or 'skids' because not only will it have moved on to a well-oiled and slippery part of the bearing but will also, in effect, be trying to roll up a hill of increasing steepness. At some position near that shown in figure 3.2, stability will be achieved, and at any moderate loads and speeds continued rotation of the journal will take place with the journal in this eccentric position. Assuming that there is ample lubricant lying around, then the rotation of the journal will continually drag fresh lubricant into the area of contact, and thus maintain the surface in a slippery and 'skiddy' condition.

3.05   This is the type of bearing operation known as *boundary lubrication*, and is acceptable only in the case of bearings that either run at relatively low speeds or carry only relatively low loads. Since the two surfaces are skidding over each other, there will always be a certain amount of wear taking place, and with heavy loads or high speeds the frictional heat would lead to trouble.

3.06   In assessing the load-carrying capacity of bearings with boundary lubrication, this frictional heat is usually the chief factor to be considered apart from the physical strength of the shaft and bearing materials. The criteria of

6

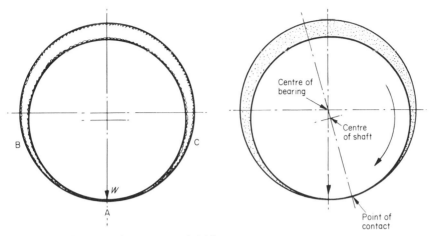

Figure 3.1 Stationary shaft (see paragraph 3.01)   *Figure 3.2* Boundary lubrication (see paragraph 3.04)

load severity in these conditions is usually measured in terms of the pV value, where p is the bearing load per unit area of projected surface (i.e. diameter × width of bearing) and V is the rubbing velocity. Various aspects of this pV factor are discussed in paragraphs 5.06 *et seq.*, but for the moment consider what happens if the journal is rotated at a much higher speed.

3.07   Provided the load is not too great, the effect will be that more and more oil will be drawn into the working area in the form of a wedge, until, at some critical speed of rotation, this wedge of oil will lift the shaft out of contact with the bearing, and the load will be carried wholly by the oil pressure generated hydrodynamically in the converging wedge of oil by the rotation of the shaft. Under these conditions, stability will be achieved by the shaft moving over to a new position of eccentricity, on the other side of the centre-line, as shown in figure 3.3.

3.08   This is the condition known as *hydrodynamic fluid-film lubrication*, and is the normal condition for a plain bearing that has to carry any substantial load at any significant speed. By comparison, all other forms of plain bearing may be regarded as less desirable alternatives, to be adopted only under the pressure of circumstances. Hydrodynamic lubrication involves no contact between the journal and the bearing, and thus gives rise to no wear so long as it persists – although some wear may take place at each start and stop.

3.09   The principles of hydrodynamic lubrication were first established by Osborne Reynolds in 1886. His equations, which can be found in most standard textbooks on lubrication, still form the basis of all calculations on the subject, even though they relate to ideal bearings having perfectly smooth surfaces made of material possessing a zero coefficient of thermal expansion, and lubricated by a fluid whose viscosity remains constant under all conditions of temperature and pressure. From a practical point of view, Reynolds' most important conclusions were that the formation of an oil wedge was essential, that a thin oil film could carry a greater load than a thick film, that the load-carrying capacity of an oil film increased directly in proportion to its dynamic viscosity, and that, under most normal conditions, it also

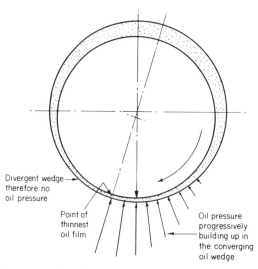

Divergent wedge therefore no oil pressure

Point of thinnest oil film

Oil pressure progressively building up in the converging oil wedge

Figure 3.3 Hydrodynamic lubrication (see paragraph 3.07)

increased directly in proportion to the relative velocity between the two opposing surfaces.

3.10   The thickness of the oil film in a bearing is determined partly by the bearing pressure (i.e. the load per unit area), and partly by the amount of clearance between the journal and the bearing bore. The combined effect of these is reflected in the *eccentricity ratio*, ε, defined as the radial clearance minus the minimum film thickness, divided by the radial clearance. This is made clearer in figure 3.4, from which it will be seen that an eccentricity ratio of

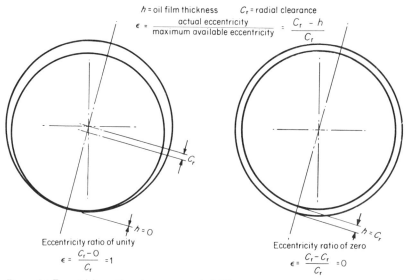

h = oil film thickness C_r = radial clearance

$$\varepsilon = \frac{\text{actual eccentricity}}{\text{maximum available eccentricity}} = \frac{C_r - h}{C_r}$$

Eccentricity ratio of unity
$$\varepsilon = \frac{C_r - 0}{C_r} = 1$$

Eccentricity ratio of zero
$$\varepsilon = \frac{C_r - C_r}{C_r} = 0$$

Figure 3.4 Eccentricity ratio, ε (see paragraph 3.10)

zero means that the journal is suspended centrally in the clearance space, whereas an eccentricity ratio of unity means that it is at its maximum eccentricity and is in metal-to-metal contact with the bearing bore.

3.11   A high eccentricity ratio obviously results in a thin oil film, but if the clearance is large, the thin part of the oil film will extend over only a short circumferential distance (as will be appreciated from an examination of the left-hand diagram in figure 3.4), so that the area benefiting from the high film pressure will be very small, and the total load carried by the bearing will not be great. Thus, the best load-carrying effect will theoretically be achieved with the smallest possible clearance, and only a moderate eccentricity ratio.

3.12   Reynolds' basic equations refer to the conditions in infinitely wide bearings, but he also produced equations dealing with the effect of oil leakage (and consequent drop of pressure) at the ends of a bearing, and showed that a high breadth/diameter ratio was theoretically always advantageous.

3.13   For a hydrodynamic bearing to operate satisfactorily, it is important to maintain an adequate thickness of oil film to prevent metal-to-metal contact between the rubbing surfaces. The relationship between oil-film thickness and the other variables is, however, extremely complex and no-one has yet found a way of expressing the facts in a simple manner.

3.14   In Britain, it is usual to estimate the oil-film thickness by a method introduced in ESDU 66023 (September 1966, amended June 1967). This employs the concept of a *non-dimensional load*, W_1, of magnitude

$$\frac{W}{\mu n b d}\left(\frac{C_d}{d}\right)^2$$

where

    W = load on bearing
    μ = dynamic viscosity of oil at operating temperature
    n = shaft speed, rpm
    b = bearing width
    d = journal diameter
    C_d = bearing diametric clearance

3.15 W_1 can be related to the oil-film thickness (at least approximately) by various empirical expressions, one of which is

$$W_1 = 19\left(\frac{\varepsilon}{1-\varepsilon}\right)^{0.707}\left[\frac{1}{1+5(1-\varepsilon)(d/b)^{1.85}}\right]$$

3.16   In the USA, the non-dimensional load is sometimes known as the *load capacity number*, defined as W_1 above but with $(C_d/b)^2$ inserted in place of $(C_d/d)^2$, and related to the oil-film thickness by the Ocvirk approximation:

$$\text{Load capacity number} = \frac{\pi}{(1-\varepsilon^2)^2}\sqrt{\pi^2(1-\varepsilon^2)+16\varepsilon^2}$$

3.17   When comparing any expressions of this kind, care must be taken with the units because $lb.sec/in^2$ is frequently used in the USA instead of cP as a measure of dynamic viscosity and, because of this, journal speeds are often expressed in rev/sec to keep the whole expression non-dimensional. There are, in any case, marked differences between the ESDU and Ocvirk approaches, particularly in their allowances for the effect of the ratio b/d. Indeed, differences of up to 600 per cent will be found in their respective values for W_1 calculated for bearings with $b/d = 1$ and $\varepsilon = 0.9$.

3.18   These divergencies are mentioned merely to underline the fact that the full interpretation of hydrodynamic theory involves a certain amount of empiricism and approximation – the accuracy of which is not necessarily related to the scientific-looking nature of the formulae in which it is expressed.

3.19   As opposed to the above, one should guard against relying on over-simplified rules, such as the old expression ZN/p, in which Z was the dynamic viscosity of the lubricant in centipoise, N the rubbing speed in ft/min, and p the bearing pressure in lbf/in^2, with an overall value for ZN/p of at least 100 regarded as essential for safety. This criterion ignored far too many of the important variables for it to be of any value except over a very limited range of sizes and speeds.

(i) Application of theory to practical design

3.20   In comparison with most other types of engineering design work, the design procedure for bearings is unusual in that the theory cannot be used to *produce* a design; only to determine whether any given design will be able to carry the required load. The reason for this is that bearing theory and practice are, in a sense, diametrically opposed to each other. Theory says that the maximum load-carrying capacity will occur with infinitesimally thin films of highly viscous oil in wide bearings having a microscopically small clearance. Practice indicates that the roughness of all real surfaces makes metal-to-metal contact unavoidable unless the oil film is of substantial thickness, that wide bearings with ultra-thin oil films would demand quite impossible standards of shaft stiffness and alignment, and that ultra-small clearances would produce seizures from differential thermal expansion – not to mention demanding uneconomic standards of tolerancing in manufacture. Too viscous an oil would also lead to overheating, as well as creating handling difficulties both outside the bearing and within its own oil grooves.

3.21   In practical bearing design, a trial-and-error cut-and-come-again approach has to be adopted, by first designing a bearing on the basis of what is reasonably practical, and then turning to theory to discover whether or not this bearing will carry the required load at the specified speed. If its load-carrying capacity is inadequate (or, alternatively, is much too great), then one must make some design changes and try again.

3.22   There are various methods by which this trial-and-error procedure can be made to converge with reasonable rapidity. One such method is fully described in ESDU 66023 and to which reference should be made for design guidance on all hydrodynamic bearings outside the range of this present book.

3.23   In ESDU 66023, the empirical relationship between W_1 and film thickness is presented in the form of a series of curves, with values of W_1 plotted against ε for a range of b/d values from 0.1 to infinity. Most of ESDU 66023 is, however, taken up with practical guidance on the calculation of minimum allowable film thicknesses, maximum permissible misalignments, surface finishes, oil flow rates, etc., together with procedures for selecting clearances, oil viscosities, etc., to achieve minimum oil drag and heat generation.

3.24   The alternative approach followed in the present book is to provide design data relating to steadily loaded bearings of one given b/d ratio, using one given lubricant, with specified clearances dependent on bearing size and rpm – and then to specify the corrections necessary for bearings that depart from these norms, or have to withstand loads that vary in either magnitude or direction. This approach is simpler and provides quicker answers, although it does not guarantee the optimum oil flow requirements or the minimum heat generation. In the vast majority of instances, however, these factors are of minor importance to the designer-draughtsman, who simply wants a trouble-free bearing that will be capable of carrying the required load.

(ii) Basis of recommendations

3.25   The values adopted for the standard bearing described above are:

$b/d = 0.6$ (The virtues of this ratio are dealt with in paragraphs 6.28 *et seq.*)

$$C_d = \left(0.0009 + \frac{\text{rev/min}}{5000000}\right)d$$

Manufacturing tolerances	BS 1916: H6/h6 or better, with nominal sizes of hole and journal differing by required C_d
Surface finish of journal and bearing	BS 1134: 0.4 μm Ra (16 μin) cla, or better
Lubricant	HVI heavy machine oil, with an assumed effective viscosity of 30 cP (30 mPa s) at the pressure and temperature existing in the bearing

3.26   A typical bearing to this specification will have a load/speed characteristic generally as shown in figure 3.5, where the caption explains the significance of the different portions of the graph.

3.27   The load-carrying capacity might in fact tend to fall off even more rapidly than shown when approaching the end marked F, because a transition from *laminar flow*, on which hydrodynamic lubrication depends, to an unsuitable *Taylor vortex flow*, takes place in bearing oil films at around

$$Q \frac{cSt}{d^2}\left(\frac{d}{C_d}\right)^{1.5} \text{ rev/min}$$

where cSt is the effective kinematic viscosity of the oil in centistokes, C_d the diametric clearance, and Q has a value of 1580 if d and C_d are measured in millimetres, or 2.45 if they are expressed in inches. This, however, need not concern anyone working within the range of speeds and sizes specified in paragraph 2.14 since, for example, a 100 mm (4 in) bearing with normal oils and clearances would have to run at about 40 000 rev/min before encountering any Taylor vortex conditions.

3.28   Figure 4.1 shows curves of the same form as figure 3.5 but relating to bearings of specific sizes, and fully calibrated in respect of rev/min and

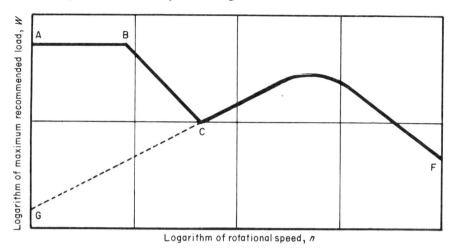

Figure 3.5 Load–speed characteristic of typical plain bearing (see paragraph 3.26)
Line A–B shows load limitation due to fatigue strength of material (about 7 MPa (70 kgf/cm^2 or 1000 lbf/in^2) for whitemetal).
Line B–C shows boundary lubrication, where frictional heat limits the load in inverse proportion to the speed.
Line C–D shows hydrodynamic lubrication in which the load carrying capacity varies approximately as $\sqrt{\text{rev/min}}$.
Line D–F shows load capacity falling off due to need for greater oil film thickness at sliding speeds above about 4 m/s (800 ft/min).
Point E represents a rubbing speed of about 16 m/s (3150 ft/min) at which the recommended maximum load reaches a peak of about 4.8 d$^{2.5}$ MN (49 d$^{2.5}$ kgf or 110 d$^{2.5}$ lbf) diameters in metre, cm, and inch respectively.
Line C–G shows the upper limit of load for hydrodynamic lubrication at low speed.
A–B–C–G shows conditions of 'mixed lubrication' such as are widely used in practice.

recommended maximum loads. This, as explained in paragraph 4.01, can be used to assist in the preliminary choice of bearing type as a prelude to detail design.

(iii) Uncertainty of predictions

3.29   Whatever design procedure is followed, there will always be some uncertainty about achieving the calculated performance in any real bearing – chiefly because the dimensional variations associated with practical manufacturing tolerances are surprisingly large when judged on the basis of their effect on load-carrying capacity. Thus, the non-dimensional concept which tells us that the load-carrying capacity of a hydrodynamic bearing will vary inversely with the square of the diametric clearance, means that in a 50 mm bearing made to H6/h6 tolerances, with a nominal clearance of 50 μm, the actual clearance can vary from 50 to 88 μm, giving a performance variation of $(88/50)^2$, or over 3:1. Similarly, on a 2 in bearing made to H6/h6 tolerances, with a nominal clearance of 0.002 in, the actual clearance can vary from 0.002 to 0.0034 in, giving a performance variation of $(34/20)^2$, or 2.89:1.

3.30   Thus with these tolerances (which are about the best one can hope to obtain in normal manufacture) the final bearing may have only about one-third of its calculated load-carrying capacity. (It is no use trying to spread the tolerance equally on each side of the nominal value, because too small a clearance would only give rise to overheating and seizure. It is always better for the clearance error, if any, to be on the 'plus' side.)

3.31   With smaller bearings, the spread of results will be even greater, although larger bearings are correspondingly more predictable. Nevertheless, even 200 mm bearings made to H6/f6 limits can still have a performance spread of more than 35 per cent from this one cause alone. In addition, of course, there are uncertainties of much the same order in respect of misalignment, surface finish, general standards of maintenance etc., all of which build up into a formidable total.

3.32   There is, fortunately, another side to this picture of uncertainty, and that arises from the effects of wear. If a whitemetal-lined bearing is inadvertantly assembled with the wrong clearance, and is thereby overloaded, then (provided the conditions are not too extreme) it may only run a little hot and experience some local 'wiping' of the whitemetal. This wiping will have two results: first, it will tend to fill the roughnesses and imperfections of the journal surfaces with smoothed-over whitemetal that will probably greatly improve the effective surface finish of the shaft – and thereby increase its load-carrying capacity; secondly, the bearing will tend to wear into a profile closely corresponding to the shaft radius, so that in way of the loaded area the form of the oil-film wedge will tend to conform closely to the ideal originally intended. Under these conditions, the load-carrying capacity may approach the designed value. Much the same thing can happen, to a lesser extent, with a hardened shaft running in a bronze bearing.

3.33   It is, unfortunately, not possible to rely on this self-correcting process being effective in every case. Many things can upset it, and even then it is also necessary to have a margin in hand for such things as dirty oil and imperfect alignment. 'Dirty oil' means, in this context, any oil that has been in use for more than a few hours, which in any ordinary application, will mean that it has acquired a substantial content of abraded metal particles plus a certain amount of such things as grinding swarf, core sand, or just ordinary factory dust.

3.34   There is, however, a further saving feature with plain bearings, namely, that most cases of marginal difficulty can be overcome by the relatively simple expedient of supplying the bearing with more oil – by increasing the oil-supply pressure, providing additional oil inlets, or taking other equivalent action. The additional flow of oil will carry away more heat and thereby reduce the bearing temperature, thus raising the effective oil viscosity, reducing the eccentricity ratio and increasing the oil-film thickness. A good flood of oil also helps to ensure the restoration of a full-width oil film in the loaded area, following the cavitation that is liable to occur in the divergent wedge region shown in figure 3.3.

3.35   In the end, therefore, although there will always be uncertainties and nothing can be guaranteed, the chances are that, with normal standards of manufacture and normal intelligent use, most hydrodynamic bearings will operate satisfactorily at the loads and speeds indicated in figure 4.1. It is, however, advisable to work as far away as possible from the limiting values of recommended loads if reliability is important.

3.36   Higher loads can certainly be carried under appropriate conditions but every marginal increase in rating introduces a greater degree of uncertainty and makes prototype testing more and more desirable.

3.37   With hydrodynamic film-lubricated bearings it may, in extreme cases, be possible to obtain satisfactory operation with loads as much as twenty times (or more) in excess of those recommended in figure 5.1. But these results will only be obtained after much experimentation with relatively exotic materials and highly specialised lubricants, as well as by going to almost unrealistic lengths in relation to such things as shaft and housing rigidity, alignment, surface finish, dimensional and geometric tolerances, oil filtration, frequency of oil changes, etc.

3.38   Some of the factors that affect the performance of highly-loaded bearings are

(a) An increase in load involves a decrease in oil-film thickness and the increased heat generated by shearing this thinner film tends to raise its temperature and lower its viscosity – thus further reducing the oil-film thickness corresponding to any given load.

(b) With ultra-thin oil films, the uniformity of their thickness becomes proportionately more sensitive to any out-of-roundness or misalignment of the journal, and any tendency towards the shaft being tapered or 'lobed' (as in a 50p piece) which can greatly increase the oil-film pressure in local areas. Even with only moderate loads, it may be necessary to impose tighter tolerances on the roundness and parallelism of the journal than on its nominal diameter.

(c) Since no journal will ever be *perfectly* round, the oil-film thickness can never show the exact pattern predicted by theory, and the rotation of the journal must consequently give rise to cyclic variations of the oil-film pressure at all points on the bearing surface. This in turn means that the compressive stress in the bearing material must also vary cyclically throughout each revolution, even if the load is nominally constant. At any reasonable rotational speed, it does not take long for the stress reversals to pass the million mark, above which the strength of any material tends to be determined by its fatigue limit. Thus, an automobile engine doing 2000 rev/min at 30 mph will complete a million revolutions every 250 miles (400 km); if its crankshaft is even microscopically 'lobed', say with seven lobes, then there will be a million load cycles on the bearings as frequently as every 36 miles (57 km) or so.

(d) Any fatigue failure will, of course, take place in the area where the stress and/or cyclic variations are greatest and – remembering that the bearing pressure over a bearing surface tends to lie in a pattern somewhat as shown in figure 3.6 – the stress at the most highly-loaded point will obviously be far above the average or nominal bearing pressure as defined by total load divided by bearing area. Thus, not only is the safe mean bearing pressure on a material far below either its uts or yield point, but is also considerably below even the nominal fatigue limit of the material.

(e) Since white metal has a relatively low fatigue limit, it is necessary to use stronger materials for highly-loaded bearings. (Some of these stronger

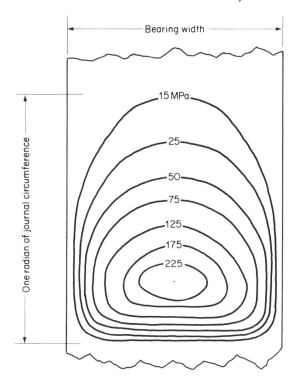

Figure 3.6 Typical pressure distribution in automobile engine bearing at instant of peak load in cycle (see paragraph 3.38d). The nominal bearing pressure is 37 MPa (5250 lbf/in²), but the pressure at one place is up to nearly 250 MPa (36 000 lbf/in²)

materials are described in chapter 7(ii).) The stronger materials are usually harder than white metal and therefore, if the danger of seizure is to be minimised, they demand a harder journal surface than would suffice for white metal. It will not always be practicable to obtain this harder surface by case-hardening, since the higher bearing loads will mean greater bending distortion in the journal (see figure 6.5), and all circumstances must be considered lest this should tend to break up the hard surface skin. See also paragraphs 7.04 to 7.08 regarding the dangers in using certain harder journal materials.

(f)   None of these harder lining materials are as good as white metal in their ability to absorb foreign particles by 'embeddability'. Yet, clearly, an ultra-thin oil film cannot give as free a passage to large abrasive dirt particles as can a thick film. This means that the question of oil filtration must be given more than normal attention.

(g)   Since the whole operation of a hydrodynamic bearing depends on the rotating journal dragging oil into the loaded region (as in figure 3.3), it is obviously important that the oil should adhere adequately to the journal surface – in other words, that the journal should 'wet' easily with the oil. Some journal materials are not as good as others in their wetting characteristics, and for most highly-rated bearings it is advisable to use oils with 'oiliness' additives to improve their wetting effect. In this connection, it may be interesting to note that the ideal conditions for a steadily loaded bearing would arise if the oil adhered strongly to the journal (and would thus be dragged into the loaded area in adequate volume), but completely refused to adhere to the bearing surface, so that there would be no friction between the rotating oil and the stationary surface. This ideal, of course, never arises in practice. In the case of a wholly-rotating centrifugal load, exactly the opposite would apply: the best condition then occurring with a non-wetting journal and a heavily-wetted bearing.

(h) Since the most heavily loaded regions of the oil film will also be the thinnest, there will tend to be considerable heat generated in them, and the oil may therefore suffer some deterioration as it passes through these areas. This means that for highly-rated bearings a complete oil change is desirable at comparatively frequent intervals – in automobile engines, for example, about every 100 to 150 hours.

(i) Cavitation erosion is another difficulty that may be encountered with highly-rated bearings. It is caused by one or more of the following conditions:

(i)   In the diverging portion of the eccentric clearance space between the journal and the bearing, the oil velocity may not fall rapidly enough to keep the clearance space full, and vapour bubbles may form. The collapse of these bubbles can readily produce erosion, similar to that which occurs on the trailing edge of marine propeller blades. The cure may lie in raising the oil pressure and/or in increasing the area of the oil-supply passages.

(ii)   Sharp discontinuities in a bearing surface (such as at a joint face relief) can also produce similar vapour bubbles and erosion.

(iii) With high-speed engines in which the connecting rod bearings are fed through a crankshaft drilling from a non-continuous circumferential groove, the sudden interruption of the oil flow at each revolution can produce a 'water-hammer' effect sufficient to damage the bearing unless the 'cut-off' edges of the drilling and groove are adequately chamfered.

(iv)   Cavitation erosion can also occur in oil grooves through sudden reversals of flow, but this is unimportant unless it is so severe as to dislodge metal particles. The cure may be to use deeper oil grooves in which the oil velocity will naturally be lower.

3.39   Few of these problems with highly-rated bearings are amenable to full mathematical analysis at the design stage, and development testing is needed to discover and correct them. Furthermore, since the performance of these thin oil films can be significantly affected by dimensional changes of less than 1 μm, one cannot be sure of any test results being reflected in the final product unless the latter has its form and sizes held to within limits of the same order. Even then it may be necessary to run extensive trials on a considerable range of production assemblies before one can feel confident about either repeatability or reliability.

3.40   The design of highly-rated bearings requires considerable experience, as well as knowledge and skill which, if one does not possess it, may best be found by consulting reputable bearing manufacturers.

Chapter 4

Assessment of bearing pressures and pV factors

(i) Simple loads

4.01 In the preliminary assessment of bearing pressures, it is an internationally accepted convention to assume that the bearing pressure *normal to the bearing surface* is uniform over the whole area of the bearing. Now, as shown in figure 4.1, if the bearing pressure is uniform over the whole curved surface of a half-bearing, then the total resultant force opposing the load will be equal to the bearing pressure multiplied by the *projected area* of the bearing (i.e. diameter × width). For this reason, the mean bearing pressure due to steady loads is always regarded as the load divided by the projected area of the bearing, the latter being corrected as necessary for any significant loss of effective bearing area arising from oil-ways, gutters, etc. In making corrections of this kind, it is, of course, not the surface area of the gutterway that has to be taken into account, but only its *projected area* in the direction of the principal load.

(ii) Corrections for non-uniform journal speed

4.02 Many types of machine (e.g. automobiles and most machine tools) have to run for long periods off their designed speed, or may indeed have no specific designed speed. The effect of this on bearing design depends on the type of lubrication involved, as explained below.

Dry-rubbing bearings

4.03 Variable speed presents no problems with these, provided the maximum load is within the manufacturer's recommended top limit. The bearing may be expected to have the same life-span as if it operated continuously at the *average pV* value it has to withstand.

Oil-impregnated porous bearings

4.04 When these are subject to variable speed, the design should be based on the highest pV^2 factor they have to meet. In other words, the load at each and every speed should be within the limits for that speed, as shown on the chain

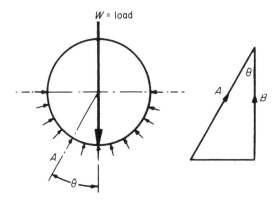

d = diameter of bearing
r = radius of bearing = $\frac{1}{2}d$
b = length of bearing
p = bearing pressure or load per unit area

A = force from bearing pressure
     acting on element of bearing
     area

  = $p\,b\,r\,\,d\theta$

B = load-supporting component of A

  = $A \cos\theta = pbr \cos\theta \; d\theta$

Total supporting force = $2\displaystyle\int_{0}^{\pi/2} pbr \cos\theta \; d\theta = 2\,pbr\,(1-0)$

$2\,rbp = dbp$ = projected area of bearing × pressure

Bearing pressure = $\dfrac{W}{\text{projected area of bearing}} = \dfrac{W}{db}$

Figure 4.1 Conventional assessment of bearing pressure, based on the assumption of uniform bearing pressure over the whole of the loaded area (see paragraph 4.01)

line marked 'porous' in figure 5.1, or as may be agreed with the manufacturer of whatever porous material it is intended to use.

Oil-film bearings

4.05   Reference to figure 3.5 shows that, provided a bearing has an adequate oil supply, if any given load lies below the value of the point C, then it will be possible for that load to be carried satisfactorily at any rotational speed from zero up to an extremely high figure.

4.06   Reference to the calibrated curves of figure 5.1 will show that, on 50 mm bearings, this range is from 0 to 30 000 rev/min, and on 200 mm bearings, from 0 to about 5000 rev/min. At the lower speeds within these ranges, the pV value will be suitable for boundary lubrication, while at the higher speeds (where boundary lubrication would give trouble) hydrodynamic conditions will prevail.

4.07   If the load is greater than that corresponding to point C in figure 3.5, and the speed range goes above that suitable for boundary lubrication, there will be a certain speed range in which operation for more than a few seconds will be impossible. If the machine can be accelerated and decelerated quickly enough through this danger area at each start and stop, then operation through a limited range of higher speeds may be quite practicable. For example, figure 4.2 (based on figure 3.5) shows that a load that could be carried safely by a film-lubricated bearing at any speed up to that represented by the point X, or at any speed between points Y and Z, would be unsafe to use at speeds between X and Y. For actual numerical examples, reference may be

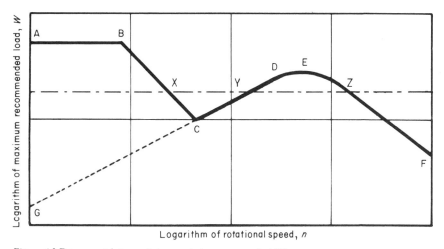

Figure 4.2 Dangers at intermediate speeds (see paragraph 4.07)

made to figure 5.1, from which it will be seen that a 50 mm film-lubricated bearing carrying a load of 200 kgf could be considered for continuous use at any speed between zero and 8000 rev/min *except* for speeds in the range between 190 and 1150 rev/min, which speed range would have to be run through quickly and/or under reduced load conditions.

4.08   Thus, for film-lubricated bearings with mean bearing pressures above about 0.7 to 1.0 MPa (100–150 psi), the load and speed characteristics must be examined for each and every speed at which the journal is required to run for any length of time, and the bearing design must be suitable for each and every one of these load–speed–time combinations.

Insidious speed variation

4.09   Designers should be on their guard against the possibility of bearing speeds not being as uniform as might at first glance appear. A striking example of this arises with the big-end bearing of any crank-and-connecting-rod mechanism. In operation, a connecting rod swings from side to side about the top-end bearing, and is thus for some time 'rotating' in one direction and for some time in the other. At bottom dead centre, the crankpin and the big-end bearing will have an *equal linear velocity*, but the pin will be rotating at engine speed around the centre of the main bearings at a radius equal to the crank throw, whereas its surrounding big-end bearing will be rotating in the same direction about the top-end bearing at a radius equal to the connecting rod length. At top-dead-centre, the conditions will be similar, except that the journal and its surrounding bearing will be rotating in opposite directions about their respective axes.

4.10   Thus, in an engine running at 5000 rev/min, if the connecting rod ratio is 4:1, the effective speed of the big-end bearing in relation to its journal will be 3750 rpm at bottom dead centre and 6125 rev/min at top dead centre. This is a wide variation that can easily have quite a marked effect on bearing performance in the cranks of presses and other machines.

4.11   If the crankpin in the foregoing 5000 rev/min engine were of, say, 50 mm diameter, then the top relative speed of 6125 rev/min would bring it beyond the peak of bearing performance, as indicated on figure 6.1, and could thus be of significance – particularly since top dead centre is around the position of highest piston loading in many types of engine.

4.12   Fortunately, in practice, the inertia effect of the piston mass helps to keep the bearing load down at top dead centre, but the interrelation of all these factors is one reason why it is so important to prepare accurate load-vector diagrams (on the lines of figures 4.4 and 4.5) for all such applications, and to examine *all* the load-speed combinations that will arise.

(iii) Corrections for non-uniform magnitude of load

4.13   A load that varies cyclically in magnitude is usually easier to carry than a continuous load of the same maximum value. This is partly because its lower average value results in less heat being generated, and partly because in a lubricated bearing the thickness of the oil film can build up during periods of lower load, and the bearing can then stand a really heavy load (limited only by the fatigue strength of the white metal or other lining) for a brief period until this thick oil film has been squeezed out. The length of time for which the heavier load can be carried depends on the magnitude of the load and on the extent to which a thick oil film has been built up in readiness for it. For normal unidirectional loads with a 360° or 720° cycle, an approximate first-order correction is to regard the equivalent steady load as being equal to the mean of the average load throughout each cycle, and the highest average load during any 15° of journal rotation. This is not an infallible rule, but it is near enough in most cases.

4.14   For slow oscillating hinge movements, such as in the arms of earth-moving equipment, it is usually better not to do any 'averaging' of the load but, instead, to design for the maximum load that has to be borne in any part of the oscillation.

(iv) Corrections for non-uniform direction of load

4.15   The load on most bearings changes direction to some extent between light load and full load. Thus, an electric motor driving through gears may have the principal working load on its bearings in an *upward* direction; although when idling, the principal load on its bearings will be the *downward* weight of the motor armature. Such incidental changes of load direction can generally be ignored, except when considering the position of oil-ways. If, however, the direction of load rotates continuously, or oscillates, then the effects may be highly significant.

4.16   In the case of dry-rubbing bearings and oil-impregnated porous bearings, a rotating load is advantageous. The frictional heating will be the same as for a steady load, but it will be spread over a greater area of the bearing, as also will the wear. The net result is that one may safely divide the actual load by a factor of 2 to find the equivalent steady-load value.

4.17   In the case of oil-film bearings, the correction for a rotating load is made by adjusting the rotational speed of the journal to an equivalent steady-load speed of $(n-2N)$, where n is the rotational speed of the shaft and N the rotational speed of the load vector. The theoretical basis for this need not concern us, but in fact $(n-2N)$ represents the rotational speed at which the mean rate of oil flow beneath the load in a steadily loaded bearing, would be the same as the mean rate of oil flow relative to the load vector in the bearing under consideration. (A negative value simply means that the 'equivalent' shaft is rotating in the opposite direction to the real shaft.)

4.18   It will be noted that, when $N = n/2$, the equivalent shaft speed is zero, which means that a hydrodynamic oil-film bearing is incapable of carrying any load under these circumstances. This is perfectly true, and is the condition known as a *half-speed load vector*.

4.19   The explanation is that when a load vector rotates at half the speed of the journal then, to an observer travelling round with the load vector, the journal would appear to be rotating in one direction and the bearing itself in another, the speed of rotation in the two directions being equal. This is illustrated in figure 4.3, from which it will be seen that the oil adjacent to the

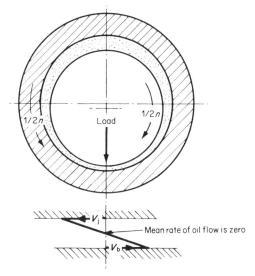

Figure 4.3 Half-speed load vector (see paragraph 4.19). The rotations and oil flows are as seen by an observer who rotates with the load. V_j is the velocity of the journal and adjacent oil adhering to it; V_b the velocity of the bearing and adjacent oil adhering to it

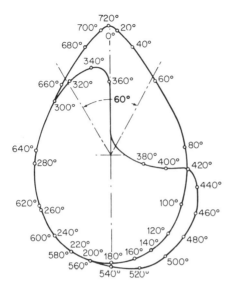

Figure 4.4 Polar load diagram for automobile engine rear-end main bearing (see paragraph 4.20); illustrating half-speed load vector between crank angles of 660° and 60° [Reproduced by permission of The Glacier Metal Co. Ltd]

journal (and partly adhering to it) will be moving in one direction in relation to the loaded area, while the oil adjacent to the bearing will be moving at an equal speed in the opposite direction. The mean rate of oil flow under the load will thus be zero, there will be no oil flowing in to replenish the oil film, and in a *very* short time the bearing will fail.

4.20   Half-speed load vectors often arise in automobile engine bearings during a portion of each cycle, and figure 4.4 is an example of a polar load diagram in which it will be noted that, between the crank positions of 660° and 60° (during which the journal rotates through 120°), the load vector moves through only about 60°. This diagram refers to a real engine of pre-1939 design; what saved it from disaster was that its bearings were very lightly loaded by modern standards. When the hydrodynamic conditions collapsed during part of each cycle, the bearing load was still well within acceptable pV limits for boundary lubrication. Modern engines are much more highly rated, and today's practice is to regard a half-speed load vector as potentially dangerous if it persists for more than about 15° of crankshaft rotation, unless this closely follows a period of load reversal, when the squeeze effect of a relatively thick oil film may intrude to ease the situation.

4.21   A half-speed load vector is, of course, dangerous only if the load vector is rotating in the same direction as the journal. If the load vector is rotating in the opposite direction, the effect is merely the same as that obtained by raising the speed of journal rotation. Thus, in using the expression $(n-2N)$, it is useful to regard clockwise rotation as positive, and anticlockwise rotation as negative. This can be of significance in automobile crankshaft bearings, where the load vector frequently changes its direction of rotation, and it would be unfortunate if anyone became alarmed at a half-speed load vector which was in fact beneficial because of its direction. Thus, figure 4.5 shows a polar load diagram typical of a main bearing in a V-engine, and it will be seen that the direction of rotation of the load vector reverses several times during each cycle, so that a half-speed load vector between, for example, points

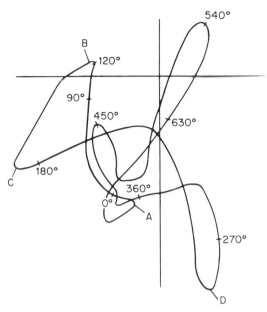

Figure 4.5 Approximate form of polar load diagram in automobile V-engine (see paragraph 4.21) [Reproduced by permission of The Glacier Metal Co. Ltd]

A and B, would have exactly the opposite effect to one occurring between points C and D.

4.22   It may further be noted that the expression $(n-2N)$ gives an equivalent shaft speed of $2N$ even if the actual shaft speed is zero. This is also true, and means that a rotating load can be carried hydrodynamically in a bearing even if the shaft is stationary. The reason for this is that the journal will be continuously squeezing out the oil from below the loaded area into the area that is about to receive the load. Thus, provided the load is not too great, or the speed of the load vector too slow, metal-to-metal contact need never occur between the journal and the bearing. This has few practical applications but it means that, under the right conditions, even at very low speeds, an oil film bearing may be capable of carrying a quite substantial rotating load.

4.23   It may also be noted that the expression $(n-2N)$ gives the equivalent shaft speed as n when the load vector rotates at the same speed as the shaft. This is explained by the fact that, under these conditions, relative to the load, there is a stationary shaft surrounded by a rotating bush, and the hydrodynamic conditions are thus exactly as they would be in the converse arrangement of a stationary bush and a rotating shaft. There is, however, an important exception to this in the case of a load vector which, although it rotates at the same speed as the shaft (arising, for example, from an unbalanced mass on the shaft), happens to have a magnitude equal to the weight of the rotating member. In such a case, the actual load on the bearing will become zero once per revolution at the instant when the rotating load vector is pointing vertically upwards. If the overall load vector situation is examined graphically, as in figure 4.6, it will be seen that one steady downward load (the

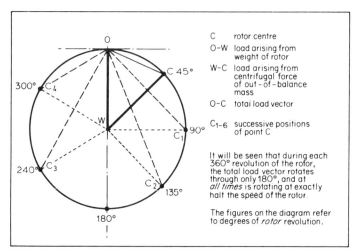

Figure 4.6 Load vector diagram for out-of-balance turbine rotor when out-of-balance force exactly equals rotor weight (see paragraph 4.23)

weight of the rotor), when combined with a rotating force of equal magnitude, produces a total resultant load vector that varies in magnitude from 0 to $2W$ but always rotates at exactly half the shaft speed. This is a true half-speed load vector, and completely unacceptable since it persists throughout 360° of each and every revolution. Even if this condition has only to be run through on the way up to normal speed, it cannot be regarded with equanimity, since even boundary lubrication will quickly break down under a half-speed load vector. If, however, the machine can be run up quickly enough through the dangerous range of rpm, then the maintenance of an adequate supply of 'jacking oil' may be sufficient to save the day.

Half-speed whirl

4.24   Instead of being steadily pressed by the load against one segment of the bearing, a shaft will sometimes whirl around inside the clearance space with an angular velocity of whirl that is exactly (or almost exactly) half the shaft rotational speed. This phenomenon, called *half-speed whirl*, is caused by a half-speed load vector being somewhat accidentally set up in one of two ways:

(a) If a shaft, such as a turbine rotor, is run close to twice its normal critical speed (deflection critical, not torsion critical), it will be subject to a harmonic excitation and will tend to vibrate vertically at its normal critical frequency (i.e. half the then shaft speed). It will thus impose on the bearing a variable load that will fluctuate with a frequency equal to half the shaft rotational speed. If the vector of this variable load is drawn, it will be found to be, in effect, a half-speed load vector which can cause the same sort of trouble as the half-speed load vectors described in paragraphs 4.19 to 4.23. The only cure is to modify the shaft in some way that will change its critical speed.

(b) A lightly loaded shaft will normally run nearly central in its clearance space (i.e. with an eccentricity ratio which is close to zero) with very little self-centring effect to hold it there. If such a shaft is subject to vibration and if this vibration chances to give the shaft a momentary accidental whirl around its clearance space, and if this whirl happens to be at a rotational speed of $\frac{1}{2}n$, then a centrifugal force will be set up. This force will also be rotating at $\frac{1}{2}n$ and will therefore be a half-speed load vector. As explained in paragraphs 4.18, 4.19 and 4.23, the bearing will offer no resistance to this force, and there will thus be nothing tending to restore the bearing to its central position. It will, therefore, be prone to continue its half-speed whirl – and in fact often does. In extreme cases, bearing failure may result. This can be puzzling, because one would naturally expect such a lightly loaded bearing to be totally safe. Relief can sometimes be obtained by employing a different value of clearance, usually larger, but the only real cures are either to increase the unit loading on the bearing by decreasing its width, or to employ a bearing with a non-circular bore, such as one of the profile bore bearings illustrated in figure 5.2.

4.25   Before closing this dissertation on load vectors it is worth emphasising that it is the resultant load vector, the summation of all the individual loads, that is of prime importance, the nature of the vectors of individual loads is of no significance. Thus, figure 4.6 shows how a dangerous result can spring from the combination of two vectors, each perfectly harmless by themselves. It is equally possible for two undesirable vectors to combine in such a way as produce quite an acceptable resultant.

(v) Corrections for ambient temperature

4.26   Strictly speaking, the assessment of bearing pressure needs no adjustment for temperature. However, the strength of bearing materials and the performance of lubricants can both be strongly influenced by the working temperature inside a bearing, and have a big effect on the load it can carry. It is therefore useful to consider temperature effects at this stage, since they may affect the choice of bearing type when this comes to be discussed in the next chapter.

4.27   With film-lubricated bearings in normal, free-standing machinery, such as machine tools, pumps and the like, there is seldom any problem with ambient temperature. In tropical climates, such bearings often tend to *feel* overheated when in operation, but this is usually only because the human hand is particularly sensitive to temperature changes in the range 40–70°C (100–160°F) – metal at the lower temperature feeling 'comfortable' to the touch, but metal at the higher temperature feeling 'too hot to hold'. This temperature range is, however, unimportant as far as bearing material strength is concerned, so that the only special precaution to take in tropical or Arctic regions is to be sure that the lubricant is of a type and grade appropriate to the local climate and other conditions.

4.28   The latter proviso is added because one epidemic of bearing failure with which the author was involved occurred in the axle bearings of old-fashioned railway goods wagons, from which the grease was regularly being stolen by the local population for cooking purposes. In the subsequent bearing failures, the affected axle-boxes reached so high a temperature that every vestige of grease drained away and thus left no clue of its having mostly been stolen and not just run off by the heat. The troubles were completely overcome by mixing a highly unpalatable additive to the grease. This is only one example of the wide range of factors that can arise in the study of bearing performance.

4.29   In the case of film-lubricated bearings that are buried in the heart of high-temperature machinery, such as steam turbines, gas turbines, or process plant, care may have to be taken in the selection of bearing lining materials and/or the provision of oil-cooling equipment, etc.

4.30   Although whitemetal does not actually melt until about 240°C (465°F), it loses strength long before that temperature is reached. Most grades have yield points of about 14 MPa (2000 lbf/in²) at 100°C (212°F), but only about 7 MPa (1000 lbf/in²) at 130°C (270°F). Thus, the maximum recommended loads for whitemetal bearings must be reduced if the bearing surface temperature is likely to exceed 100°C, and the vertical lines on figure 6.2, corresponding to the fatigue strength limit of whitemetal, must be moved to the right to increase the bearing sizes to which they refer. This increase in size is about 1.4 per cent for each degree Celsius by which the bearing surface temperature is expected to exceed 100°C (or 0.8 per cent for each degree Fahrenheit above 212°F).

4.31   At first glance, it might be thought that the maximum recommended loads of figure 6.2 are too high, since a quick calculation will show that they are based on a mean bearing pressure of about 7 MPa or 1000 lbf/in². Since figure 3.6 shows that the peak bearing pressures may, in places, be nearly seven times as much as the mean or average pressure, this means that bearings loaded in accordance with figure 6.2 may, in places, be subject to bearing pressures of nearly 49 MPa or 7000 lbf/in², which is more than 25 per cent above the fatigue rating for whitemetal at 100°C (212°F), as shown in figure 7.1.

4.32   There is, however, a redeeming feature in that the peak pressures in a bearing tend to act on relatively small areas that are surrounded by what might be called a 'solid wall' of less highly-stressed material. Thus, in an extreme case, it would be possible to imagine a small pocket of highly-stressed whitemetal that had gone plastic somewhere in the middle of a bearing area, yet could still carry a high load because, being trapped and surrounded by solid metal, it would simply pass the applied pressure on its surface to the bearing shell behind it and to the solid wall of metal surrounding it. This would be true even if the small area of lining was so plastic as to be almost fluid.

4.33   It is always hoped that such extreme conditions would never arise, but, even if they did, bearing failure would be by no means inevitable. It really all comes back to what was said in paragraph 1.08, and an appreciation which confirms that figure 6.2 is no more than a statement of what it is usually possible to 'get away with'.

4.34   Expert metallurgical advice is highly desirable for high-temperature applications that involve any risk of the whitemetal temperatures at any point approaching 120°C (250°F).

4.35   In engines with aluminium crankcases, or other machinery with aluminium bearing housing, the higher coefficient of thermal expansion of these (and consequent greater difference between the 'hot' and 'cold' dimensions, as well as the effect of differential expansion between the housing and the journal and/or between the housing and the backing of a thin-walled bearing) may produce a disadvantageously higher running clearance in the bearing. It may be necessary to make some provision in the design to meet this contingency, and/or accept some compensatory reduction in the performance rating of the bearing in tropical climates.

4.36   It also has to be remembered that, for really high-temperature process applications (in any climate), it may be necessary to rule out the use of mineral oils and, instead, employ synthetic lubricants, or else use special types of bearing that can employ the actual process fluid as their lubricant. For comments on these special bearings, see paragraphs 5.15 to 5.27.

4.37   With *dry-rubbing bearings* the life decreases with rising temperature. To calculate this effect, the actual load on the bearing should be multiplied by a factor to bring it to a higher 'equivalent' value. The correct factor to use depends on the exact type of bearing, and must thus be obtained from the manufacturer – although the following values are specified by one maker and may be used as a guide:

Ambient temperature, °C	25	60	100	150	200	280
Factor	1.0	1.25	1.7	2.5	5	10

4.38   *Oil-impregnated porous bearings* are not particularly well adapted for high-temperature use since, unlike dry-rubbing bearings, they contain oil which needs relatively frequent replenishment if it is not kept cool. Unlike oil-film bearings, they have no external oil circulation which can be passed through a cooler. Any proposed use of this type of bearing in ambient temperatures either higher or lower than normal ought to be discussed fully with the bearing manufacturers.

Chapter 5

Selection of bearing type

(i) Types of plain bearing

5.01   The first step in the selection of bearing type must be to decide whether to use a plain bearing – or whether a ball or roller bearing might not be more suitable. Guidance on this will be found in appendix 1, to which reference should be made if the designer has any doubts at all on the subject.

5.02   If the decision is to employ a plain bearing, then the types from which he can choose are:

Film-lubricated bearings

5.03   These are the normal plain bearings found in steam turbines, marine engines, automobile engines, machine tools, lawn mowers, and many other everyday applications. They are normally lubricated by either oil or grease, and operate on either the principle of boundary lubrication (see paragraph 3.5) or the principle of hydrodynamic fluid flow (see paragraph 3.7), depending on the load as well as on the rotational speed and other conditions. Many of the materials employed as bearing surfaces are described in chapter 7.

Dry-rubbing bearings

5.04   These are bearings with an essentially slippery working surface, usually formed from ptfe or graphite. They can be run without other lubricant provided the loads and rotational speeds are not too great. (See paragraphs 7.33 *et seq.*)

Graphite-impregnated bearings

5.05   These are made from a matrix of metal and/or carbon, impregnated with graphite. They are used both as dry bearings for relatively high-temperature applications, and as 'wet' bearings in situations where they have to run in water, or in some process fluid, which acts as the only lubricant. (See paragraphs 7.34, 7.37, 7.40 to 7.46.)

Pre-lubricated bearings

5.06   Sometimes called 'marginally lubricated bearings', these come in two forms, one being made from a porous bronze matrix impregnated with lubricating oil, and the other being virtually a dry-rubbing bearing that is provided with surface pockets for pre-loading with grease on assembly. Both types are capable of running for long periods without attention. (See paragraphs 6.23 to 6.25.)

(ii) Preliminary selection

5.07   The first tentative consideration of which bearing types to employ should be made with reference to figure 5.1, which indicates the range of loads that can readily be carried by 25 to 200 mm (1–8 in) diameter bearings of each type.

5.08   From the specified requirements already entered in the check list (figure 2.1), the designer can use figure 5.1 to discover what options are likely to be open to him. Thus, if the bearing load and speed are already fixed, he can mark the point on figure 5.1 where these two values intersect, and observe where this lies in relation to bearing types and sizes. Alternatively, where, as frequently happens, the load, speed *and* bearing diameter all form part of the designer's brief, then figure 5.1 will show him which type(s) (if any) of plain bearing will be capable of fulfilling the duty.

5.09   For example, if the load to be carried was 2000 N (450 lbf) at 1000 rev/min, it would be clear from figure 5.1 that no dry-rubbing bearing or porous impregnated bearing would meet the specification, and that a hydrodynamic bearing of at least 55 mm ($2\frac{1}{4}$ in) would be necessary. On the other hand, if the specified duty had been 1000 N (250 lbf) at 250 rev/min and the size of the bearing had not yet been determined, there could be a choice between using a 25 mm (1 in) diameter boundary-lubricated bearing or having full hydrodynamic lubrication with a bearing of about 65 mm ($2\frac{1}{2}$ in) diameter. Even a porous bearing of 25 mm (1 in) diameter might be considered.

5.10   In most cases, there will be little difficulty in making a final choice between such options, and the designer will be guided mostly by commonsense engineering practice, taking into account the general nature of the machine in which the bearing is to be used. For example, with relatively high-performance equipment, such as automobiles, the users accept the need to employ a variety of lubrication systems and lubricants: engine oil for the engine, brake fluid for the brakes, Vaseline for the distributor, light oil for the throttle linkage, WD40 for the window mechanism, and two different grades of gear oil for the differential and automatic gear-box, respectively – and perhaps an odd place needing the application of a grease gun. The users of ordinary industrial machinery normally expect much greater simplicity. Thus, unless it were absolutely essential, it would be stupid to have one bearing that needed periodic attention with a grease gun if all the other bearings in the machine were fed from a continuously-running oil pump. The designer must also consider the relative priorities of reliability, cost, frequency of replacement, etc., appropriate to the job in hand.

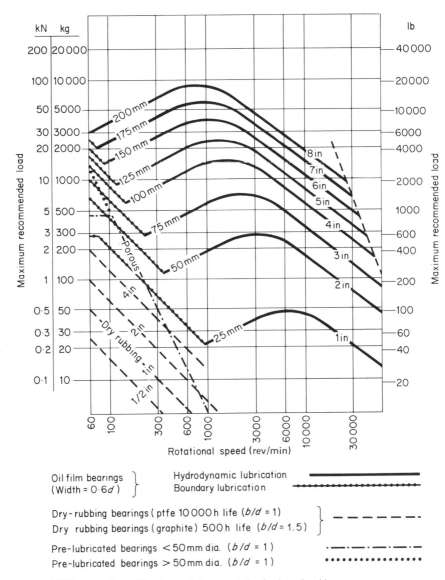

Figure 5.1 Selection chart of load–speed characteristics for broad guidance
(see paragraph 5.08)

5.11   In using figure 5.1, the following points should be borne in mind:

(a) It is intended only for preliminary guidance in the selection of a bearing
type. For oil-film bearings, all the information is repeated in a more useful
form, and on a larger scale, in figures 6.1 and 6.2. The details on dry
bearings and porous bearings are given in paragraphs 6.18 to 6.26, with
further comments in paragraphs 7.33 *et. seq.*

(b) Its values relate only to *steady-load conditions.* If the load is variable in
magnitude and/or direction (as in reciprocating machinery), figure 5.1

must be read in conjunction with section 4, where factors are given for converting non-uniform loads to equivalent steady-load values. In general, reciprocating machinery can carry heavier bearing loads than purely rotating machines, and chapter 15 on engine bearings explores an extreme example of this property. Rotating loads in turbomachinery are dealt with in section 4(iv).

(c) Figure 5.1 recommends different b/d ratios for the different types of bearing. These recommendations should not be ignored without first reading chapter 6, where the reasons for them are fully explained and the dangers of departing from them underlined.

(d) Hydrodynamic lubrication can be maintained at lower speeds than those shown in figure 5.1, since the heavy lines on this diagram ought actually to extend down to the left, as shown dotted in figure 3.5. These portions of the lines have been omitted from figure 5.1 merely to make it more readable. Thus, for example, a load of 500 N could be carried hydrodynamically at 100 rev/min by a 50 mm bearing. The upper limit of hydrodynamic lubrication represented by the line $A-G$ on figure 3.5 is, incidentally, relatively safe to approach much more closely than most limiting values, since loads below this line would normally provide the ideal of hydrodynamic lubrication; in the event of accidental overload, the bearing would not automatically fail, since the conditions would probably move into an acceptable state of boundary lubrication.

(e) It will be noted that the load–speed characteristics of all sizes of porous and pre-lubricated bearings are (for the reasons given in paragraph 6.24) shown on a single line, except for their limiting values at low speeds, which are based on a maximum bearing pressure of the order of 7 MPa (70 kp/cm^2, 1000 lbf/in^2). This is a conservative presentation that is felt desirable since porous bearings are made in a wide range of porosities with an equally wide range of performances. Many authorities quote higher figures. For example, BS 1131, Part 5, suggests, in dealing with porous bearings, maximum loads that correspond closely to those shown for boundary lubrication in figure 5.1, with a low-speed maximum of 14 MPa (2000 lbf/in^2). Such values, however, tend to be on the limit of what can be carried without supplementary lubrication and/or hardened shafts and other refinements. These higher values ought thus to be avoided unless one is very sure of one's ground *and* has confirmation from the bearing supplier. Porous bearings can give very satisfactory service indeed with light loads, and it is pointless to risk disappointment by pushing them to their limits.

(iii) Mixed lubrication

5.12 It is sometimes helpful to operate a bearing partly on one type of lubrication and partly on another. This applies not only to accidental excursions into the A-B-C-G area described in paragraph 4.07, but also to a mixture of dry-rubbing and hydrodynamic, or dry-rubbing and boundary lubrication. Thus, many types of dry-rubbing bearing material (especially those with a ptfe base) will operate even better than normal if they are occasionally provided with a smear of oil or grease, and will be almost everlasting if they are run submerged in some form of liquid lubricant.

5.13   Provided the bearing pressure is within the top limit for the dry material, dry-rubbing linings may be used in place of whitemetal for normal oil-lubricated hydrodynamic bearings. Such a lining will provide a hydrodynamic bearing with low-friction conditions and low wear when starting and stopping, which are virtually the only times when a normal hydrodynamic bearing is subject to any wear. For maximum pressure limits, see paragraph 6.22.

5.14   Some types of 'dry' bearing will benefit from water as a lubricant. When the load on a bearing with $b/d = 1$ is less than $d^3n/5000$ (mm, rpm, N) or $d^3n/80$ (in, rev/min, lbf), such bearings will normally operate under full hydrodynamic conditions. Care must, however, be taken before using water in this way – firstly to ensure that it does not result in corrosion of any constituent of the bearing material or its backing; and secondly because some 'dry'-bearing materials will work satisfactorily either permanently wet or permanently dry, but may break down rapidly if these conditions alternate. For wet–dry usage, consult the material suppliers.

(iv) Special applications

5.15   For special applications, there are several types of plain bearing (or pseudo-plain bearing) other than those dealt with in figure 5.1. All of them lie outside the scope of the present book in so far as their detail design is concerned, but a general description of six is given in the following paragraphs.

Pressurized gas bearings

5.16   In appearance, these are normal plain bearings, but their bearing load is carried by gas pressure acting directly on an appropriate area of the journal. The bearing has a number (at least three) of separate shallow pockets recessed into its surface, and each of these pockets is supplied with gas through its own calibrated orifice connected to a common supply under pressure, as shown in figure 5.2. When the journal is displaced from its central position (shown exaggerated in figure 5.2), one pocket has its outlet to atmosphere restricted, while the diagonally opposite pocket has a freer-than-normal vent to the open air. The result is that gas pressure will build up in pocket A to a higher pressure than in pocket B, and this pressure acting on the journal will generate an upward force to carry the load.

5.17   In actual bearings of this type, the clearances employed are extremely small in relation to those in normal bearings. Indeed, it is fairly standard practice for the journal to be a slight interference fit in the bearing when idle, the required running clearance being generated by the expansion of the bearing when subjected to the internal pressure of the 'lubricating' gas. To avoid damage during its original assembly, the housing has to be warmed, or the journal cooled (with solid CO_2), to provide sufficient difference in their dimensions for one to be inserted in the other. Such bearings must *never* be rotated except when under the influence of the full gas pressure. Since the viscosity of most gases is virtually zero, there is no frictional heat generated when running, and there is thus no objection to the use of such ultra-tight

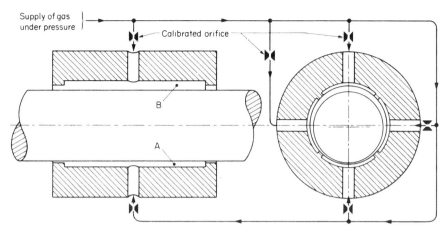

Figure 5.2 Pressurised gas bearing (see paragraph 5.16)

clearances, which result in the journal rotating continuously in one very well-defined position.

5.18   Air is the gas most commonly employed since its escape into the atmosphere presents no difficulty, but in process plant applications there are sometimes other gases available that are more convenient to use for the avoidance of leakage and contamination problems.

5.19   The unusually steady running of pressurised air bearings is of immense benefit in fine grinding machines and roll-grinders, where the precision of the journal position enables much finer finishes to be more easily obtained, and in which the absence of journal friction means that the grinding wheel can be balanced on the machine more accurately than on the usual knife-edge balancing stand. Even the weight of a postage stamp stuck on the side of a 12 in grinding wheel is enough 'out-of-balance' to show up quite clearly on an air-bearing wheelhead. For this class of work, ordinary shop air at 7 bar (100 lbf/in²) is frequently used as the supply so that no special compressors are needed.

5.20   The chief disadvantage of pressurised gas bearings is their first cost, since this must include superb air-filtering and air-drying equipment as well as fool-proof trip gear to sense any incipient failure of the air supply and stop the machine completely before the supply pressure has fallen too far.

5.21   In Britain, the practical development of gas bearings has been carried out mostly by the National Engineering Laboratory at East Kilbride, working in conjunction with the firms incorporating such bearings in their products. The National Tribology Centre at Risley is also understood to have a team experienced in this field.

Liquid pressurised bearings

5.22   These are essentially the same as pressurised gas bearings except that they employ liquid instead of gas, and can thus use higher pressures since there is no cooling effect from expansion when the fluid pressure drops on its way through the bearing. They can be designed to carry almost any load at any desired speed, using virtually any liquid as the 'lubricant'. They are particularly useful for process plant where the bearing liquid can be the process liquid itself, thereby avoiding all risk of lubricant contamination. Their principal disadvantages are their first cost and the complication of fail-safe equipment.

Self-acting gas bearings

5.23   Very lightly loaded, high-speed bearings can sometimes be built to run satisfactorily on normal hydrodynamic principles using atmospheric air or other gases as their sole lubricant, which they pick up from the environment in which they are immersed. On the face of it, nothing could be simpler, but in fact such bearings are completely impractical for everyday applications, and are mentioned here merely to say that any information on them should be sought from government research establishments such as the National Engineering Laboratory, East Kilbride, or from universities actively engaged in tribological research. This type of bearing has been used in gyros for inertial navigation systems and in circulators submerged in a gas circuit such as used in some nuclear reactors.

Profile-bore bearings

5.24   With high-speed machinery, the practical need for large running clearances, see chapter 6(iii), means that the bearing pressures in high speed work cannot be more than moderate. This combination of light loads and large clearances results in a lack of hydraulic stiffness in the bearing, so that relatively little force is required to produce significant radial displacement of the journal from its normal running position. The overall effect is one of considerable shaft movement and vibration, usually described as *instability*, which may or may not be associated with half-speed whirl as described in paragraph 4.24. With these high-speed journals, the stiffness and damping characteristics of ordinary cylindrical plain bearings tend to be unsatisfactory, and various forms of profile bore have been developed to give improved results. Some of the better known of these are illustrated in figure 5.3 but no hints on their design are given because, except on vertical shafts, the need for them arises comparatively rarely and, when it does, the problem can seldom be solved without trial-and-error procedures based on first-hand experience with comparable applications. Some comments on offset halves are, however, made in paragraphs 8.13 to 8.16, and comments on floating bushes will be found in paragraphs 10.19 to 10.24. If further information is required, it should be sought from one or other of the advisory services on bearing design.

Camella restricted-clearance bearing

5.25   This is a proprietary profiled bore bearing that operates as shown diagrammatically in figure 5.4. It consists essentially of a series of completely plain round bores each placed eccentrically to its adjacent partners. Figure 5.4 shows three such sections, but more may be used and all need not be of the same length. The individual bores need not be in separate parts as shown, but may all be machined in the one bearing. The construction, as will be seen,

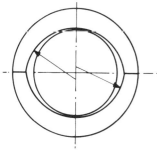

(a) Lemon bore bearing

Moderate cost and widely used,
but liable to half-speed whirl

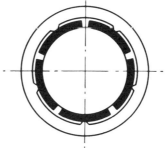

(b) Tilting-pad bearing
Expensive but excellent, except
for rotating loads, for which
a proprietary form is available
with hydrostatically pivoted pads

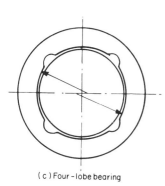

(c) Four-lobe bearing

Moderate cost and widely used,
although cannot carry much load

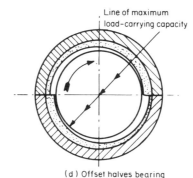

(d) Offset halves bearing

Highly effective modern procedure,
but useless if rotation reverses.

Offset halves may be helpful or
harmful, depending on circumstances
– see paragraphs 8.13 to 8.15

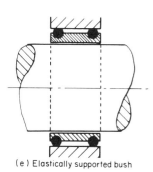

(e) Elastically supported bush

Moderate cost, but mostly used in
small machines with light loads

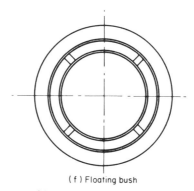

(f) Floating bush
(See paragraphs 10.19 to 10.24)

Figure 5.3 Profiled bearings: diagrammatic (and hence somewhat
exaggerated) illustrations of six special forms of bearing for high-speed
shafts (see paragraph 5.24)

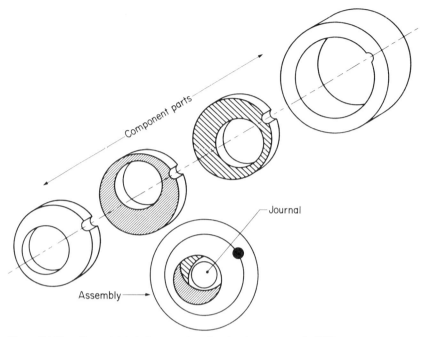

Figure 5.4 Camella: restricted-clearance-type bearing (see paragraph 5.25)

results in each section of the bearing having a high eccentricity ratio and thus considerable stability, yet the overall effect is one of small clearance with considerable self-centring effect and anti-whirl properties. Since, within any given overall length, such an assembly can have a reasonable load-carrying area in only one direction, it is best suited to very light loads unless the load is unidirectional. If the direction of load is variable, as between one condition and another (e.g. in reversing gear-boxes), there may be problems with the overhang of the load beyond the load-carrying part of the bearing, and the question of lateral shaft stiffness and vibration must be carefully examined in such circumstances – although with loads that vary in direction cyclically, there can be an advantage in that the natural period of vibration will vary throughout each load cycle, and thus tend to assist damping.

Rubber-lined bearings

5.26   When rubber is wet, it can become very slippery, and is thus often used as a bearing material in ships' sterntubes and other similar applications where the bearings are relatively lightly loaded and can be continuously flooded with water. Rubber-lined bearings tend to be proprietary articles, and no design guidance on them is given in this book.

Magnetic bearings

5.27   These are quite different from conventional bearings in that the rotating shaft (or other moving component) is held in suspension in, or near, the desired

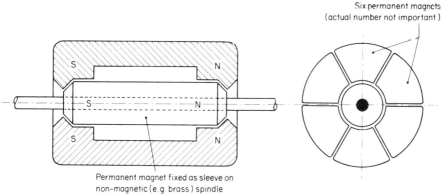

Six permanent magnets
(actual number not important)

Permanent magnet fixed as sleeve on
non-magnetic (e.g. brass) spindle

The main illustration above shows the ends of the spindle sleeve
and the matching faces of the surrounding magnets chamfered to an
angle of 45° to the spindle axis. This angle is not mandatory
and may be varied to suit the relative proportions of the radial
and axial loads involved. Thus, the scrap view alongside shows
a case where the angle at one end has been made 60° to suit a
vertical application with a downward axial load substantially
greater than the radial load

Figure 5.5 Magnetic bearing (see paragraph 5.27)

position by forces from magnetic fields, instead of by fluid pressure, or rubbing or rolling contact. The general principle is shown in figure 5.5 which illustrates how the effect could be achieved with permanent magnets, although in all known applications it has been necessary to use electromagnetic fields to obtain adequate forces and to control the position of the shaft by means of positional sensors arranged to regulate the forces and maintain stability. A practical application is in high vacuum pumps where the absence of any form of lubricant is very advantageous.

Chapter 6

Principal dimensions

(i) Journal diameter

6.01   Once the type of plain bearing to employ has been decided from a study of chapter 5 (and, where appropriate, any unusual load conditions have been converted to an equivalent steady load and speed in accordance with chapter 4), then consideration can be given to the precise journal diameter to employ. This is dealt with in paragraphs 6.03 to 6.27, according to type of bearing.

6.02   In cases where the journal diameter has already been specified, paragraphs 6.03 to 6.27 ought still to be studied to confirm that a bearing of the specified diameter will adequately fulfil the desired function.

Hydrodynamic fluid-film bearings

6.03   The information presented in figure 5.1 is repeated in a different form and on a larger scale is figure 6.1a (SI) and figure 6.1b (Imperial units), from which the minimum recommended diameter for any given steady load and uniform speed may be read off directly.

6.04   Journal diameters are usually fixed from considerations of shaft strength and stiffness, with the result that, in most machinery, the bearings are considerably larger than the minima stipulated in figure 6.1, which merely represents the minimum to which it is wise to go if there are restrictions on diameter.

6.05   *Note*: These minimum recommended diameters refer to bearings that conform to the standards laid down in paragraph 3.25, and in all other respects are also built and used in accordance with the recommendations of the present book. If it is necessary to depart from these standard dimensions and/or recommendations, then guidance on the probable effect will be found in the paragraph(s) that deal(s) with the condition or property being varied. The appropriate paragraph(s) may be located from the index. For example, guidance on the effect of changes in the b/d ratio will be found under 'Bearing width' in paragraphs 6.28 to 6.51.

Boundary-lubricated bearings

6.06   The friction of these is greater than that in hydrodynamic film bearings. Consequently, they tend to run with higher surface temperatures, making the limits of their performance dependent mainly on the frictional heat input per unit bearing surface. This is usually described in terms of the 'pV factor', i.e. bearing projected pressure multiplied by rubbing velocity, which is expressed in Imperial units as $lbf/in^2 \times ft/min$, and in metric as $kgf/cm^2 \times m/s$.

6.07   In SI, the pV factor is probably best worked out in terms of $kPa \times m/s$, because if this is written out in full it becomes

$$1000 \times \frac{newtons}{square\ metres} \times \frac{metres}{seconds}$$

which is the same as 1000 N m/s per m^2, or kW/m^2. In other words, it represents the rate of heat input to the bearing if the coefficient of friction were unity.

6.08   Thus, when thinking in SI units, it is useful to regard pV factors as a measure of the heat input that the bearing has to withstand, and to visualise them in terms of kW/m^2 although *calculating* them as $kPa \times m/s$. Sometimes one sees pV expressed in terms of $N/mm^2 \times m/s$ but this is really too large a unit for its purpose: for example, a common pV rating such as 20 000 $lbf/in^2 \times ft/min$ (700 kPa $\times$ m/s or 700 kW/m^2) represents less than one unit of $N/mm^2 \times m/s$.

6.09   There is a wide range of applications for boundary-lubricated bearings in industry, but the conditions of usage vary so enormously that it is virtually impossible to lay down hard-and-fast rules for safe loads, speeds, etc. All guidance must necessarily be of a somewhat general nature.

6.10   The information on boundary-lubricated bearings in figure 5.1 is repeated in a different form and on a larger scale in figure 6.2a (SI and metric units) and figure 6.2b (Imperial units), from which the minimum size for any given steady load and speed may be read off directly. These recommendations refer to bearings of the proportions described in paragraph 3.25.

6.11   Figures 6.2a, b are based on a pV of 700 kW/m^2 (700 daN/cm$^2 \times$ cm/s, 7.2 kgf/cm$^2 \times$ m/s or 20 000 lbf/in$^2 \times$ ft/min), which is usually found satisfactory for bearings that have to run more or less continuously with reasonable reliability. There are, however, many applications, such as cranes and elevators, where the usage is relatively intermittent, and many others where the combination of full load and full speed is infrequent and even then has to be sustained for only very brief periods. In such cases, it may well be appropriate to use a higher pV factor, although the extent to which a designer cares to take a chance on this must be left to his own judgement and discretion.

6.12   In considering this question, the designer must take account of the form of the oil supply. The pV factor previously mentioned of 700 kW/m^2 was selected on the assumption that the bearing would always be completely flooded with splash lubrication or the like, but that there would be no positive flow of oil to carry away the heat. If the lubrication arrangements are less generous than this then the safe pV factor will tend to be lower, whereas if there is a substantial oil flow to keep the bearing cool, a higher pV value may be quite acceptable.

6.13   A somewhat unusual characteristic of any bearing rated on a pV basis is that its load-carrying capacity is influenced not by its diameter, but only by its length. This is because, at any given load and speed, although an increase in diameter will provide a greater bearing area, nothing will be gained from this because the lower value of p will be exactly offset by the corresponding increase in V. Thus, the curves show different load values for boundary-lubricated

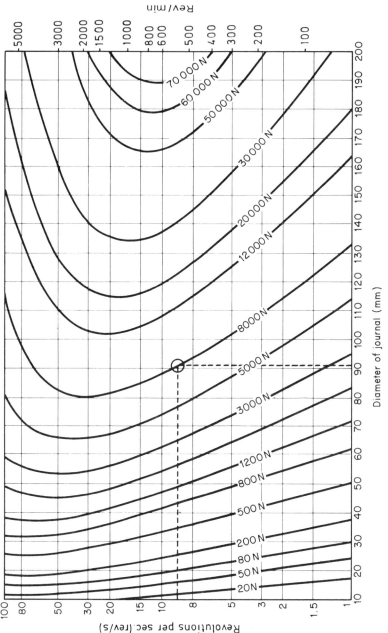

Figure 6.1a Hydrodynamic film-lubricated bearings (SI units): minimum recommended journal diameters with steady load, constant rotational speed and $b/d = 0.6$ (see paragraph 6.03). An example of how to use this diagram is shown by the dashed lines. These refer to a bearing required to carry a load of 8000 N at 10 rev/s, and indicate that the minimum safe journal diameter for these conditions would be a little over 90 mm

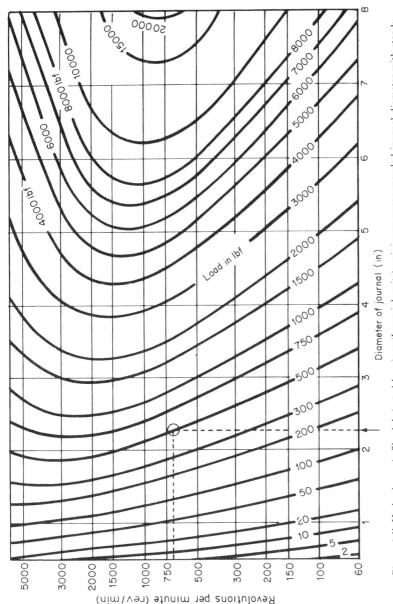

Figure 6.1b Hydrodynamic film-lubricated bearings (Imperial units): minimum recommended journal diameter with steady load, constant rotational speed and $b/d = 0.6$ (see paragraph 6.03). An example of how to use this diagram is shown by the dashed lines. These refer to a bearing required to carry a load of 500 lbf at 700 rpm, and indicate that the minimum safe journal diameter for these conditions would be about $2\frac{1}{4}$ in

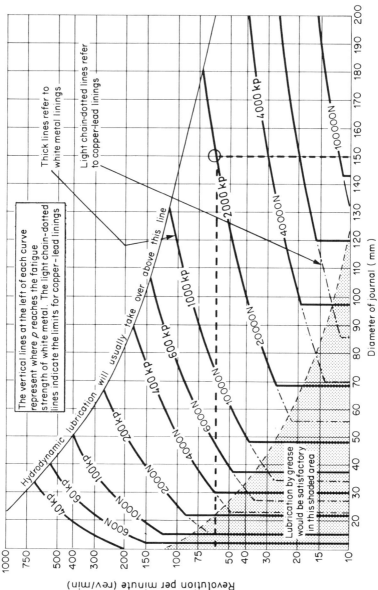

Figure 6.2a Boundary-lubricated bearings (SI units): minimum recommended journal diameters with steady load, constant rotational speed and b/d = 0.6 (see paragraph 6.10). An example of how to use this diagram is shown by the dashed lines. These refer to a bearing required to carry a load of 20000 N at 60 rev/min, and indicate that for these conditions the smallest satisfactory bearing would have a diameter of about 150 mm

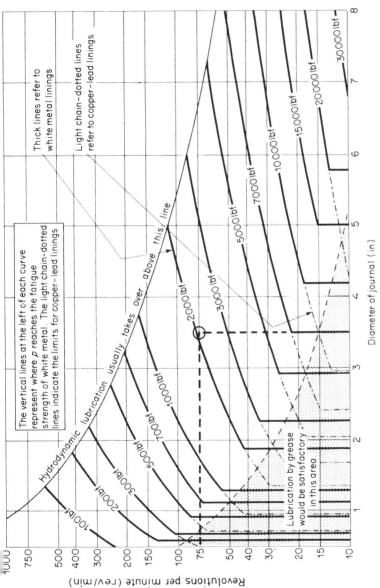

The vertical lines at the left of each curve represent where p reaches the fatigue strength of white metal. The light chain-dotted lines indicate the limits for copper-lead linings

Thick lines refer to white metal linings

Light chain-dotted lines refer to copper-lead linings

Hydrodynamic lubrication usually takes over above this line

Lubrication by grease would be satisfactory in this area.

Diameter of journal (in)

Revolutions per minute (rev/min)

Figure 6.1b Boundary-lubricated bearings (Imperial units): minimum recommended journal diameters with steady load, constant rotational speed and $b/d = 0.6$ (see paragraph 6.10). An example of how to use this diagram is shown by the heavy dashed lines. These refer to a bearing required to carry a load of 2000 lbf at 750 rev/min, and indicate that under these conditions a journal diameter of $3\frac{1}{4}$ inches is the minimum that could be recommended

bearings of different diameters solely because they refer to bearings of constant b/d ratio, in which the larger diameter bearings also have a greater length. In other words, the recommended load is simply varying with the length, and is limited at any given diameter only by the need to keep within a sensible b/d ratio, as explained in paragraphs 6.29 *et seq.*

6.14   Boundary lubrication is now used mostly in applications with rather ill-defined loads and speeds, in which it is difficult to produce well-documented guidance data. Earlier practice, however, was different, and one can instance the case of railway wagon and carriage axles, which for many decades had plain grease-lubricated journals. The practice with these (at least in Britain) was to true up the journals and re-metal the bearings at intervals of about 18 000 miles, which probably corresponded to a running life of about 500 h. If the pV value of 700 kW/m^2 adopted for figure 6.2 were to be adjusted for this application, it first ought to be halved to allow for grease lubrication instead of splash, and halved again to allow for the undesirably high b/d ratios of these old bearings – thus bringing it to 175 kW/m^2. If this value is regarded as applying to an anticipated life of 10 000 hours, and one goes by the analogy of dry rubbing bearings in which the anticipated life varies inversely with the pV factor, then an equivalent pV value for a 500 hour life would be 3500 kW/m^2. Remembering that the average speed of carriages and wagons was substantially less than their maximum, as also was their average load, one might perhaps raise this figure by, say, 25 per cent, thus concluding that if one were to be satisfied with a performance equal to that of these old axle-boxes, one could employ a pV factor of about 4375 kW/m^2. This lines up very well with the facts, which are that the nominal pV value on these carriage and wagon bearings was from 100 000 to 150 000 lbf/in^2 × ft/min. The mean of these figures is to be precise for once, equal to 4378.163 kW/m^2.

6.15   The snags with this high pV value, apart from the short life already mentioned, were that to make it practicable the bearings had to be re-loaded with grease every evening and, on anything other than short-run local services or slow goods trains, were checked by hand for bearing temperature at every stop – which in practice meant about once per hour. Even so, 'hot boxes' were not uncommon. This kind of 'nursing', unreliability and short life would not be acceptable in ordinary industrial usage today, so that a pV factor of over 100000 may now be regarded as quite impracticable.

6.16   The fact is that boundary lubrication is an inherently unscientific procedure that produces continuous wear. The appropriate loads and speeds depend greatly on how much wear and other trouble the user is prepared to accept as normal. This is obviously difficult to quantify, or to express in terms that would satisfactorily cover everything from the spindles of jig-borers to the axles of yard wheelbarrows.

6.17   It is hoped, however, that the figures and comments in paragraphs 6.06 to 6.16 may be of some help to the inexperienced designer in finding his way through this largely uncharted maze.

Dry-rubbing bearings

6.18   These are proprietary components, therefore, although figure 5.1 gives general guidance, it must be remembered that each manufacturer produces an individual range of bearings with its own set of characteristics. Thus, manufacturers and their technical catalogues should always be consulted before a final decision is taken on the dimensions, clearances, tolerances, surface finishes, etc., of any application in which it is proposed to use dry bearings.

6.19   The load and speed values shown in figure 5.1 correspond to those given by one leading manufacturer for a 10000 hour bearing life. These load values would be more than trebled (at the same rpm) for a bearing life of 3000 hours – which in many applications would probably be quite acceptable. It is certainly well above the total life-span, in actual running hours, of the average automobile.

6.20   It should be noted that the values in figure 5.1 relating to dry-rubbing bearings are for bearings in which the length is the same as the diameter (and *not* with the b/d ratio of 0.6 used for the oil-film bearings).

6.21   It should also be noted that, although with a constant b/d ratio the bearing area will go up as d^2, the load-carrying capacity of dry bearings at any given rpm rises only proportionately with d. This is because, in these bearings, the performance is measured in terms of pV and is therefore affected by the anomaly described in paragraph 6.13.

6.22   Quite apart from watching the pV rating to ensure an adequate bearing life, the designer must (particularly on slow-speed or intermittent applications) make sure that the load per unit area does not exceed that recommended by the manufacturers – usually about 28 MPa (290 kgf/cm^2 or 4000 lbf/in^2).

Oil-impregnated porous bearings

6.23   These again are proprietary components that should be used only in strict accordance with their maker's recommendations. These recommendations vary enormously, and each manufacturer can usually offer quite a range of different material specifications and porosities (each with its own load–speed characteristics) to suit different applications.

6.24   A simple pV factor tends to be meaningless with most types of porous bearing, because their effectiveness diminishes with any rise in bearing temperature. The allowable unit load thus tends to vary in inverse proportion to V^2. Under these conditions, the permissible load at any given rpm is independent of bearing size, because with bearings of constant width/diameter ratio both the bearing area and the permissible bearing pressure vary as the square of the shaft diameter, but in opposite directions, thus exactly cancelling each other. This explains why there is only one line for all sizes of porous bearings in figure 5.1, except at the lower rpm where the load limit is set by the fatigue strength of the material.

6.25   It will be found that the load—speed characteristics in figure 5.1 are fair average values for porous bearings, better than some specifications but not as good as others.

Graphite-impregnated bearings

6.26   Designed for use in a dry state, these show to best advantage in relatively lightly-loaded, high-temperature intermittent motion applications, or for work in which there is a danger of product contamination. Each maker produces a variety of grades with different supporting matrices (which may be metal, carbon, or metal–carbon mixtures) and in every case the manufacturers ought to be consulted for advice regarding the most suitable grade for any given application.

6.27   At normal temperatures, graphite-impregnated bearings tend to have a higher rate of wear than ptfe-impregnated structures. Although both types are shown to have the same load–speed characteristics on figure 5.1, this is only because in the case of graphite-impregnated bearings the characteristics refer to normal practice with such bearings, in that they relate to $b/d = 1.5$ and to a wear rate of about 50 μm (0.002 in) in 500 h. With intermittent motion, this may correspond to many years of life; for example, there are the instances of a chain grate stoker in which this type of bearing has worn by only 75 μm (0.003 in) in seven years, and of another in which similar bearings have run for seventeen years without replacement.

(ii) Bearing width

6.28   The width/diameter ratio of a bearing (the b/d ratio) tends to have a very marked effect on its performance. The effect varies from one type of bearing to another, and is also influenced by the type of load, namely, steady or fluctuating.

Hydrodynamic film-lubricated bearings

(See paragraphs 6.47 *et seq.* for notes on other types of bearing.)

6.29   The recommended 'norm' for these is $b/d = 0.6$, and it is on this basis that figures 5.1, 6.1 and 6.2 have been prepared.

6.30   Hydrodynamic bearings are, unfortunately, very sensitive indeed to changes in their b/d ratio, and it is therefore important to understand why 0.6 is regarded as a preferred value. A brief answer is that any lesser width tends to result in too great a loss of oil pressure from the ends of the bearing, while any greater width tends to create difficulties with alignment and edge loading.

6.31   These two conflicting limitations are worth examining in detail.

6.32   Because of the exposed ends of the clearance-space annulus, an oil film cannot have any pressure in it at the extreme edge of a bearing, and the distribution of pressure across a narrow bearing is consequently of the general form shown in figure 6.3. By reference to figure 6.4, it will be seen that an increase in bearing width not only enables a higher pressure to be sustained in the centre of a bearing, but also increases the proportion of the bearing width that is capable of carrying a significant load. In bearings with b/d ratios

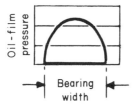

Figure 6.3 Oil-pressure distribution across narrow bearing (see paragraph 6.32)

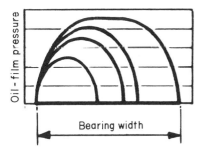

Figure 6.4 Change of oil-pressure distribution with variation in width of bearing (see paragraph 6.32)

between 0.5 and 1.5, and eccentricity ratios around 0.6, the theoretical load-carrying capacity varies roughly as $(b/d)^{1.5}$, so that doubling the width of a bearing of given diameter (allowing also for the resultant increase in area) would apparently allow it to carry more than five times as much load.

6.33   This theoretical advantage of wide bearings is, however, based on the assumption of a constant oil-film thickness across the whole bearing width but very minor misalignments or shaft deflections can badly upset this assumption and lead to most of the load being concentrated at one end. Indeed, the situation in a wide bearing can be far from happy even if one ignores all the normal bending and misalignment of the shaft, and considers only what happens within the bearing length itself.

6.34   Take as an example a 50 mm diameter bearing running at 1000 rev/min for which figure 6.1 indicates a maximum recommended load of about 1800 N (400 lbf) on a bearing 30 mm long. If the length of this were to be increased to 75 mm, giving it $b/d = 1.5$, its theoretical load-carrying capacity (allowing for both the increased area and improved b/d ratio) would be increased by a factor of about 10, bringing its apparent safe load up to 18 kN (4000 lbf). Reference to figure 6.5, however, will show that a uniformly spread

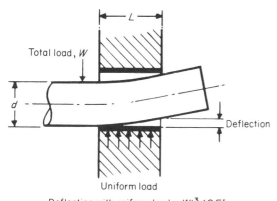

Deflection with uniform load = $WL^3 / 8EI$

d	L	W	E	Deflection
50mm	75mm	18kN	210GPa	0·0147mm
2in	3in	4000lbf	30×10^6 lbf/in²	0·000573 in

Figure 6.5 Deflection of journal within length of bearing (see paragraph 6.34)

load of this magnitude could not be carried in a cylindrical bearing, because it would bend the shaft by an amount much greater than the oil-film thickness in the loaded area. In other words, the uniformity of oil-film thickness needed to produce a uniformly spread load would be incompatible with the bent shape of the shaft that would arise from such a load – even assuming the external alignment to be perfect and the shaft beyond the bearing to be infinitely stiff.

6.35   The effect of any assembly misalignment or external shaft flexure will naturally increase directly in proportion to the bearing width, while the alignment could be further upset by the higher load (or hoped-for higher load) of a wide bearing increasing the structural deflection, as shown exaggerated in figure 6.6.

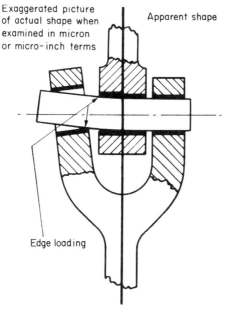

Exaggerated picture of actual shape when examined in micron or micro-inch terms

Apparent shape

Edge loading

Figure 6.6 Edge loading due to bending of shaft and deflection of structure (see paragraph 6.35)

6.36   All shaft or housing deflections affecting oil film thickness must be judged in relation to the μm or 'tenths-of-thou' dimensions of the oil films themselves. When examined on this scale, it will be found that all shafts and structures are remarkably flexible, as may have been gathered from figure 6.5.

6.37   With dynamic loads, the line of the shaft will be continually varying, and within each cycle may change from being straight to having the form shown exaggerated in figure 6.5. There is no simple way of coping with this, other than by the use of relatively narrow bearings, i.e. with a b/d ratio of certainly no more than 0.6. In dynamic applications where the problem is serious, it can sometimes be alleviated by adopting the principle of 'conjugate deflection', which means arranging for both the bearing support and the shaft to deflect along similar lines, thereby keeping the two substantially co-axial. One example of this principle applied to a marine crosshead bearing is shown

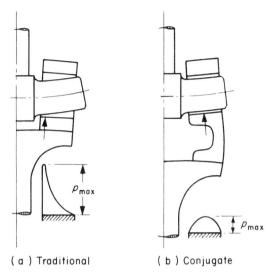

(a) Traditional   (b) Conjugate

Figure 6.7 Conjugate deflection (see paragraph 6.37) [After Hill, *Trans. Inst. Mar. Eng.*, vol. 88, 1976 (Discussion)]

in figure 6.7. Here, the journal is fixed, and the bearings are in the connecting rod top-end.

6.38   The disadvantages of wide bearings are, therefore, just as serious as the disadvantages of narrow bearings, and it is a question of deciding where the optimum solution lies.

6.39   Throughout the years, it has become increasingly realised that most bearings have been too wide, and the practice of the 1980s tends to regard a b/d ratio of about 0.6 as the optimum for normal industrial bearings. Considerably narrower bearings are often used quite successfully, even in highly-rated applications such as modern automobile big-ends, where lack of space would make $b/d = 0.6$ impracticable. (See also paragraph 6.70.)

6.40   The trend is well illustrated by Talbot/Hillman cars, whose 1937 Minx had a big-end with $b/d = 0.56$, whereas their 1963 Imp had a big-end with $b/d = 0.46$, and the 1982 Talbot Solara figure is 0.41. In the same high-rating class, there are the Perkins V8-510 big-end with $b/d = 0.386$ and the E-type Jaguar big-end with a ratio of 0.467.

6.41   Turning to industrial applications, the diagram of load–speed characteristics in ESDU 65007 (November 1965) is based on $b/d = 1$, but the later ESDU 66023 gives three worked examples, one of which results in a recommended b/d ratio of 0.35, whilst the other two give 0.5. Paragraphs 16.09 to 16.11 deal with another industrial example showing the same trend.

6.42   From a wide review of current practice and experience, it is believed that $b/d = 0.6$ is probably the best for the vast majority of applications. It is therefore recommended that, wherever possible, bearings for normal film-lubricated industrial usage should be designed with these proportions.

6.43   If it is necessary to use narrower bearings, then, for any given bearing diameter and speed shown in figures 6.1 and 6.2, the corresponding maximum load value should be reduced by multiplying it by a factor of $2.8(b/d)^2$. If bearings wider than $0.6d$ have to be employed, a corresponding correction

Film lubricated bearings		b/d	Dry rubbing bearings	
pV factor	Safe load		pV factor	Safe load
0.50	0.25	0.3	1.0	0.3
0.54	0.45	0.4	1.0	0.4
0.84	0.70	0.5	1.0	0.5
1.00	1.00	0.6	1.0	0.6
0.90	1.05	0.7	1.0	0.7
0.81	1.07	0.8	1.0	0.8
0.70	1.05	0.9	1.0	0.9
0.60	1.00	1.0	1.0	1.0
0.40	0.80	1.2	0.97	1.16
0.21	0.50	1.4	0.88	1.24
0.10	0.25	1.5	0.83	1.25
		1.6	0.77	1.24
		1.8	0.64	1.16
		2.0	0.50	1.00

Figure 6.8 Loads on bearings of non-standard widths (see paragraphs 6.43 and 6.48): the amounts by which to multiply recommended loads and *pV* factors of figures 5.1, 6.1 and 6.2 when applied to bearings of non-standard widths. The above tables are based on somewhat arbitrary assumptions regarding shaft and structural rigidity. They are thus no more than broad guidelines to the results it would be reasonable to expect in average circumstances.

factor of $[2.67b/d - 1.67(b/d)^2]$ is recommended. These correction factors have been worked out and tabulated in figure 6.8, in which it will be seen that the maximum recommended bearing *pressure* occurs with a *b/d* ratio of 0.6. Thus, increasing the length by one third, thereby bringing the *b/d* ratio up to 0.8, raises the theoretical load-carrying capacity by only 7 per cent – which is a poor recompense for all the edge-loading trouble that the longer bearing will be risking.

6.44   Although really a matter for discussion under oil grooves in chapter 13(ii), it may be noted that just about the worst conditions of bearing length arise in a bearing with a central circumferential oil groove. This is because, from the view-point of end leakage of oil pressure, a central groove converts any one bearing into two separate bearings, each of less than half the original width, and thus reduces the theoretical load-carrying capacity by a factor of about three. In addition, for any given bearing diameter and area, the length of the bearing is increased by the width of the oil groove, and the longer bearing has a correspondingly greater sensitivity to misalignment. One is therefore losing effectiveness on two fronts, and a central circumferential oil groove is consequently something to be strenuously avoided unless it is essential as a duct for the transfer of oil to other places – and even then it is often advisable to reduce the adverse effect by the means suggested in paragraph 13.12.

6.45   In the case of steady loads (arising from gravity or the like), it is possible to compensate in part for shaft deflections by mounting the bearings 'out-of-line', as described in paragraph 6.69. Under such conditions, a *b/d* ratio as high as unity may be quite satisfactory.

6.46   With purely centrifugal loads (i.e. of constant force but rotating with the shaft), conditions may be helped by the bearing wearing into a slightly conical form, as shown exaggerated in figure 6.9. Here again, therefore, a

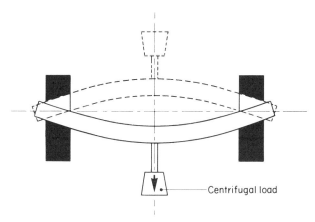

Figure 6.9 Bearing worn into conical form by effect of centrifugal load (see paragraph 6.46)

bearing slightly longer than $0.6d$ may be employed, although not without some reluctance.

Dry-rubbing bearings

6.47 Since these have no oil film to assist in spreading the load, they are more susceptible to edge loading than are hydrodynamic bearings. On the other hand, their loadings are normally very much lighter and therefore they rarely suffer from bending of the shaft within the bearing length. They also wear more during the bedding-in process than do hydrodynamic bearings, and thus tend to adjust themselves better to the required conditions of alignment. Furthermore, being subject to pV limitation, they are affected by the considerations described in paragraph 6.13 and thus gain nothing from an increase in diameter, thereby tending to make a high b/d ratio attractive. As a result, the 'norm' for these bearings is usually regarded as $b/d = 1$ and it is on this basis that the recommended loads and speeds of figure 5.1 have been chosen.

6.48 With wider bearings, there is always some uncertainty about whether the stiffness and alignment of the assembly will be sufficient to ensure an even spread of load. For this reason, the safe load for wider bearings should be regarded as the unity b/d safe load, multiplied by $[3b/d - (b/d)^2 - 1]$. This formula indicates (with some truth) that the maximum load capacity for dry rubbing bearings occurs with $b/d = 1.5$, and that any wider bearing will carry *less* load, not more. The use of dry bearings wider than $1.5d$ is therefore not recommended.

Oil-impregnated porous bearings

6.49 The recommended 'norm' for these is $b/d = 1$, and it is for these proportions that the load and speed relationships shown in figure 5.1 apply.

6.50 For porous bearings either narrower or wider than their diameter, the recommended loads should be adjusted exactly as described in paragraph 6.48 for dry rubbing bearings.

Carbon graphite bearuings

6.51   In the interests of reduced wear, $b/d = 1.5$ is regarded as normal for carbon bearings and even a value of 2 is not uncommon. A b/d ratio of less than unity can seldom be recommended for them when used in a dry condition, although lengths of only one quarter the diameter are often employed with wet bearings, i.e. those running in water.

(iii) Clearance of journal in bearing

Oil-film lubricated bearings (hydrodynamic and/or boundary)

6.52   If too large a clearance is provided around the journal of a film-lubricated bearing, its load-carrying capacity will be reduced, as explained in paragraph 3.11, but the effect of minor upward variations from the optimum value is seldom serious. On the other hand, too small a clearance can rapidly lead to overheating and failure. In any design, it is thus important to determine first a minimum acceptable clearance, and then to select dimensions and tolerances that will ensure that this minimum figure is always exceeded – aiming, if possible, for the mid-point of tolerance spread to coincide with the optimum clearance value.

6.53   For bearings that have to operate near to the limit of their capacity, it may be necessary to define both the maximum and minimum permissible clearances with considerable precision, but for many everyday slow-speed applications the old-fashioned workshop rule of one thou' clearance per inch of diameter (1 μm/mm) has much to commend it. Even for bearings operating within the load–speed characteristics of figures 6.1 and 6.2, the load-carrying capacity will not be greatly affected if the clearance is up to 100 per cent more than the minimum specified in paragraph 3.25, namely, $0.0009d + dn/5\,000\,000$. For lightly-loaded bearings, see paragraph 5.24.

6.54   For more heavily-loaded bearings, it may be necessary to operate nearer to the optimum clearance, i.e. rather less than that mentioned in paragraph 6.53, but the actual optimum value will depend on whether the load and speed are relatively steady, or are violently fluctuating as in a diesel engine. For steady loads, it will seldom be wise to use lower clearances than those indicated in figure 6.10 (under the worst conditions of tolerance summation).

6.55   For diesel engine bearings and the like, an equivalent set of values is plotted on figure 6.11, which is based on the formula shown thereon.

6.56   Clearances of less than 25 μm (0.001 in) ought never to be used unless unusual care is taken and particularly close limits are set on journal circularity and other tolerances. Even then they tend to be rather too impractical to recommend. The very finest Swiss watches usually have far greater bearing clearances than this.

6.57   The slightly lower minimum clearances acceptable for the diesel bearings are explained partly by the fact that their average load is well below the peak value, so that for the same nominal speed and load their pV factor is lower. They also benefit from reverse loads and the consequent squeeze effect described in paragraphs 12.17b and c.

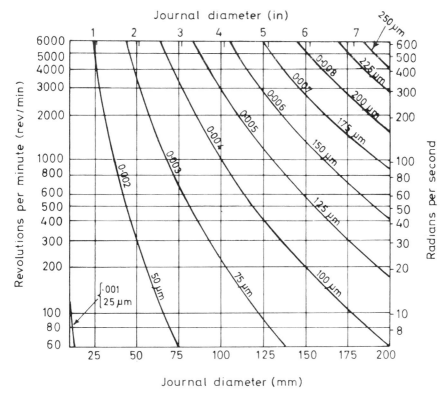

Figure 6.10 Minimum diametral clearance advisable for steadily loaded
bearings (see paragraph 6.54)

Dry-rubbing bearings

6.58 The required running clearance varies with the type of dry-rubbing
material, and must therefore be taken from the manufacturer's instructions. It
will usually be relatively high in comparison with other types of bearing, and
may be of the order of $0.005d$.

Pre-lubricated bearings

6.59 In view of the variety of porosities and material compositions in which
this type of bearing is available, it is always advisable to obtain guidance on the
optimum clearance and its appropriate tolerances from the supplier's
literature.

6.60   BS 1131, Part 5, recommends clearances corresponding to H8/f7 fits
for oil-impregnated bearings, but other authorities recommend slightly larger
clearances for higher speeds. One maker has a chart of recommended mean
clearances (i.e. the mid-point of the tolerance spread) which corresponds
closely to the value

$$K \times 7\sqrt{d^6} \times 12\sqrt{n}$$

where $K = 1/750$ (mm and rev/min), or $1/1200$ (in and rev/min).

6.61   Incidentally, the correct clearance in bushes or bearings of this type should never be obtained by machining their bore, since this may seriously affect their performance. The correct bore size should be obtained by the use of an appropriate size of fitting pin, as described in paragraph 10.18.

Graphite-impregnated bearings

6.62   Since these are made from any one of a wide range of proprietary materials, it is essential to obtain guidance on clearances from the suppliers. In general, it may be found that optimum clearances are of the order of D9/f8 fits for dry applications, and E9/f8 for lubricated applications.

General

6.63   It has to be remembered that the required clearance must be obtained under working conditions, so that if the shaft and housing are of materials having different coefficients of thermal expansion, an appropriate allowance must be made for their dimensional changes between static and running temperatures.

6.64   This differential thermal expansion will tend to be more pronounced immediately after starting, because a journal nearly always warms up more rapidly than the surrounding housing. Several cases have arisen where this has given trouble even when the clearance under the final running conditions would have been ideal. This particular trouble is not very common, but it should nevertheless always be borne in mind at the design stage.

6.65   To summarise: for film-lubricated bearings having the loads, speeds and sizes specified in figure 6.1 or 6.2, a good working rule for the minimum running clearance is

$$C_d = \left(0.0009 + \frac{\text{rev/min}}{5\,000\,000}\right)d$$

This, of course, assumes that the surface finishes, manufacturing tolerances, alignments, lubricating oil supplies, etc. are all generally in accordance with the appropriate recommendations. Other conditions and/or other types of bearing need special consideration, as discussed in the preceding paragraphs.

(iv) Accuracy of alignment

6.66   For film-lubricated bearings in the range of loads, speeds and sizes dealt with in figures 6.1 and 6.2, a total misalignment of 1 part in 5000 (i.e. a shaft slope through the bearing of 0.0002 rad) is about the maximum that can be accepted without undue risk of trouble from edge loading. This figure refers to the sum of the misalignments arising from shaft and/or housing deflections under load, and those arising during assembly or as a result of subsequent wear. (For an explanation of edge loading, see figure 6.6.)

6.67   For misalignments *greater than 0.0002 rad*, the load-carrying capacity of a bearing will be reduced, and for all sizes of bearing the load figures shown in figures 6.1 and 6.2 should be divided by a factor of

$$\frac{0.0008}{(0.001 - m)}$$

where m is the actual misalignment in radians.

6.68   When thinking of shaft deflections on the scale of oil-film thicknesses (tenths of thousandths of an inch or hundredths of a millimetre), it will be found that all shafts are remarkably flexible – as already commented upon in paragraph 6.36 – and if one is working anywhere near to the maximum load values of figure 6.1 or 6.2, it is advisable to calculate the deflection in every case as a precaution.

6.69   When the deflection is due to gravity, and is thus always in the same direction, it can be overcome by mounting the bearings parallel to the local shaft centre-line instead of truly in line with each other. This is illustrated in figure 6.12, which is not so exaggerated as might be supposed. With large turbo-alternators, there is often a sag of several centimetres in the natural lie of a long in-line assembly of several turbines (hp, ip, dual-flow multiple lp, etc.) and their associated alternator. The arrangement in figure 6.12 is, broadly speaking, adopted in most British turbo-alternators, although there have been considerable changes in detail throughout the years. Comments on these changes, together with comments on an alternative approach to turbo-alternator alignment favoured in the USA, will be found in example 2 of chapter 16.

6.70   In engine bearings, cyclic variations in load which recur every 360° or

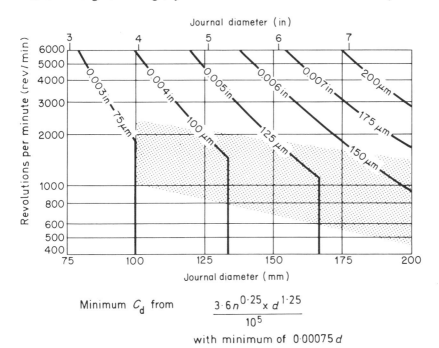

Minimum C_d from

$$\frac{3 \cdot 6 n^{0 \cdot 25} \times d^{1 \cdot 25}}{10^5}$$

with minimum of $0 \cdot 00075 d$

Figure 6.11 The shaded area shows the range in which most medium-speed diesel engines operate [After Warriner, 'Thin-shell bearings for medium-speed diesel engines', *Proc. Diesel Eng. Users' Ass.*, 1974] (See paragraph 6.55.)

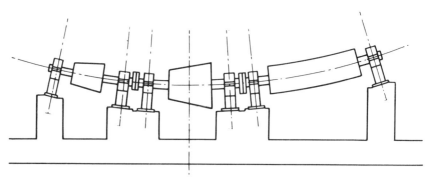

Figure 6.12 Exaggerated diagram of turbo-alternator assembly having bearings aligned with local centre-line of shafting (see paragraphs 6.69 and 16.05)

720° naturally change the shaft deflection and the bearing alignment throughout each cycle, but the harmful effect of this (in so far as edge loading is concerned) is largely offset by the 'squeeze' lubrication effect of the relatively thick oil film that tends to build up when the deflection is, as it were, 'the other way' elsewhere in the cycle.

6.71   Dry-rubbing bearings and oil-impregnated porous bearings will also give trouble if badly aligned with their journals, but their tolerance of minor misalignments is greater than that of oil-film bearings and, in general, one does not have to go to extreme lengths to avoid difficulty. An alignment accuracy of 1 part in 5000 ought, however, still to be the aim. If alignment to this standard presents problems, then there is at least one range of proprietary self-aligning dry bearings available in sizes up to 100 mm (4 in) diameter.

6.72   In addition to misalignment between a journal and its bearing, there may be equally undesirable mismatching or misalignments between parts of a bearing itself. Figures 6.13, 6.14 and 6.15 (taken from *Steam, diesel and electric traction engineer*, 1979) show examples of this. There is not much a designer can do to avoid these particular faults, but it can be useful for him to bear them

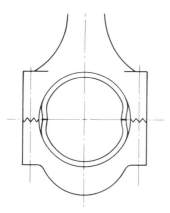

Figure 6.13 Excessive crush [Reproduced by permission of The Glacier Metal Co. Ltd]

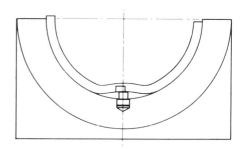

Figure 6.14 Fouled dowel [Reproduced by permission of The Glacier Metal Co. Ltd]

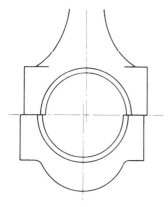

Figure 6.15 Offset cap [Reproduced by permission of The Glacier Metal Co. Ltd]

in mind if one of his bearings runs into trouble, and to check them out before trying to improve matters by changes to his design.

(v) Manufacturing tolerances

6.73   Information on manufacturing tolerances will be found in the paragraphs indicated below

Journal diameter	3.23, 3.29, 3.30
Bearing bores	3.25, 3.30, 6.52, 14.8 to 14.14
Bearing width	11.04
Bush bores and o.ds	10.08 to 10.14
Shell diameters	8.01
Housing bores	6.52, 8.01, 10.08 to 10.14
Cap and keep bores	9.03
Misalignment	6.66 *et seq.*
Thrust washers	11.04 and figure 11.1
Wall thickness	6.52, 8.01

The effect of manufacturing tolerances on bearing performance is dealt with in paragraphs 3.29, 3.30, and 8.13.

6.74   It is important to note that the tolerance deviation from the nominal value will seldom be the same on all the main-bearing journals (or on all the big-end journals) of an engine, and that the individual bearing shells may also differ from each other tolerance-wise. When an engine (or any machine) is dismantled for repair or maintenance, it is thus *essential* to identify each bearing in turn as it is removed, and to *make certain* that all are re-assembled in the same positions as they originally occupied (and the same way round if they are symmetrical), otherwise the re-assembled machine will be highly liable to serious bearing trouble if bearings that have become nicely 'run in' against a particular journal have to run under a new set of conditions in a situation where variations of a fraction of a micro-inch can be significant. To those unaccustomed to this type of work, it is prudent to issue a warning against identifying the shells with chalk marks or paint marks which, if not thoroughly cleaned off, could come between the shell and its housing, thereby perhaps upsetting the running conditions even more than they might have been by random assembly.

Chapter 7

Materials and finishes

(i) Journal materials

7.01 For recommended journal hardnesses, see paragraphs 7.02 to 7.09.
For recommended surface finishes, see paragraphs 7.49 to 7.53.
For recommended dimensional tolerances, see paragraph 6.73.

7.02 Plain mild steel journals will usually be found perfectly adequate for use with whitemetal linings under the load and speed conditions corresponding to the values in figures 5.1 and 5.2.

7.03 As loads and/or speeds are increased, however, the maintenance of a truly circular journal will become more important. Yet the rate of journal wear will obviously be greater, so that, to achieve a satisfactory longevity in the bearing, it soon becomes necessary to adopt some form of surface hardening for the journal – although a mild or low-alloy steel will still be preferable for the basic material. Care must, however, be taken lest the relative brittleness of the hard surface leads to its cracking and breaking off under the high-strain conditions that may obtain on the surface of heavily-loaded journals. (See paragraph 3.38e and figure 6.4.)

7.04 If the peak loads are such as to call for the use of copper–lead or other stronger linings, then a harder journal becomes almost mandatory – even if the 'equivalent steady load' is well within the limits of figure 5.1 or 5.2. If a suitable surface cannot be obtained by case hardening, then some harder shaft material may be necessary, but care has to be exercised here because of the danger of what is known as a 'wire-wool' failure. There is some disagreement about the precise cause of such failures, the symptoms of which are that the journal (or thrust face) appears as though it had been machined with a very fine, sharp tool, producing swarf that resembles wire-wool.

7.05 These failures are mostly (but not always) associated with chrome steel journals which have a relatively high chromium content. They frequently occur within a few seconds of the plant being started up in service for the first time, but they do sometimes occur only after long periods of satisfactory service.

7.06 The usual mechanism of such failures is thought to be that a contaminant steel particle bridges the oil film and becomes partially embedded in the bearing lining. Because of its rubbing against the journal the exposed tip of the particle may reach a high temperature causing carburisation in the presence of the oil. This produces a hard carbide tip that transforms the steel particle into a highly-effective cutting tool that can machine away the journal surface to a catastrophic extent in seconds rather than minutes.

7.07   Wire-wool failures have occurred with chromium contents as low as 2 per cent, and appear to depend to some extent on the particular combination of chromium content, oil specification and/or e.p additive employed. No reliable guidance can be given on how to avoid these troubles, other than to be very chary about using any combinations of chromium steel and oil specifications that are not backed by direct experience on which the designer is fully aware of all the details.

7.08   A 'last ditch' expedient in difficult cases may be to fit the journal with a mild steel sleeve, that if it wears can be renewed at much less cost than would be involved in replacing the whole component of which the journal forms a part.

7.09   Some grades of cast iron make excellent journals, but suitable applications for these tend to be too specialised to deal with here, except to note that at least one car manufacturer of world-wide reputation has used cast-iron crankshafts for many years.

(ii) Types and grades of lining materials

General

7.10   For applications where the equivalent steady load conditions are within the limits of figures 5.1, 6.1 or 6.2, it may often be quite practicable, and perhaps very convenient, to dispense with a bearing lining and simply run the shaft direct in the normal material of the stationary part. This applies particularly if the part concerned is of brass, Duralumin, or some polymer (or even, occasionally when of steel – see paragraphs 12.01 to 12.03).

7.11   The performance of an unlined bearing will obviously be precisely the same as if the bearing were lined with the same material, but a disadvantage is that the bearing becomes more expensive to replace if damaged or worn. In general, this form of construction is used only for lightly-loaded bearings subject to somewhat intermittent operation in relatively low-cost applications, such as domestic appliances and the like.

7.12   The various types of true 'lining' material have been divided into five groups in the following paragraphs, with cross-reference in the case of those materials which fall into more than one group. The groups are: oil-film bearings; dry-rubbing bearings; oil-impregnated porous bearings; water-lubricated bearings; and bearings for high-temperature applications.

Oil-film bearings

7.13   The first choice here is almost inevitably whitemetal, unless its use is ruled out by considerations of high peak loads or high ambient temperature, in which case the alternative materials in descending order of preference are as table 7.1.

7.14   Everyone may not agree with the order of preference indicated in table 7.1, but it is based on the fact that the softer bearing materials are much more tolerant of misalignment, dirty oil, and general rough usage, than are the harder materials. It is thus best to employ the softest material that it is safe to use under the prevailing conditions of peak load, temperature, or other limiting factors.

TABLE 7.1.

Material	Recommendations					
	Maximum peak load			Minimum shaft hardness (HV_{30})		Characteristics (paragraph numbers)
	MPa	kg cm²	lbf/in²	plain	o'lay	
White metal	7	70	1000	180	—	7:16 to 7:21
40% SnAl	10	100	1500	180	—	7:27 to 7:29
20% SnAl	18	180	2500	250	250	7:27 to 7:29
6% SnAl	19	190	2700	300	250	7:27 to 7:29
CuPb	19	190	2700	300	250	7:22
Pb bronze	20	200	3000	300	250	7:22
AlSi	20	200	3000	300	250	7:30

7.15   The maximum recommended peak loads in table 7.1 refer to bearings carrying constant loads at constant speeds. For such, they represent good working limits for everyday use, and ought not to be exceeded without arranging for development testing to prove (or disprove) the practicability of higher values for any particular application under consideration. For the higher loads that can be carried satisfactorily under variable load–speed conditions (as in i.c engines, for example), see paragraph 2.03 together with sections 4(ii) to 4(iv), and especially paragraphs 3.36 to 3.40 inclusive.

Whitemetal

7.16   Suitable specifications for whitemetal are

For thin-walled bearings
BS3332/1 (Sn89, Sb7.5, Cu3.5) or
SAE13 (Pb83.25, Sb7.5, Sn6, Cu0.5, As0.25)

For thick-walled bearings
BS3332/2 (Sn87, Sb9, Cu4) or
BS3332/7 (Pb74, Sn12, Sb13, Cu1)

7.17   The advantages of whitemetal as a lining material are

(a) When properly applied, it has excellent adhesion to steel, cast-iron and bronze.
(b) It 'wets' well with all ordinary oils, and thus tends to remain oily after standing for some time.
(c) Its embeddability is better than that of any other present-day lining material, which means that foreign matter readily embeds or sinks into it, especially under the near metal-to-metal contact conditions at starting, instead of remaining proud and scoring the journal.
(d) It is better than harder materials at withstanding lubrication starvation at starting, or when subjected to momentary overloads.
(e) It is better than harder materials at tolerating slight journal misalignment or deflection under load.
(f) It is not corroded by the acid introduced into the lubricants of i.c engines by combustion of the sulphur content in the fuel.

(g) Thick-walled bearings lined with whitemetal can be relined when worn or damaged.

(h) In the case of thick-walled bearings, it can be 'scraped' to adjust for slight misalignment or the like.

7.18   A disadvantage of whitemetal is that not only is its compressive fatigue strength low but that this falls off rapidly with rising temperature as shown in figure 7.1. For example it decreases by about 6 per cent between

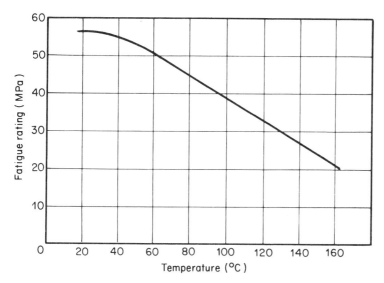

Figure 7.1 Reduction in fatigue strength with rising temperature of a typical thin whitemetal lining on steel backing (see paragraph 7.18) [Reproduced by permission of The Glacier Metal Co. Ltd]

winter and summer air temperatures (0 and 40°C, 32 and 100°F), and it falls even more steeply as the temperature increases further. Ultra-thin whitemetal linings, 120 μm (0.005 in) instead of 800 μm (0.03 in), can be up to 40 per cent stronger, but obviously have less 'wear' in them and are less tolerant of dirt and/or misalignment.

7.19   Tin-based whitemetal has always been regarded as superior to lead-based, and has traditionally been used as standard in Britain for over a hundred years, although engineers in the USA and Germany have been equally consistent in their use of lead-based compositions – probably because tin is regarded as a strategic material in these countries. Also, since the automobile industry is something of a trend-setter in these matters, it may be of significance that American automobile engines tended to be relatively large and 'woolly' in comparison with the more compact, high-revving and 'hotted-up' engines typical of British cars.

7.20   In the last decade or so, however, the price of tin has soared so drastically that lead-based whitemetal is now being much more widely used and it is being discovered that in practice its alleged inferiority to its tin-based rival is scarcely noticeable. The automobile industry is now mostly using aluminium alloy or copper–lead linings, so that its influence on whitemetal specifications is much less than formerly.

7.21   The importance of all this is that, although most British textbooks imply that only tin-based whitemetal is of any use, a designer need not be afraid of using a good lead-based equivalent – and in the normal course of things need introduce no special allowances for this change.

Copper–lead and lead bronze

7.22   In copper–lead (Cu 70, Pb 30) and lead bronze (Cu 74, Pb 22, Sn 4), the tin (if any) combines with the copper to form bronze, but the lead remains in a free state dispersed in the copper or bronze matrix. This free lead provides the bearing properties, while the copper-based matrix supplies the strength. These materials are suitable for peak loads up to the values shown in table 7.1, but are otherwise less attractive than whitemetal because

(a)   They are suitable only for steel shells.
(b)   They require a harder journal (see paragraph 7.13).
(c)   They are less tolerant of misalignment.
(d)   They are less resistant to attack from the acids in engine oils (see paragraph 7.23).
(e)   They are less tolerant of dirty oil.
(f)   They are less tolerant of momentary overloads.
(g)   They are more expensive.

Lead–tin or lead–indium plating

7.23   The copper-based bearing materials (paragraph 7.22) should, wherever possible, be used with an overlay plating of either lead–tin or lead–indium (say, Pb 89, Sn 11) or lead–indium (say, Pb 93, In 7) of from 0.025 to 0.040 mm (0.001–0.0015 in) thick. This makes little difference to the safe peak load capacity but permits the use of somewhat softer journals (500 HV_{30} minimum unplated or 230 HV_{30} when plated), and makes the bearing slightly more tolerant of dirty oil. Copper-based bearing material in i.c engines must always be plated, otherwise the acids in the lubricant (see paragraph 7.17f) will strongly attack the lead phase in the material. The tin content of the overlay inhibits the lead in it against similar attack, whereas any tin added to copper–lead or lead bronze would combine preferentially with the copper, and leave the lead unprotected.

7.24   The chief disadvantage of overlay plating is that the bearing cannot be scraped or bored out in place, but must be fitted pre-finished.

7.25   For many years, overlay-plated copper–lead was the most widely used lining material for heavy-duty bearings, but it is now being supplanted by aluminium alloys.

Lead–tin or lead–indium flashing

7.26   In applications other than i.c engines, there is not the same need to protect the free lead in copper–lead or bronze–lead bearings, and in such cases, a light 'flash' (about 5 μm or 0.0002 in thick) of lead–tin or lead–indium (of the same compositions as in paragraph 7.23) is sometimes applied to copper-based bearing materials to facilitate the 'running-in' process. This flash soon wears away after it has served its purpose, and the characteristics of the bearing are then exactly the same as those for the unflashed material – except, of course, that the helpful running-in conditions will probably have improved the surface finish of the journal. (See paragraph 3.32.)

Aluminium alloys

7.27   The normal aluminium alloys used for machinery construction (such as Duralumin) do not make particularly good bearing materials, and in general are now used only for relatively lightly-loaded applications. The alloys suitable for bearings have been developed specially for this purpose and consist essentially of aluminium and tin. The tin does not combine with the aluminium but, in effect, forms a 'mechanical' mixture of lumps of tin held in a matrix of aluminium. This free tin provides the 'bearing properties' of the material – exactly like the lead in copper–lead or lead–bronze alloys. The aluminium matrix is, however, much softer than bronze and can thus absorb foreign matter to a greater extent than copper–lead or lead–bronze alloys of comparable strength. It can also be run successfully with softer shafts than the copper-based alloys.

7.28   Another advantage over copper-based alloys is that both the free tin and the aluminium matrix are highly resistant to most of the acidic and other impurities that form rapidly in i.c engine lubricants. A further advantage of aluminium is that it is cheaper and more plentiful than copper.

7.29   Aluminium–tin mixtures have, unfortunately, a relatively high coefficient of thermal expansion and are thus awkward to use in the solid state if held in ferrous housings. The aluminium matrix in the as-cast condition also tends to be rather weak if the tin content is much in excess of 5 per cent. These disadvantages have, however, been overcome in recent years by the introduction of cold-working and heat-treatment processes that produce a reticular or net-like aluminium structure of relatively high strength, in which both the aluminium and tin phases are continuous, and also by the introduction of economic methods of bonding these alloys to steel backing strips for the production of thin-walled bearings.

7.30   The strongest aluminium-based bearing material available at present is aluminium–silicon (Al 99, Si 11, Cu 1). This, like the others, is available in steel-backed thin-walled form with a reticular matrix structure.

7.31   Since aluminium–tin bearings are immune to the corrosion described in paragraph 7.17, they do not need a protective overlay. It is, nevertheless, normal practice to provide them with an ultra-thin overlay (13–20 μm, 0.0005–0.0008 in) of lead–tin or lead–indium to assist in the initial running in. (See paragraph 3.32.)

Heavily-loaded bearings

7.32   If the equivalent steady-load and uniform-speed conditions are above the upper limits of figure 6.1 or 6.2, then the whole design falls outside the strict scope of this book, and becomes a project better handled by an experienced team rather than by someone seeking answers in a book.

Nevertheless, anyone engaged on the design of heavily-loaded bearings for the first time may find it useful to read paragraphs 7.13 to 7.31 on the higher-fatigue-strength materials, and to consider them in conjunction with the comments in paragraphs 3.35 to 3.40, as well as those in chapter 15.

Dry-rubbing bearings

7.33   Suitable lining materials for these are made by several specialist suppliers, a comparison of whose products would be out of place here. Very broad guidance on loads and speeds is given in figure 5.1, and further broad guidance will be found in BS 4480, but in any final design it is the recommendations and instructions of the specialist supplier that ought to be followed.

7.34   Apart from the plastics bearings mentioned in paragraph 7.10. The 'active' part of most dry bearing materials is either ptfe (polytetrafluoro-ethylene) or graphite, either of which can act as a dry, solid lubricant but lack the necessary strength to provide a useful bearing surface. They are thus usually employed as the 'filling' for a matrix of some stronger material, such as carbon or bronze produced in a porous form. The graphite or ptfe oozes slowly in minute quantities from the matrix during the operation of the bearing. This occurs partly because the frictional heat causes the filling to expand, and partly because more and more of the filling becomes exposed as a result of surface wear.

7.35   Since there is no flow of lubricant to carry away the frictional heat, the latter has to be dissipated by conduction through the housing and adjacent parts. The performance of the bearing is thus limited by the maximum temperature it is safe to have in the bearing material, and in a continuously running bearing, this is related to the pV value (bearing pressure × rubbing speed – see paragraphs 6.6 to 6.8), while the 'life' of the bearing can best be expressed as a given number of pV hours. In other words the life is inversely proportional to the conditions of usage. The maximum peak load is, of course, determined by the nature of the supporting matrix.

7.36   If the operation is intermittent, with very short periods of use followed by long idle periods, then the bearing temperature will, of course, not reach that corresponding to steady running and the use of a higher pV value may be quite acceptable.

7.37   Since carbon and graphite retain their strength and general charac-teristics at much higher temperatures than bronze or ptfe, the effect is that carbon-based graphitic materials will withstand higher pV values than will those with metal matrices, although the use of a carbon matrix will do nothing to reduce the wear associated with high pV values. Indeed, at normal temperatures, bronze may be expected to wear less than carbon.

7.38   In general, therefore, carbon-based materials are particularly suitable for highly-rated bearings subject to intermittent use, where relatively few hours of actual *wear* may correspond to many years of service. They are also very suitable for use in conditions of high ambient temperature and/or conditions subject to chemical attack. (See also paragraph 7.43.)

7.39   For most normal dry-bearing applications not troubled with unusual temperature or corrosive influences, it is difficult to do better than to use one of the ptfe-in-bronze materials that are available in the form of wrapped bushes over a wide range of sizes as well as in strip form for use on flat surfaces.

7.40   In addition to the wholly carbon and wholly bronze matrices, there is also a wide range of materials (mostly with graphite filling) in which the matrix is formed from various carbon–metal mixtures. The metals used include copper, lead–bronze, iron, nickel, and iron–nickel alloys, each of which are available in a range of different porosities and graphite contents. With any given material as a matrix, the mechanical properties are very dependent on the graphite content. As the graphite is reduced, the material becomes harder, more ductile and stronger in tension, but the friction and wear characteristics become less favourable. The nature of the application dictates the graphite content at which the optimum properties are obtained; the general rule is to use the maximum graphite content compatible with adequate mechanical strength.

7.41   It is normal practice for bearings of carbon-based materials to be specially made to order (often by pressing techniques) and supplied fully machined ready for fitting. This is economic for large quantities but not for one-off jobs. Many grades of metal–carbon graphited materials are, however, available in the form of standard blanks of ring-like, cylindrical or rectangular shape suitable for machining to form bushes. Some can also be obtained as metal-backed flat strip for straight-line sliding applications.

Oil-impregnated porous bearings

7.42   These have no linings as such, but are made throughout from the same material, which is usually a sintered bronze, manufactured in a range of porosities from about 3 to 30 per cent, the higher porosities tending to have lower friction and less wear, whereas the denser material has the greater strength. Since these are proprietary products, only the manufacturers can supply accurate data on characteristics and performance of their various grades, but general guidance will be found in paragraphs 5.11d and 6.23, as well as in figure 5.1.

Water-lubricated bearings

7.43   Almost any dry-rubbing bearing material will operate extremely well with water lubrication provided it does not have any iron in its composition, and provided the journal is of stainless steel or other suitable non-rusting substance. In some applications, there may be process fluids in or around the bearing location that can usefully be employed as the lubricant – always provided, of course, that the journal and bearing materials are selected with care to avoid any chemical or electrolytic corrosion by the process fluid. The water, or other fluid, can reduce the friction and mechanical wear, as well as help to dissipate the frictional heat.

7.44   Because of the difference in viscosity between lubricating oil and water, the load-carrying capacity ought theoretically to be reduced by a factor of about ten as compared with normal oil-film operation. In fact, however, the ability of dry-bearing material to approach much more closely to boundary lubrication conditions without risk of seizure enables proportionately higher loads and speeds to be used in safety. Indeed, there have been successful applications in which the loads and speeds have been virtually identical with those shown in figure 6.1. Such results are, however, unusual and can be achieved only after development testing has removed the 'bugs'.

7.45   For off-the-board designs, it would be unwise to reckon on water lubrication increasing the performance by more than 100 per cent above the 'dry' values for the same materials and/or approaching 50 per cent of the equivalent oil-lubricated values, whichever may be the lower.

7.46   An important point to note with most of these dry-bearing materials is that, although they operate well with a copious supply of water, they can soon give trouble if allowed to dry out, even when stationary. The transition stage from wet to dry running will generally create temporary high friction and undesirable wear. Continuous immersion in the water is therefore essential.

Bearings for high-temperature applications

7.47   For bearing surface temperatures above about 200°C (400°F), the material to use is normally one or other of the graphite-impregnated sintered types described in paragraphs 7.34 to 7.41. Almost any of these are suitable for use up to about 350°C (650°F), although most suppliers can offer special grades capable of operating satisfactorily at up to 500°C or even 600°C (abt. 1000°F to 1110°F).

7.48   Even with the best materials, the wear rate at these elevated temperatures tends to be rather high, and any mechanism which has to operate under these conditions should be so designed that its satisfactory performance is not too dependent on the maintenance of a good fit between any bearing and its journal. Care must also be taken that differential thermal expansion between different materials used in the construction of the mechanism does not either crush the bearing, loosen it in its housing, or cause it to grip the shaft.

(iii) Surface finishes of journals and bearings

7.49   To comply with the load–speed characteristics of figures 6.1 and 6.2, the surfaces of journals and bearings should always be finished to something better than 0.4 μm Ra (16 μin per BS 1134).

7.50   If the final finish is obtained by grinding, then with some materials such as cast-iron, it is important that the direction of motion of the grinding wheel relative to the surface should be the same as the direction in which the opposing part will ultimately slide against the same surface. With any material on which the direction of grinding is readily detectable, a bearing that has to run 'against the grain' of the grinding will be liable to give trouble. With such materials, the finish should be around 0.2 to 0.25 μm (8–10 μin) if the journal has to carry load in either direction of rotation.

7.51   An improved finish ought also to be given to any bearing that has to work close to the limit of its capacity. Thus, automobile engine journals are commonly given a 0.2–0.25 μm Ra (8–10 μin) finish, and automobile gudgeon pins 0.1–0.15 μm Ra (4–6 μin).

7.52   With dry-rubbing bearings, a 0.4 μm Ra (16 μin) finish on the journals is necessary to achieve the rated life, but there is seldom anything to be gained by using a finer finish than this.

7.53   With oil-impregnated porous bearings, a 0.4 μm Ra (16 μin) finish is also desirable, although some suppliers specify 0.8 μm Ra (32 μin) as acceptable.

Chapter 8

Bearing shells

(i) Thin-walled bearings

8.01   These have a wall thickness of the order of $d/30$, or less, with a minimum of about 1.5 mm ($\frac{1}{16}$ in), of which about 0.12 mm (0.005 in) is whitemetal or other lining material, while the remainder is steel. Thin-walled bearings are essentially flexible and in use are clamped firmly in their housings, the shape of which maintains their circularity, and the size of which determines the finished bore of the assembly. The housings must be bored with accuracy, the usual limits being size to minus 13 μm (–0.0005 in) in the smaller sizes, and size to minus 25 μm (–0.001 in) in bores above 125 mm (5 in). The shells themselves are usually made with a wall-thickness tolerance of about $+7$ μm (0.00025 in) in the smaller sizes, and $+13$ μm (0.0005 in) in sizes above 125 mm (5 in).

8.02   Thin-walled bearings were first introduced to meet the demands for high load-carrying capacity and light weight in automobile engines, but their success has led to their wider adoption, and now they are regarded as the preferred type for nearly all engine applications. Their chief disadvantage is that the tooling and set-up required to produce them are expensive and can seldom be justified for order batches of less than about 10000 half-bearings of small size or 500 of the larger sizes up to, say, 300 mm (12 in) diameter – which probably represents an order size of between about a quarter and half a ton of bearings in each case.

8.03   For applications where the volume of demand makes ordering on this scale appropriate, the basic advantages of thin-walled bearings may be summarised as:

(a) They are made with high precision, and consequently tend to have a higher load-carrying capacity than other types.

(b) They employ a slightly smaller bore housing and bearing cap than thicker bearings, which can lead to marginal savings in the weight and cost of the complete machine, which can be important for volume production.

(c) When they become worn, they can be cheaply and quickly replaced, whereas normal bearings are time-consuming to re-metal and re-bore.

(d) Compared with medium-walled shells, they do not require such high clamping forces to produce accurate conformity with the housing bore, and they can thus be used with smaller bolts.

(e) Their construction makes available a wider range of bearing materials.

(f) Replacements may be held in a variety of standard metric or inch undersizes, such as 0.25, 0.50 and 0.75 mm (0.010, 0.020 and 0.030 in), so that worn shafts may be retained in service by grinding the journal down to an appropriate undersize.

(g) Their method of production ensures a more consistent quality of material and strength of bonding between lining and shell than can be obtained with individually-produced bearings.

(h) They are extremely cheap when manufactured in large quantities.

8.04   If it is particularly desired to use thin-walled bearings in a new product for which the volume of demand is too small to justify a mass-produced bearing, then it will sometimes be possible, provided the bearing is less than 75 mm (3 in) or so in diameter, to design the product to incorporate standard thin-walled automobile or truck-engine bearings that are readily available in sets of four or six as replacements – and are probably stocked all over the world. Care must, of course, be taken to see that the standard oil holes and oil grooves (if any) in such bearings are appropriate to the new application. Care must also be taken not to choose bearings from an engine that is soon to be withdrawn from production, otherwise the bearings will have to be bought and stocked in sufficient quantity to meet all foreseeable demands for both production and servicing. Guidance on these points may be obtained from automobile replacement-bearing manufacturers.

8.05   It is vital to have the correct interference fit between the housing and shell of a thin-walled bearing. But since such bearings must always be obtained from specialist suppliers, it is advisable to consult the actual makers for detail guidance on fitting procedures, and to accept their recommendations in respect of 'free spread' – the amount by which the dimension d (figure 8.1) of each half-bearing *prior* to assembly is greater than its intended value *after* assembly. The 'spring' effect provided by this free spread acts to hold each half bearing in place during the assembly process.

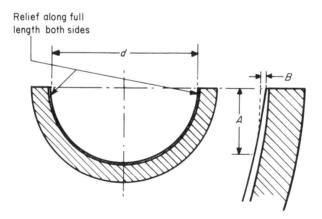

$A = 1.2$ mm $(0.05$ in$) + d/10$

Up to 150 mm (6 in) bore, $B = 0.012$ to 0.025 mm ($0.0005 - 0.0001$ in)
Over 150 mm (6 in) bore, $B = 0.025$ to 0.050 mm ($0.001 - 0.002$ in)

Figure 8.1 Bore relief (see paragraphs 8.05, 8.11 and 8.12)

(ii) Thick-walled bearings

8.06   The only real alternative to the thin-walled bearing is the thick-walled bearing, which may be defined as one with a wall thickness in excess of $d/10$, and preferably nearer $d/8$. Even a shell of this thickness could, of course, be distorted if it were heavily tightened into a mis-shapen housing, but in all ordinary circumstances it would maintain its original free shape and size to within acceptable limits. The function of the housing is then merely to support the load and to locate the shell in the machine. The shell may be of steel, cast-iron, or bronze.

8.07   Thick-walled bearings are the traditional and well tried type with which most people are still familiar. Their advantages over thin-walled bearings may be summarised as:

(a) They are not specialised products, and can be made or repaired in any reasonably well-equipped workshop.
(b) They can be bored in situ, which is sometimes useful with complicated assemblies. The boring of a bearing involves no more than the removal of relatively soft and easily machined metal, and can thus often be done quite effectively with a portable boring rig that would not be suitable for boring out the corresponding housings.
(c) If the shell is of bronze and the bearing is flanged, then normal location end-thrusts can usually be carried without any special arrangements other than the provision of one or two oil grooves in the thrust face, as in figure 8.2. (For dimensions of suitable oil groove, see paragraph 13.14.)

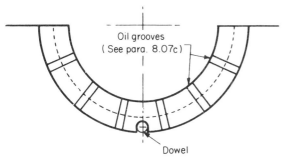

Figure 8.2 Location dowel for thick-walled flanged bearings (see paragraph 8.16)

(d) In emergency, they may be 'let down' by the removal of metal (or shims) from the joint faces of both the bearing and the keep. This is not a recommended practice but can be very useful on occasions, for example, at sea, when an urgent repair may be needed far from home and with no spare parts available.

8.08   The material of the bearing shell is preferably of mild steel, although cast-iron is sometimes used in special applications – mostly on the grounds of lower cost. Bronze shells were popular at one time because it was believed that, in the event of bearing failure, less damage would be caused to the shaft by relatively soft bronze shells than by mild steel shells. However, experience, coupled with the results of carefully monitored and deliberately planned experimental bearing failures, has shown that in fact the surface damage to the shaft tends to be markedly worse with bronze shells than with steel. Furthermore, with bronze shells there can be significant danger of shaft failure through the phenomenon of 'copper penetration'. It is also recognized that whitemetal has a stronger bond to steel shells than to bronze.

(iii) Location of shells in housings

8.09   Thin-walled bearing shells are held against rotation by the interference fit resulting from the tightening down of the bearing caps (see paragraph 9.03f, i), but additional location devices are needed to hold the shell in the right place during assembly.

8.10   During assembly, thin-walled bearings are normally located by lugs bent back from near one corner of each half-bearing, as shown in figure 8.3.

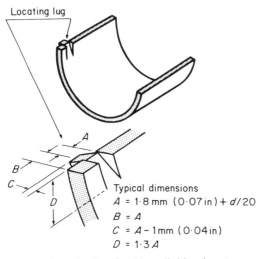

Locating lug

Typical dimensions
$A = 1·8$ mm $(0·07$ in$) + d/20$
$B = A$
$C = A - 1$ mm $(0·04$ in$)$
$D = 1·3 A$

Figure 8.3 Location lugs for thin-walled bearings (see paragraph 8.10). The matching notch in the housing should have corresponding dimensions, but with the value of C doubled, the value of D increased by 2.5 mm (0.10 in), and a clearance of 0.25 mm (0.01 in) allowed on A. (See BS 1131, Part 4)

These lugs engage in corresponding notches in the housing. The notches in a pair of half-bearings should be arranged to lie on the same side (i.e. on the same joint face), otherwise any slight error in the exact diametric splitting of the bearing may result in one or both lugs slightly fouling its notch, thus tending to distort the shell.

8.11   When thin-walled bearings are tightened down by their caps, or keeps, they are obviously compressed circumferentially, which implies a slight circumferential sliding between the shell and the housing throughout the tightening operation. In fact, however, the tightening of the bolts results in a

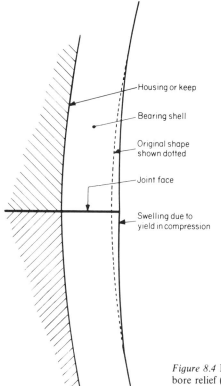

Housing or keep

Bearing shell

Original shape
shown dotted

Joint face

Swelling due to
yield in compression

Figure 8.4 Effect that would arise in absence of
bore relief (see paragraph 8.11). To assist clarity,
no lining material is shown on this diagram

progressive increase in friction between the shell and the housing, so that the
final stages of tightening usually cause the shell to yield slightly in compression
along the joint faces, as shown exaggerated in figure 8.4. To overcome this, the
bearings must always be provided with a 'bore relief', as shown in figure 8.1.

8.12   This bore relief also helps to minimize any trouble that might occur
from tolerance spread in wall thickness, as shown exaggerated in figure 5.3d.
The tolerance spread of all bearing thicknesses is relatively large in relation to
the running clearance and the oil-film thickness (especially the latter), so that
very little offset between the two halves can significantly change the
hydrodynamic pattern of the oil flow. With thin-walled bearings, this
particular danger is perhaps less than that with thick-walled, because bearing
sets are always made up from halves produced from the same piece of bimetal
strip, and are thus virtually identical in thickness. Thus, bore relief is necessary
with all bearings, although not necessarily for the same reasons on each.

8.13   An offset halves bearing, as in figure 5.3d, can clearly never run with a
uniform film thickness around its circumference, and each half must thus
always have, in effect, an eccentricity ratio greater than zero. This means that
the bearing generates its own inherent internal loading which, when superimposed on the normal load, may well lead to bearing trouble by bringing the
total load above its safe limit.

8.14   The same effect can, however, sometimes be employed to advantage

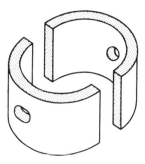

Figure 8.5 Location dowels for thick-walled bearings (see paragraph 8.16). The diagram shows a round clearance hole in one half-liner, and a circumferentially elongated hole in the other half-liner

by using it to generate enough artificial load to stabilize a lightly-loaded high-speed bearing that might otherwise tend to rattle around in its clearance space (as described in paragraph 4.24b). The exact amount of offset required to give the best results in any given application will generally need to be determined by ad hoc tests, but will typically lie between one-half and two-thirds of the radial clearance.

8.15   When deliberately using offset halves in this manner, it has to be

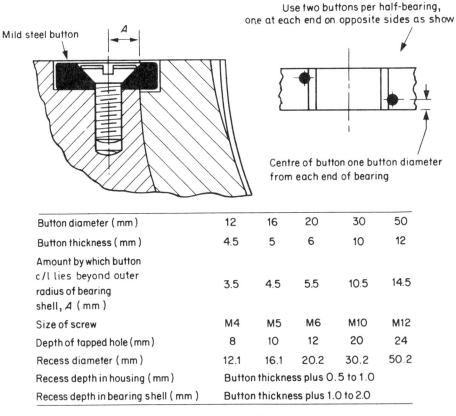

	12	16	20	30	50
Button diameter (mm)	12	16	20	30	50
Button thickness (mm)	4.5	5	6	10	12
Amount by which button c/l lies beyond outer radius of bearing shell, A (mm)	3.5	4.5	5.5	10.5	14.5
Size of screw	M4	M5	M6	M10	M12
Depth of tapped hole (mm)	8	10	12	20	24
Recess diameter (mm)	12.1	16.1	20.2	30.2	50.2
Recess depth in housing (mm)	Button thickness plus 0.5 to 1.0				
Recess depth in bearing shell (mm)	Button thickness plus 1.0 to 2.0				

Figure 8.6 Button stops (see paragraph 8.18). The table shows typical dimensions

remembered that such bearings are essentially unidirectional, and that care must be taken with the detail design to ensure that the bearing halves cannot be assembled incorrectly in relation to the shaft rotation.

8.16   Thick-walled bearings with flanges are located endwise by the flanges themselves, and are usually located rotationally by a dowel in one housing face, as shown in figure 8.2. Thick-walled bearings without flanges are usually located by dowels fixed in the housing and bearing cap, and engaging with clearance holes in the shells. To avoid interference between the two dowels, the hole in one of each pair of half-liners is usually slotted or elongated circumferentially, as shown in figure 8.5.

8.17   A reasonable dowel diameter is $d/16$ with a minimum of 4 mm ($\frac{5}{32}$ in).

8.18   A difficulty with the arrangement shown in figure 8.5 is that the lower half of the bearing cannot be removed for inspection without first lifting the shaft, which is often highly inconvenient in large machines. This problem can be overcome by using the alternative method of bearing location shown in figure 8.6, which has the additional advantage that, if applied to both halves, it enables the top half to be lifted off in safety with the bearing cap. Bearings not secured in this way can be dangerous if they adhere to the bearing cap firmly enough to be lifted with it, but do not adhere firmly enough for the shell to remain in place for more than a few seconds thereafter, with the result that it falls and can cause incidental damage and/or injury.

8.19   For the size of bearing covered by this book, these buttons tend to range in diameter between about $d/15 + 4$ and $d/9.2 + 8.5$ mm (where d is the journal diameter), but in general it is wise to use the largest size of button consistent with the space available.

Chapter 9

Bearing caps or keeps

9.01   The optimum design of a bearing cap, or keep, depends on the nature of the application, and particularly on the direction of load. If the load is always away from the cap (i.e. if the cap bolts never have to carry any of the load), the construction can be relatively light and the main function of the cap may be merely to hold the various items in place during erection, shipment and installation. In the extreme case, there may be no top half-bearing at all, and the cap (if any) may be there only as a protection against dirt and water, as with many marine intermediate-shaft bearings. On the other hand, if the load rotates and is thus applied cyclically to the cap, then the design must be relatively strong and rigid.

9.02   Figure 9.1 shows a typical design of cap for a highly-rated bearing carrying a rotating load. The dimensions of this design may be scaled up or down directly in proportion to the journal diameter, since, if the b/d ratio remains constant, the loads will vary as d^2 (where d is the journal diameter), as also will the area of the cap bolts. Similarly, the bending moment in the centre of the cap will vary as d^3, as also will the modulus of the section. The extent to which this design may be simplified, lightened and cheapened, will depend on the details of the application, and must be judged by the designer himself on the basis of his knowledge of the circumstances.

9.03   Any bearing cap that has to carry a significant load should be designed with the following factors in mind:

(a)   For bearings with $b/d = 0.6$, it is usual to have only two bolts, but for wider bearings there are more often four bolts.

(b)   To reduce the type of distortion shown exaggerated in figure 9.2, the bolt centres should be kept as close together as possible. With thick-walled bearings, it is often helpful to recess the shells in order to permit closer bolt centres, as shown in figure 9.3.

(c)   The 'bolts' may be ordinary bolts with nuts, or may be studs, or tap-bolts, but in any case they should be as long as is reasonably practicable, since this reduces the risk of fatigue failure.

(d)   Bolts may be 'fitted bolts', but with studs or tap-bolts other means must be provided to maintain the two halves of the housing and the bearing bores concentric. A simple and satisfactory cap-locating device is the stepped cap, as in figure 11.1. A more sophisticated arrangement is the serrated cap shown in figure 9.4. This can be made narrower than a stepped cap, and is useful for big-end bearings which are intended to be drawn up through the

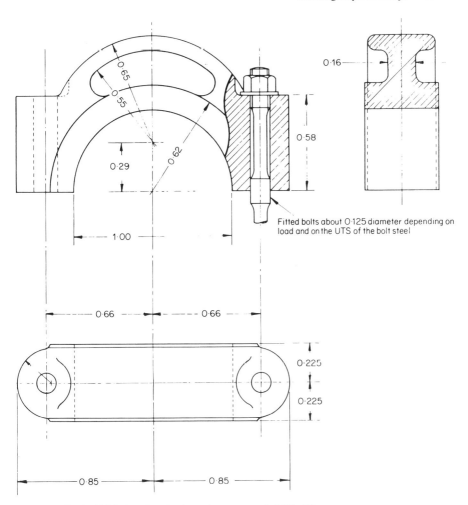

0·16

0·58

Fitted bolts about 0·125 diameter depending on
load and on the UTS of the bolt steel

Figure 9.1 Typical high-duty bearing keep (see paragraph 9.02). All
dimensions are proportions of the housing bore, which is therefore shown
as 1.00

engine cylinder. To help this operation, these serrated caps are often
combined with an obliquely split rod, as also shown in figure 9.4.

(e) To prevent the setting up of localised high stresses, all bolts, studs or tap-
bolts ought to be reduced to their bottom-of-thread diameter throughout
their length, except in way of the threads themselves and in way of any
sections where a fit is required, as in figure 9.1.

(f) The bore of the bearing housing and keep should be an H7/m6 fit on the
shell for thick-walled bearings in cast-iron or steel housings. For thin-
walled bearings, the amount of nip should be agreed with the bearing
manufacturers. The machining sequence ought to be:

(i)   Assemble caps in position and bore, leaving just enough material for
finishing to size later.

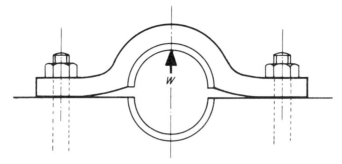

Figure 9.2 Exaggerated diagram of keep distortion under load (see paragraph 9.03b)

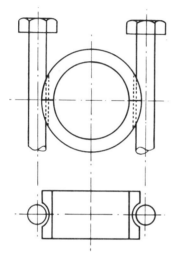

Figure 9.3 Thick-walled bearing recessed to permit close bolt centres (see paragraph 9.03b)

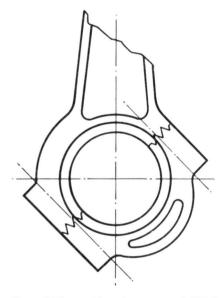

Figure 9.4 Serrated keep (see paragraph 9.03d)

(ii)   Remove caps and machine all notches, grooves, dowel holes, etc.

(iii) Re-assemble and tighten cap bolts to the same tightness as will be used in service.

(iv)   Finish bore or grind to size. This should be the final machining operation, since if any further machining were done in that region it might cause some stress-relieving and consequent distortion of the finished bore.

(g) With steel-backed bearings in non-ferrous housings, it may be necessary to provide a greater interference between the shell and the housing in order to allow for the effect of differential thermal expansion. The precise allowance must depend on the general design of the machine, the coefficient of thermal expansion of the actual material(s), and on the maximum temperature at which the bearing has to run. Guidance figures often quoted for housings of different materials are

Brass of other copper-based alloys	reduce housing bore by 0.05 per cent
Aluminium alloys	0.10 per cent
Zinc-based alloys	0.15 per cent

(h) With skilled erectors, it is often best to allow them to judge the correct degree of bolt-tightening to employ. But for the majority of present-day assembly shops, it is safer to insist on the use of torque spanners, and to specify a torque that will correspond to a stress of about 140 MPa (20000 lbf/in^2) at the bottom of a well-oiled thread on a bolt of, say, EN3 mild steel, and about 60 per cent more for a bolt of EN8.

(i) With thin-walled bearings, the load on the keep bolts does not usually arise primarily from the bearing load, but rather from the securing of the shells into the housing. As may be inferred from paragraph 8.11, the latter load will be equal to the yield strength of the steel multiplied by the area of the liner cross-section (i.e. bearing length × twice the wall thickness of the steel backing).

(j) In most cases, bearing cap bolts are locked by one or other of the usual means, such as split pins, tab washers, nylon-insert nuts, but this practice is by no means universal, and some cars have their engine-bearing caps held up by plain tap-bolts with no locking provisions whatever. The tendency to loosen with vibration is less with fine threads than with Whitworth threads.

(k) In applications where the bearings have to be dismantled frequently, as in the roll necks of steel mills, it is beneficial to use helical insert wires in the housing threads. These eliminate wear in the tapped holes and make reconditioning of the thread much easier.

Chapter 10

Bushes

(i) Bushes

10.01   In all calculations for load-carrying capacity, oil flow rates, etc., the performance of bushes and split bearings may be regarded as identical, and the choice between them in any given application will usually depend only on their relative convenience of manufacture, supply and assembly.

10.02   Bushes are widely employed in general engineering practice, particularly in the smaller sizes (up to, say, 50 mm diameter) and for the less exacting load conditions (loads up to, say, one quarter of those in figures 6.1 and 6.2), often under relatively poor lubrication conditions. (The top-end bearings of internal combustion engines are a non-typical exception – see paragraph 12.17b(iii).)

10.03   BS 1131, Parts 1 and 2, lists ranges of solid bushes from $\frac{1}{4}$ to 5 in diameter, and wrapped bushes from 10 to 58 mm and $\frac{3}{8}$ to 3 in diameter – most of which standard sizes are still stocked by the leading bearing manufacturers.

Wrapped bushes

10.04   These consist of thin-walled, steel shells lined with lead bronze or other comparable material. They have a higher fatigue strength than solid bushes and are generally preferable to the latter, although they must be obtained from specialist suppliers and must therefore be one of the maker's stock sizes unless ordered in batches of 5000 or so.

Solid bushes

10.05   These are normally of lead bronze, and are more readily adaptable to suit non-standard journal sizes than are wrapped bushes, since the extent to which they can be bored out is not limited by any fear of breaking through into the steel shell. For this same reason, solid bushes are the better for applications demanding deep oil or grease grooves. They are also more suitable for emergency repair jobs, since they can be made quickly from stock bronze bar in any normal workshop.

Wall thickness

10.06   BS 1131 gives the recommended thickness of wrapped bushes as:

Nominal bore (mm)	Thickness (mm) +0.05-0	Nominal bore (in)	Thickness (mm) +0.002-0
10 to 18	1.01	$\frac{3}{8}$ to $\frac{11}{16}$	0.047
20 to 25	1.51	$\frac{3}{4}$ to 1	0.0625
26 to 44	2.01	$1\frac{1}{8}$ to $1\frac{5}{8}$	0.078
45 to 58	2.51	$1\frac{3}{4}$ to 3	0.094

The lining thickness is usually of the order of about 0.3 mm (0.012 in).

10.07   The wall thickness of standard solid bronze bushes is specified in BS 1131, Part 1, but those who wish to make their own bushes may care to note that these BS thicknesses correspond roughly to $0.63\sqrt{d}$ mm or $0.125\sqrt{d}$ inches.

Interference fit in housing

10.08   Standard proprietary bushes (both wrapped and solid) are usually supplied with an o.d having a suitable interference fit in a housing machined to H8 limits on a nominal bore equal to the nominal journal diameter plus twice the nominal wall thickness.

10.09   For those making their own bushes, it will usually be found satisfactory to have an H8/p7 fit between the bush and the housing, assuming the latter to be of steel or cast iron.

10.10   If the housing is of aluminium or other non-ferrous material, fits of greater interference will almost certainly be needed to allow for the effect of differential thermal expansion, but the exact figure will depend on the actual material as well as on the maximum temperature anticipated in service. For aluminium housings in i.e engines, the *extra* interference required is likely to be of the order of $0.001d$.

Running clearances

10.11   The running clearance required in bushes is exactly the same as for any other type of plain bearing, and is discussed in chapter 8. For non-critical general industrial usage, an H8/e8 fit is common for wrapped bushes, and an H8/f8 fit for solid bronze bushes.

10.12   With any type of bush, because of the inevitable bore distortion occurring on insertion into the housing, the standard and highly desirable practice is to leave a finishing allowance in the original bore, and for this to be finally removed by burnishing, broaching, reaming or boring, whichever may be convenient or appropriate. If the wall thickness tolerance is $-0+0.50$ mm (0.002 in) and the housing bore is to H8 limits on a nominal size equal to the nominal journal diameter plus twice the nominal wall thickness, then the closing-in of the bush bore on assembly (plus the fact that a running clearance is required) will normally provide adequate metal for reaming the bore to size after assembly.

Surface finish

10.13   The recommended surface finish for wrapped bushes is 0.2–0.3 μm Ra (8–10 μin) Ra, and for solid bushes 0.2–0.4 μm Ra (8–16 μin).

Pre-finished bushes

10.14   BS 1131 makes reference to pre-finished bushes, by which is meant bushes with a bore that is originally oversize but closes in to the correct size when the bush is inserted into its housing. The use of such bushes simplifies assembly procedures but gives rise to performance uncertainties because of the tendency for there to be a considerable spread of tolerance on the final bore. This tolerance will be the sum of the tolerances on the housing bore, the o.d of the bush, and the pre-finished bore of the bush. Even if all the bores are held to H6 limits and the bush o.d to p6 limits, the finished bore will not be up to H8 standards. Machining to H6 and p6 limits can often be more expensive than the additional operation of final reaming. In general, therefore, unless the application is lightly-loaded and unimportant, pre-finished bushes should not be employed in production without carrying out preliminary tests to justify their adoption.

End thrust

10.15   With plain bushes (either wrapped or solid), any end thrust associated with endwise location of the shaft must be taken care of by separate thrust washers. These may, in effect, be plain bronze washers with a few shallow oil grooves, $0.5 - 1$ mm ($0.02 - 0.04$ in) deep, cut on the face that is to run against the journal collar. Appropriate groove forms are shown in figure 11.1. Recommended washer outside diameters are $\frac{4}{3}(d + 12)$ mm, or $\frac{4}{3}(d + 5)$ in, where d is the shaft diameter. The recommended thickness is about $\frac{1}{16}$(o.d of washer), and the depth of recess in the housing 1.6 mm ($\frac{1}{16}$ in). The hole in the centre of the washer should be about 5.5 mm ($\frac{7}{32}$ in) larger than the journal diameter.

10.16   Flanged bushes are sometimes used, but these have only a limited application because they can take thrust in only one direction, and for normal location purposes it is still necessary to have a separate thrust washer at the other end of the bush.

Assembly

10.17   Bushes can easily be damaged on assembly and must never be driven into the housing with a hammer – not even a soft-faced hammer. They should be pulled in with a draw-bolt and washers, or inserted by some appropriate form of hydraulic ram. Prior to insertion, all burrs should be removed from the edge of the hole in the housing, and to ease its entry the bush (and preferably also the housing) should be provided with a suitable chamfer on their appropriate edges. A 15° chamfer about 0.4 mm ($\frac{1}{64}$ in) wide is sufficient.

10.18   Porous metal bushes require especial care with their insertion, because they are more plastic and their bores are therefore more liable to be closed in by the forces involved in assembly. To avoid this difficulty, the required bore should be preserved by the use of a fitting mandrel as, for

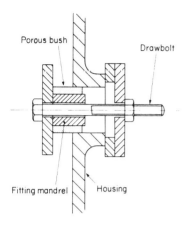

Porous bush

Drawbolt

Fitting mandrel

Housing

Figure 10.1 Insertion of porous bushes (see paragraph 10.18)

example, in the method of insertion shown in figure 10.1. The diameter of this mandrel should be about 0.012 mm (0.0005 in) larger than the nominal bore size of the bearing.

(ii) Floating bushes

10.19   A low-speed journal carrying a heavy rotating load in a fixed bush will often act like a mangle, in that (speaking in terms of μm or micro-inch) it will tend to roll a wave of bush material in front of it, with the result that the bush may creep around in its housing. This rotational creep can be inconvenient in many ways and, for example, will shut off the oil supply if this reaches the bearing through matching radial holes in the bush and housing. Also, since the creep may take a spiral form, it can give rise to trouble with endwise locations. Since this creep is often difficult to prevent, the alternative is simply to accept it as a fact of life, and design the bush deliberately as a floating bush with clearance on both its o.d and i.d (see figure 5.3f).

10.20   At the other end of the scale, with very lightly-loaded bearings running at high speed, a floating bush is sometimes adopted because, under these conditions, the outer oil film can be useful in damping down the half-speed whirls and other vibrations to which such bearings are prone. (See paragraphs 4.24 and 5.24.)

10.21   It is difficult to forecast the performance of any floating bush, since the hydrodynamic conditions in the two clearance spaces depend on the speed of revolution of the loose bush, which tends to be both variable and indeterminate. It may be necessary to experiment with several different combinations of clearances and oil feed arrangements before achieving a satisfactory result.

10.22   For heavily-loaded applications, a good starting point is to allow 0.001 of the bore as clearance, both in the housing and the bush. If the movement is rotational, then a first trial may be made with oil fed from a hole in the housing, which leads into a shallow circumferential groove in the bush o.d, with four radial holes leading from this groove to the inside of the bush – but with no grooves of any kind in the bush bore. This may not be the final answer, but its performance will probably give a few clues as to what is actually needed.

10.23   For heavily-loaded oscillating applications, it *may* be better to have a few (say, six) axial grooves in the bush bore, together with an equal number of axial grooves on the bush o.d, the two sets of grooves being staggered so that the bush thickness is not doubly reduced anywhere by two grooves backing against each other. The outer grooves should be interconnected by a central circumferential groove opposite the incoming oil hole, and each internal groove should be connected by a hole to this external circumferential groove.

10.24   For high-speed, lightly-loaded type applications, an equivalent design starting-point would be to have an external clearance in accordance with the *Tribology Handbook* (Butterworth 1975) recommendation in paragraph 6.60, and an internal clearance in line with the bearing manufacturer's clearance quoted in that same paragraph – with an external circumferential oil groove, four radial holes, and no other grooving. For these high-speed applications, it is sometimes found helpful to provide a peg to prevent the bush rotating, which makes the internal hydrodynamic conditions more stable, and leaves the outer oil film to act merely as a vibration damper.

Chapter 11

End thrust

11.01   Although thrust bearings, as such, lie outside the scope of this book, it nas to be accepted that the shafts of most machines need some form of restraint to prevent unwanted axial movement, and arrangements to absorb a limited amount of end thrust are, therefore, obligatory in nearly every application.

11.02   In the majority of heavy machinery running today, the bearings are probably of the flanged thick-walled type, in which end location is effected by collars on the shaft acting against the sides of the bearing flanges. The flanges usually have a few shallow grooves across their surface (as shown in figure 8.2) and are lubricated by oil overflowing from the journal.

11.03   Machines of more recent design are likely to have plain (i.e. unflanged) journal liners, and to be provided with loose thrust washers of the type illustrated in figure 11.1. These are usually of steel faced with whitemetal, and, although described as 'loose', are in fact held effectively trapped in a recess, as shown. If two half thrust washers are used together, then one must have a locating lug to prevent rotation.

11.04   In theory, there is little to choose between flanged bearings and separate thrust washers, but in practice the latter are preferable since they can be made in various oversize thicknesses and thus used to control the end float of the shaft. An end float of 0.1–0.25 mm (0.004–0.010 in) is normal practice. Plain liners and separate thrust washers also tend to be much cheaper than flanged bearings.

11.05   BS 1131, Part 4, gives detail dimensions for a range of thrust washers for journals of from 1 to 6 in diameter. The BS 1131 dimensions agree closely with those given in figure 11.1.

11.06   The actual thrust load that has to be borne by a locating collar tends to be rather indeterminate, although in applications where there may also be a small but measurable specific end load, it is useful to know the safe load limit of these simple thrust washers. This figure is difficult to state with precision because it depends on the rate of oil flow through the journal and can also be badly upset if the parts are carelessly assembled, since any particles of dirt under the thrust washer can seriously affect its flatness relative to the rotating collar, and consequently lead to local overload. In general, however, each half of a well lubricated thrust washer of the dimensions given in figure 11.1 will – assuming clean assembly and a good flat presentation to a thrust collar of 0.4 μm Ra (16 μin), or better, surface finish % accept a thrust load of $0.0035d^2$ N ($12.5d^2$ lb), where d is the journal diameter in millimetres and inches, respectively.

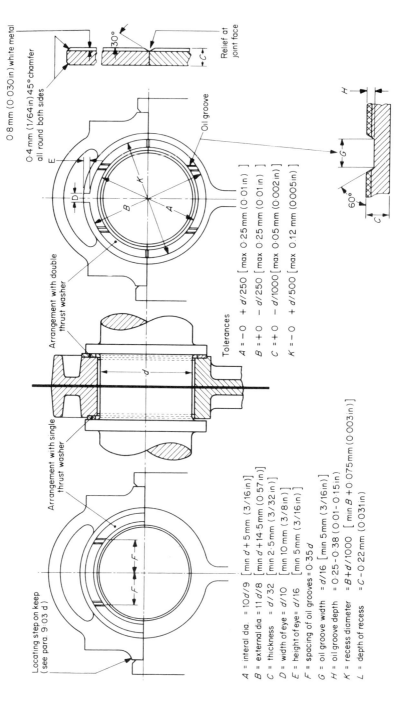

Figure 11.1 Typical thrust-washer dimensions (see paragraph 11.03)

A = internal dia.   = 10d/9   [min d + 5 mm (3/16in)]
B = external dia = 11 d/8   [min d +14·5 mm (0·57 in)]
C = thickness   = d/32   [min 2·5 mm (3/32 in)]
D = width of eye = d/10   [min 10 mm (3/8 in)]
E = height of eye = d/16   [min 5 mm (3/16 in)]
F = spacing of oil grooves = 0·35 d
G = oil groove width   = d/16 [min 5 mm (3/16 in)]
H = oil groove depth   = 0·25-0·38 (0·01-0·15 in)
K = recess diameter   = B+d/1000 [min B + 0·075 mm (0·003 in)]
L = depth of recess   = C − 0·22 mm (0·031in)

Locating step on keep
(see para. 9·03 d)

Arrangement with single
thrust washer

Arrangement with double
thrust washer

Oil groove

Tolerances

A = −0   + d/250 [max 0·25 mm (0·01 in)]
B = +0   − d/250 [max 0·25 mm (0·01 in)]
C = +0   − d/1000 [max 0·05 mm (0·002 in)]
K = −0   + d/500 [max 0·12 mm (0·005 in)]

0·8 mm (0·030 in) white metal

0·4 mm (1/64 in) 45° chamfer
all round both sides

Relief at
joint face

30°

11.07   The foregoing value should be treated with respect, and if the anticipated load even remotely approaches it, then this aspect of the design must be treated with the utmost caution. Any known thrust load should, of course, be added to the journal load when calculating the required oil flow from $\frac{1}{6}Wdn \times 10^{-6}$ (see formulae in paragraph 13.07).

11.08   Loose thrust washers should always be located against a fixed face. They are not suitable for applications such as connecting-rod big ends, where both faces are in motion. Connecting rods on all sizes of engine normally have their location end thrusts (which are usually very light) taken simply by steel against steel lubricated by the flow and splash of the escaping lubricant. The finish of the abutting faces ought not to be rougher than 0.4 μm (16 μin) cla.

Fillet radii

11.09   When designing thrust collars, or indeed any change in shaft diameter close to a bearing, care must be taken that the bearing surface is kept well clear of the fillet radius. This naturally applies to the bearing surfaces on the journal as well as to those on the thrust washer. A safety margin of around 0.5 mm (0.02 in) is advisable, rising to half as much again in the larger sizes. Similarly, when using thrust washers, the width of the bearing liner should be at least 0.5 mm (0.02 in), narrower than the housing, otherwise there may be a danger of the thrust washer not being able to lie flat in its recess.

Chapter 12

Oscillating bearings

12.01   These cover a very wide range, in the majority of which the loads, speeds and amounts of usage are so low that no significant design problems arise. In other applications, however, oscillating bearings may be subject to such severe loading conditions as to present very real problems. Examples of both extremes may often be found in the same product.

12.02   Thus, the carburettor throttle linkage in automobiles usually includes several oscillating bearings in the form of hinges, often of the ball-and-socket type, where mild steel bears against mild steel and lubrication is by the occasional drop from an oil-can when the user feels like it. Yet, the system is virtually trouble-free because the loads are low and the movement slight. Much the same may be said of the other oscillating bearings in the hand-brake mechanism, the window mechanism, the door-closing mechanism – yet in the same car the connecting-rod top-end bearings are probably examples of the most highly-rated oscillating bearings to be found in any machinery.

12.03   Thus, in the design of oscillating bearings, one of the first things to determine is whether there is any problem at all. It may be that the need is simply for parts capable of carrying and transmitting the various loads and forces, and that their performance as a *bearing* can almost be ignored.

(i) Lowly-rated oscillating bearings

12.04   Examples of lowly-rated applications outside the automobile industry include domestic door hinges, door locks and keys, control levers for machine tools, typewriter mechanisms, piano keys, gas taps, and hundreds of other devices where it would be quite out of place to let the theories of plain bearings play any major part in the design.

12.05   If the kitchen door in an ordinary house is opened through 90° in one second then the pV factor will be of the order of 2 kW/m² (50 lbf/in² × ft/min). If the door is opened and shut every 15 minutes, the usage will be 2 revolutions per hour, and almost certainly less than 20 revolutions per day. These door hinges will last indefinitely with a minimum of lubrication.

12.06   The spring shackles used on private cars prior to the general adoption of flexible rubber bushes for this application (see paragraph 12.19) were another example where the journal dimensions were fixed primarily from considerations of stress, and the bearing aspect fell into place automatically. The pV factor on car spring shackles was of about the same order as that on kitchen doors, but was made up of a higher pressure multiplied by a lower speed   typically 60 lbf/in^2 × 0.75 ft/min for the doors, and 250 lbf/in^2 × 0.2 ft/min for spring shackles – both being equivalent to 1.6 kW/m^2.

12.07   The outstanding difference between the two applications is, however, the much more intensive usage of the spring shackle, which in a typical case may have a cumulative movement equivalent to a steady 2 rev/min. For a car averaging 25 mph overall, this would amount to about 24 000 revolutions in a 5000 mile interval between servicing, which would make greasing at every 5000 miles roughly equivalent to oiling a household door once every 4 years. These periods are reasonably representative of what would constitute good maintenance practice in the two cases (being rather better than either of them usually gets), and thus serve to emphasize the similarity of the two applications.

12.08   Because of the higher usage (100000 rev/year as against 6000 rev/year for the door), the spring shackle is usually a hardened pin in a bronze bush – a construction that is also common for the more frequently used doors in offices and public buildings.

12.09   These two everyday applications have been examined in some detail because they form a good basis of comparison on which to judge the probable performance of any new design of lightly-loaded oscillating bearing.

12.10   In so far as one can quantify these lessons, it would appear that, for bearing pressures of the order of 350 kPa (3.5 kgf/cm^2, 50 lbf/in^2), there is no need to exercise any special care in the choice of materials if the usage is low, and that for pV values of the order of 2 kW/m^2 (50 lbf/in^2 × ft/min) lubrication is likely to be required at intervals of about 25 000 revolutions. If a life expectancy of more than 100 000 revolutions is required at this pV factor, then a bronze bush should be used.

12.11   These bald figures, however, fail to convey anything like as complete a picture of the facts, as a designer is likely to get by simply mentally comparing his prescribed conditions with those obtaining in the two cases instanced – or by comparing them with any other application of which he has an equal understanding.

(ii) Highly-rated applications

12.12   In the case of heavily-loaded oscillating bearings, even more than in the case of rotary bearings, designs have evolved more from practical experience than from any basis of pure theory.

12.13   Leading authorities are still far from agreement on many fundamental issues, such as the arrangement of oil grooves. There are, for example, those who maintain that, in oscillating bearings with small rubbing velocities, it is necessary to have axial oil grooves in the loaded zone, particularly if the load is unidirectional. Figure 13.5a shows a typical arrangement of this type.

12.14   Others, however, maintain that it is only the lower pair of grooves that can serve any useful purpose, and that the others are better omitted, but that an oil feed should also be provided at the top of the bearing, if only to keep it thoroughly wetted with oil and thus avoid corrosion.

12.15   Yet another school of thought insists that oil escapes fast enough from the end of a bearing without any axial groove to help it, and that the best results are obtained with no grooves – but merely with oil supply holes in the centre of the bearing length and displaced from each side of the load centre-line by an angle equal to the angle of oscillation.

12.16   There are also divergent views on the subject of bearing profile. Some claim that the best results are obtained with perfectly round, hardened journals having a clearance of 0.001d running in perfectly round bearings. Others claim that the bearing metal ought to be formed or to conform exactly to the journal profile over an arc of 120° *plus* half of any angle of displacement *of the load* in relation to the bearing surface, as in figure 12.1. Those who favour

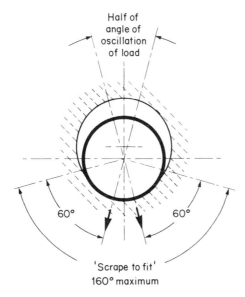

Half of
angle of
oscillation
of load

60°   60°

'Scrape to fit'
160° maximum

Figure 12.1 Oscillating bearings: suggested bedding-in procedure (see paragraph 12.16)

the scraping-in procedure maintain that, even if this is not done deliberately, the same shape will ultimately be acquired as a result of wear, and that reliable operation on heavy loads will not be obtained until the bearing has been run in sufficiently to produce this form. The words 'of the load' are italicized above because the angle of oscillation of the load is not necessarily the same as the angle of oscillation of the journal in the bearing, and it is the former angle that is significant in the 'scraping' context.

12.17   Points on which there appears to be general agreement include the following:

(a)   When the angle of oscillation is small (of the order of 5°) and the oscillation is sufficiently infrequent or erratic, the chief limitation to the acceptable load is the fatigue strength of the bearing material – although the method and frequency of lubrication may also enter into it. Commonly accepted values for the fatigue strength in the case of oscillating bearings are: whitemetal 7 MPa (70 kgf/cm^2, 1000 lbf/in^2); copper–lead 14 MPa (140 kgf/cm^2, 2000 lbf/in^2); lead–bronze 24 MPa (24.5 kgf/cm^2, 3500 lbf/in^2). All these pressures are, however, limiting values that should not even be remotely approached except when it is absolutely unavoidable.

(b) Oscillating bearings can often carry a much higher pV factor than rotating bearings operating under boundary lubrication conditions. This is because

(i) At each reversal of rotation, the journal makes part of its rotation by *rolling* from the eccentric position shown in figure 3.2 to the equivalent position on the other side of the centre-line corresponding to the opposite direction of rotation. The true *sliding* is thus less than the nominal figure.

(ii) If the application is one in which the load falls to zero at any point in the oscillation, then the pressure of the oil supply will tend to centralise the journal in the bearing and thus provide a relatively thick oil film that will take a little time to squeeze out when the load is re-applied. Metal-to-metal contact can thus be avoided with loads heavier than those that could be sustained continuously.

(iii) If the load reverses at any point in the cycle (as in the top-end bearings of many four-stroke connecting rods), then the oil film built up under the journal will be even thicker, and the subsequent brief load-carrying capacity even greater. Incidentally, if the oil film thickness is artificially increased in this way, then the bearing becomes slightly more prone to end leakage of oil pressure, and also becomes slightly more tolerant of misalignment. Both these effects point to a higher optimum b/d ratio. For this reason, $b/d = 1$ is common in modern gudgeon pins.

(iv) As a result of (ii) and (iii), oscillating bearings frequently operate virtually under hydrodynamic conditions, even though their apparent loads and speeds may be well outside the theoretical hydrodynamic range.

(v) When these quasi-hydrodynamic conditions prevail, as in four-stroke automobile gudgeon pins, then loads up to the limits specified in sub-paragraph (a) above are often carried successfully.

(c) For bearings with small angles of oscillation, and on which the load is both heavy and continuous, grease is often a more satisfactory lubricant than oil.

(iii) Applications with intermediate ratings

12.18 For applications between the extremes already dealt with, there is a variety of different bearing types from which to choose. In some cases, the choice is easy. Thus, the rocker arm bearings in an automobile engine are naturally bronze bushes lubricated by a pressure feed from the main oil system. They are relatively low-rated bearings because they carry no load for most of each cycle and thus benefit markedly from thhe squeeze effect already mentioned. In any case, the load they have to carry is not high in relation to the space available for the bearing – although they *are* somewhat highly rated in that they are in rapid and continuous use all the time that the engine is running.

12.19 In other applications, the choice is wider. Thus, the hinges in the arms of earth-moving equipment, grabs and the like, may be normal oil-lubricated bearings with seals to retain the oil and exclude water and dirt, or they may be grease-lubricated, or they may be dry rubbing bearings. Each manufacturer has his own preference based on experience, but dry rubbing bearings are undoubtedly attractive for this class of work. Another solution, not strictly a 'plain bearing' in the usual sense, is to use flexible rubber bushes which distort under torsion and thus have no rubbing surfaces or lubrication problems whatever. These are obtainable as proprietary items in a wide range of sizes and various angles of movement.

12.20   The flexible rubber bush and the dry rubbing bearing are both worthy of serious consideration for many industrial oscillating-bearing applications commonly dealt with at present by more conventional means.

(iv) Other forms of oscillating bearing

Kromhout top-end

12.21   Figure 12.2 shows a form of connecting rod top-end bearing employed in the Kromhout semi-diesel engine formerly used in many canal boats. This, being a two-stroke engine, always had the connecting rod in compression, and

Figure 12.2 Kromhout diesel top-end (see paragraph 12.21)

the load on the top-end bearing was consequently always in the one direction. Because of this, and because the unit was relatively lowly rated, it was found thoroughly satisfactory to carry the load on a hardened gudgeon pin that merely rolled on a flat, hardened-steel pad held in the connecting rod as shown. Splash lubrication was found to be adequate to deal with the slight amount of slipping that took place. This slipping occurred mostly at starting, until the pin had settled into a central position, from which it could roll freely to either side with the swing of the connecting rod.

GMT–Fiat eccentric-type crosshead pin and bearing

12.22   This is a more sophisticated top end for a modern, highly-rated engine. As shown in figure 12.3, each pin section and its bearing are divided so that the extreme sections of the pin are eccentric to the inner sections. Due to the oscillatory motion of the connecting rod, the contact between the bearings and

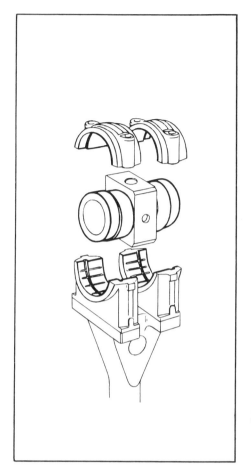

Figure 12.3 GMT–Fiat eccentric-type
crosshead pin and bearing, patented
December 1974 (see paragraph 12.22)
[Reproduced from *Marine
Engineering Review*, October 1975]

the pin alternates between the inner and outer sections. An alternating lift is
thus obtained, allowing lubricating oil to flow into the sections not under load.
In this way, an effective film of oil is built up to carry the load, and at the same
time the increased flow of oil helps to cool the sliding surfaces. This principle of
operation may be seen more clearly in the simplified and exaggerated line
diagram of figure 12.4.

Fully-floating gudgeon pin

12.23   In motor-cycle and automobile connecting rod top-end bearings, it is
not unusual to find that the gudgeon pin is free to turn in both the connecting
rod and the piston. The total angle of oscillation is thus split between two sets
of bearing surfaces, with the pin (at least nominally) oscillating through half
the angle of the connecting rod. This immediately halves the value of V in the
pV factor, and consequently improves the load-carrying capacity as compared
with a pin that is fixed in either the connecting rod or the piston. In fact, the pin
seldom swings at exactly half the angular speed of the connecting rod, and the

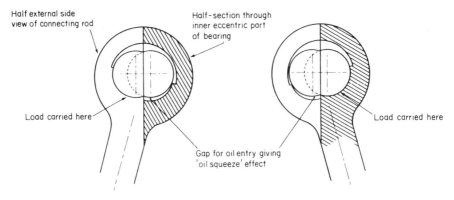

Figure 12.4 Exaggerated diagram showing how the double eccentric pin on the GMT–Fiat crosshead provides clearance for oil entry to bearings on alternate strokes (see paragraph 12.22)

gain in performance is thus considerably less than 100 per cent – although still significant.

12.24   A new range of special applications for oscillating bearings is arising in the development of artificial hip-joints, knees, etc., for the medical profession, but these lie outside the scope of this book.

(v) General comments

12.25   The factors governing the performance of oscillating bearings are so involved that the only safe way to design them is to be guided by the nearest equivalent application of which one has full particulars. This is not quite such a counsel of despair as it might sound, because there is usually an earlier model of the product which can be used as a yardstick and which can be improved in the light of experience with it.

12.26   In designing a completely new product, it may be more difficult to find a reasonably equivalent application to use as a model, and some development testing may be needed. This, however, is unlikely to be necessary unless one is particularly pressed for space or for some other reason has to adopt loadings that are approaching dangerous limits. In most cases of oscillating bearing design, as mentioned in paragraph 12.3, there is really no problem at all. Rarely does one have to be more scientific than the Victorian engineers' rule-of-thumb 'Use a gunmetal bush and keep the pressure below 200 pounds per square inch'.

12.27   If one wishes to use figures 5.1, 6.1 or 6.2 for broad guidance on oscillating bearings, then the equivalent 'rotation speed' of such bearings may be regarded as their average angular velocity throughout each cycle. Thus, if the angle of oscillation is $\theta°$ each side of the mid-position, then

rotational speed $= n^\theta/90$ rev/min

where $n =$ cycles/min.

Chapter 13

Lubrication

(i) Lubricating oil requirements

Type of oil

13.01   Unless the conditions are particularly severe or unusual, most industrial bearings have to make use of the oil that is bought for many other bearings, and no great choice exists. For the purposes of calculating the values in figures 6.1 and 6.2, therefore, it has been assumed that (except for grease-lubricated bearings) the oil will always be some reputable brand of heavy-machine oil, not necessarily containing any particular additives. A nominal viscosity of 30 cP (30 mPa.s) at the average oil outlet temperature from the bearing, say, 60°C (140°F), has been assumed in all cases.

13.02   For overseas installations in either high or unusually low ambient temperatures, there may be trouble if one attempts to use the same lubricants as in Europe – since these may well be either much too fluid or much too viscous for satisfactory operation under the local conditions. In some countries it may be necessary to use a different grade of oil in summer as compared with winter. Advice is best obtained on the spot from the nearest branch of any reputable oil company.

13.03   For some applications, it may be advisable to use oil containing one or more of the following additives:

Anti-acid
Anti-foam
Anti-oxidant
Anti-rust
Anti-wear
Corrosion inhibitors
Pour-point depressants
Water emulsifiers
Extreme-pressure (e.p) additives (for very high bearing loads)
Oiliness additives (to reduce stick–slip effects)
Tackiness additives (to reduce loss of oil from vertical surfaces)
Viscosity index improvers (to reduce temperature-dependent viscosity changes)

13.04   If there is a need for any of these additives, it will normally be apparent to the designer from his knowledge of the bearing environment. Detailed information on additives may be obtained from the oil suppliers.

Grease

13.05   All comments about grease lubrication in this book refer to the use of normal, general-purpose grease to BS 3223.

Quantity or flow rate of oil

13.06   Many complicated formulae exist for estimating the amount of oil that needs to be supplied to a bearing (a) to maintain the lubricating oil film, and (b) to remove the frictional heat and keep the bearing cool. In practice, there is little point in trying to calculate the flow with any great accuracy, because the resulting values ought always to be increased to allow for ultimate wear in the pump and in the bearings – wear in the former tending to reduce the flow, and in the latter to increase the amount required. A margin is also needed to take care of partially choked filters, sludgy oil, and other contingencies.

13.07   A useful rule-of-thumb approach is to assume (quite wrongly) that the only purpose of the oil is to carry away heat, and that the amount of heat to be removed will correspond to that which would be produced by a coefficient of friction of 0.005 in the bearing. Using average values for the density and heat capacity (specific heat) of lubricating oil, and a reasonable oil temperature rise, this results in the very simple expressions

$$\text{required oil flow} = \frac{\text{load (N)} \times \text{diameter (m)} \times \text{rad/s}}{6} \times 10^{-6} \text{ litre/s}$$

$$\text{or} = \frac{\text{load (kgf)} \times \text{diameter (mm)} \times \text{rev/min}}{6000} \times 10^{-6} \text{ litre/s}$$

$$\text{or} = \frac{\text{load (lbf)} \times \text{diameter (in)} \times \text{rev/min}}{40} \times 10^{-6} \text{ gal/min}$$

13.08   The foregoing will be found to be reasonably accurate within the range of loads, speeds and sizes covered in figures 6.1 and 6.2, although the pump itself ought to be designed for about double that flow when new, for the reasons mentioned in paragraph 13.6.

13.09   If these expressions are applied to the three worked examples given at the end of ESDU60023, which gives very precise values for the *minimum* required flow rate, it will be seen that paragraph 13.07 would provide margins of 75, 27 and 85 per cent respectively. The case with the 27 per cent margin is, incidentally, outside the load and speed range of figures 6.1 and 6.2.

(ii) Oil grooves

13.10   At one time, nearly all bearings were provided with a pattern of oil grooves, ostensibly to spread the oil around. These grooves usually did more harm than good, since they provided easy paths over part of the distance from the oil hole to the bearing edge, and thus allowed much of the oil to short-circuit the loaded area in which it was primarily required.

13.11   In most cases, and certainly with the bearing proportions recommended in this book, the best results are obtained with no grooves. Practically the only exceptions to this are some types of oscillating bearing (see paragraph 12.13), and the cases where circumferential grooves are needed to act as channels for the supply of oil to other parts, such as from main to big-end bearings in i.c engines, or from big-end bearings to top-end bearings *and* piston cooling passages in large diesels.

13.12   These circumferential grooves considerably reduce the effectiveness of any bearing (see paragraph 6.44), and care must therefore be exercised in using them. With large diesels, the bearing loads are often low enough for the harmful effects of the groove to pass unnoticed, but in more highly-rated engines it is common practice to use an interrupted circumferential groove, and thereby leave the bearing surface intact throughout its full length in the loaded area. In such cases, the crankshaft is provided with multiple drillings, as shown in figure 13.1, so that at least one drilling is always opposite a grooved part of the circumference. Since highly-rated engines tend to use thin-walled bearings, in which the provision of deep grooves is impracticable, the construction is often of the 'gashed' half-liner form shown. A more detailed view of the gashed half-liner is given in figure 13.2.

13.13   For normal applications having no oil grooves, the oil is usually supplied through a single hole placed centrally in the bearing length and positioned at approximately 180° from the area of maximum load. A hole diameter of $d/16$ is customary, where d is the diameter of the journal with a minimum of 3 mm ($\frac{1}{8}$ in). Remembering, however, that an ordinary stop valve has to be lifted off its seat by one quarter of its diameter before it can be regarded as fully open, it will be appreciated that, in a bearing with a running clearance of $0.001d$, an eccentricity ratio of 0.60, and an oil hole of $d/16$ diameter, the escape path for the oil (as shown in figure 13.3) would be

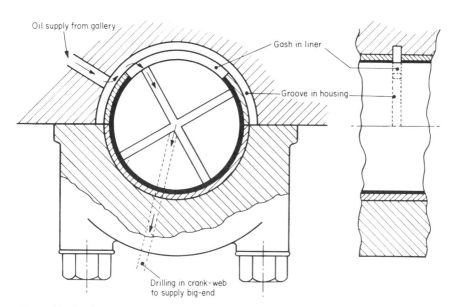

Figure 13.1 Partial circumferential groove: gashed half-liner construction (see paragraph 13.12)

Figure 13.2 Gashed half-liner (see paragraph 13.12)

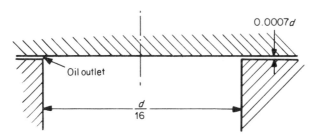

Figure 13.3 Limited oil-flow area from an oil hole (see paragraph 13.13). The assumed conditions are: diametrical clearance $0.001d$; eccentricity ratio 0.60; oil-hole diameter $d/16$

equivalent to less than one-twentieth of the area of the oil hole. It is, therefore, advisable to countersink the hole and de-burr the edges to something approaching double the hole diameter, as shown in figure 13.4. In applications where the oil supply pressure is low, as when the supply is from splash or ring oiling, an even larger inlet area will be needed. In such instances, a short axial oil groove can sometimes be helpful, not to spread the oil, but just to get it into the bearing.

13.14   If grooves are employed, whether for oil or grease, their width ought to be about $d/16$, chamfered as shown in figure 13.4.

Oil grooves in oscillating bearings

13.15   Grooving arrangements in oil-lubricated oscillating bearings are dealt with in paragraphs 12.13 to 12.15. When bushes are used for oscillating bearings, grooves of the form shown in figure 13.5a are commonly employed in

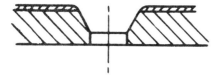

Figure 13.4 Countersinking of oil hole and chamfering of oil groove (see paragraph 13.14)

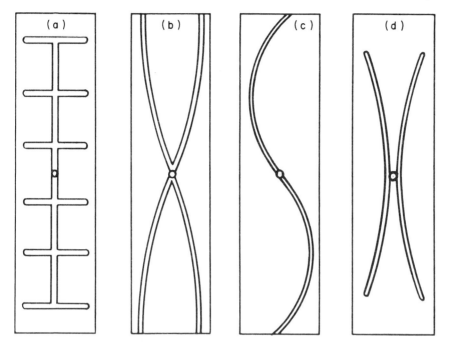

Figure 13.5 Various patterns of oil grooves (see paragraphs 13.15–13.17)

the actual loaded zone, the grooves being spaced at intervals of 60° irrespective of the angle of oscillation. With heavier loads, the spacing may be the angle of swing less one groove width.

13.16   For small grease-lubricated oscillating bearings (which are usually bushes), the best results are normally obtained with either elliptical or figure-of-eight grooves, as shown in figures 13.5b or 13.5c.

13.17   The groove form shown in figure 13.5d is suitable for larger grease-lubricated bearings.

Oil grooves in journals

13.18   In cases where a steadily-loaded revolving bearing receives its oil from a hollow, fixed journal, as in the track wheels of dock cranes, and the like, any axial groove in the bearing will lie right in the middle of the most heavily loaded area at least once per revolution. In such cases, it is particularly desirable to use ungrooved bearings. If a groove *has* to be employed, then it should be cut axially in a relatively lightly-loaded part of the journal, say about 180° from the point of maximum load. There may be difficulty in following this advice if the load rotates, and the choice may then lie between using a simple oil hole at some carefully selected position, or of reluctantly accepting the need for a circumferential oil groove. Choosing a good place for the oil hole can be complicated, and the reader will find an excellent article on this in the August 1973 issue of the *Journal of Automotive Engineering*, pp. 23–27.

General

13.19   Oil holes and oil grooves require careful planning only if the loading on a bearing is approaching the limit of its capacity or the conditions of loading are complex. If the loading conditions are moderate and straightforward then the exercise of common sense is all that is necessary.

(iii) Lubrication systems

13.20   Lubricant may be supplied to a bearing in various ways, the more common being pressure feed, splash, ring oiling, disc oiling, drip feed, mechanical drip feed, wick feed, hand oil-can, grease gun, and screw-down grease cups.

13.21   For heavily-loaded high-speed bearings, pressure feed is virtually essential, but in other applications the choice of lubrication system tends to be influenced mostly by what the customer will expect in a machine of the type involved. If there is any difficulty in making a choice, then the following comments may be of help.

Pressure feed

13.22   Pressure feed from an oil pump is the preferred system unless it is ruled out on the score of first cost. The size of pump may be assessed from paragraph 13.06, and the pump delivery pressure may be chosen by starting with the bearing most remote in the oil system from the pump, deciding what oil pressure is needed at that point to give the bearing enough oil to carry away its heat, and then working back through calculations of pipe sizes and hydraulic flow losses to determine the pressure needed at the pump itself. Pump delivery pressures used in practice vary from about 0.1 MPa (1 bar, 15 lbf/in^2), found mostly in applications such as process plant and machine tools, up to over 0.5 MPa (5 bar, 75 lbf/in^2), in many internal combustion engine lubrication systems. Oil pressures intermediate between these extremes are found in applications such as large steam turbo alternators.

13.23   The oil pum p is usually driven from the machine to be lubricated. If the machine has to operate at more than one speed, care must be taken to ensure that the pump is large enough to deliver the required volume at the lowest speed, and that a by-pass relief valve is provided to prevent too great a pressure being generated at the top speed. With large and expensive plant, it is often useful to supplement the built-in pump with an independent motor-driven oil pump that can be run to establish full pressure throughout the system before the machine is started, and also sometimes to circulate oil for cooling after it has stopped. See paragraphs 14.03, 14.04 and 14.05(c). The oil pump should always be mounted below the sump oil-level, so that it lies continuously primed and ready to deliver full flow immediately on starting.

13.24   A pressure-feed system lends itself to added features such as oil filtration (see section 13 (v)) and oil cooling (see section 13 (vi)), but if either of these are employed, the pump delivery pressure should be raised to compensate for the pressure drop through the filter and/or cooler. If by-pass filtration is employed, the pump capacity should be increased to compensate for the additional flow required.

13.25   Lubrication oil pumps, whether built-in or independently driven, are usually bought-out items.

Splash lubrication

13.26   It is virtually impossible to give any firm guidance on the design parameters for splash lubrication systems, which are usually developed on an ad hoc basis by trial-and-error on prototype testing. Automobile gudgeon pins are examples of highly-rated bearings that are satisfactorily lubricated with splash, although in general this form of lubrication is confined to more lightly-loaded applications.

13.27   In situations such as gear-boxes, it is sometimes possible to arrange ducts that will catch localised oil splash from one part of the mechanism and conduct it by gravity to some other part needing lubrication. In most cases, the use of some form of deflector(s) and/or funnel(s) will be found helpful. Even the car gudgeon pin needs a fairly large diameter countersunk recess around the oil hole to make it successful.

Ring oiling

13.28   This is suitable for bearings carrying moderate unidirectional loads at relatively low speeds – generally not more than about 1.7 MPa (17 kgf/cm^2, 250 lbf/in^2) nor above 10 m/s (2000 ft/min). Ring dimensions are not critical, but values for a typical design are given in figure 13.6.

13.29   Ring oiling is unsuitable for portable machinery, or for shipboard use, since a relatively slight inclination of the assembly will cause the ring to drag against the side of its slot – and probably stop.

13.30   Ring oiling is also unsuitable for rotating loads, since the slots for the rings considerably reduce the bearing area of the top half-liner. Indeed, many ring-oiled applications have no top half-bearing at all, the inside of the bearing keep being machined to the equivalent of the bearing bore, and the existence of a unidirectional load being relied upon to prevent the top half ever experiencing any load or wear.

Disc oiling

13.31   A simple disc mounted on a rotating horizontal shaft will pick up a considerable amount of oil in each revolution. This can be wiped off by a close-fitting deflector plate (as shown in figure 13.7), and the oil can then be directed into an adjacent bearing. This somewhat crude device is surprisingly effective, although at speeds above about

$$\frac{11}{\sqrt{\text{shaft diameter (mm)}}} \quad \text{or} \quad \frac{125}{\sqrt{\text{shaft diameter (in)}}}$$

the oil might be expected to be thrown off the disc by centrifugal force, the worst condition arising at an angle of about 45° above the horizontal. This

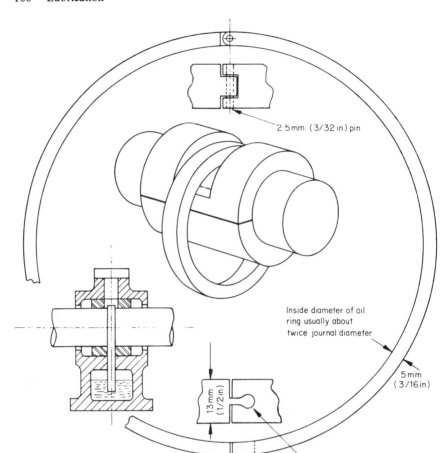

2·5mm (3/32 in) pin

Inside diameter of oil
ring usually about
twice journal diameter

5mm
(3/16in)

13mm
(1/2in)

Interlocking portions of 'key' joint may be comparatively loose fit since they cannot separate unless
ring is forcibly distorted from circularity. The joint can easily be sprung apart by hand for assembly
around the journal

Figure 13.6 Ring oiling (see paragraphs 13.28 to 13.30)

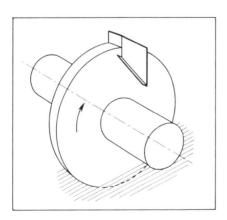

Figure 13.7 Diagrammatic arrangement of
disc oiling (see paragraph 13.31)

would seem to restrict the speed to as little as 10 rev/min on a 50 mm (2 in) bearing, but in fact the oil slips around the disc and does not attain the full rotational speed of the latter. Operation at much higher speeds is thus quite practicable, but much depends on the surface finish of the disc and its wettability in relation to the lubricant employed. It is almost impossible to 'design' a disc lubrication system, other than from experience or trial-and-error, but it is nevertheless a useful expedient in many simple applications.

Scoop oiling

13.32   This is a more sophisticated and effective form of disc oiling. The general principle is shown in figure 13.8. The dimensions shown on this diagram are typical but in no way obligatory. Anything that looks right will usually provide more oil than is needed for loads up to 1.7 MPa (175 kgf/cm^2, 250 lbf/in^2) and speeds up to 10 m/s (2000 ft/min). In some applications, prototype testing and development have enabled satisfactory results to be obtained up to 20 m/s (4000 ft/min) or more.

13.33   Disc and/or scoop lubrication is impractical in many applications simply because it takes up considerable space in the vicinity of the bearing, and cannot supply oil at a pressure of more than a few centimetres head. Furthermore, when starting from rest, the lubrication does not become effective until the shaft has made several revolutions, and the oil has had time to find its way down the deflector and into the bearing.

Drip feed

13.34   The old-fashioned drip feed lubricators of the type illustrated in figure 13.9 have little to commend them other than cheapness, and tend to be employed only in situations where their use has been traditional.

Multiple drip-feed mechanical lubricators

13.35   These are useful for bearings in which it is impossible to have any form of oil recirculation, and in which all the oil supplied is consequently lost (see figure 13.10). In such total-loss systems, it is important to control the oil flow with some precision, since too much would be expensive and too little might be disastrous. This type of lubricator is suitable for flow rates of up to about 19 mm^3/s (0.0003 gal/min) per feed, which corresponds to about one drop of oil each $2\frac{1}{2}$ s. They may be driven through reduction gearing from any rotating shaft, or through a ratchet from any reciprocating part of the machine. The oil rises up in a sight-glass through water, guided by a wire, and can be delivered at pressures up to about 0.20 MPa (30 lbf/in^2) or more, if necessary.

Other systems

13.36   Wick feed, hand oil-can, grease gun and screw-down grease caps are all traditional systems. Those who wish to use them will have their own reasons for doing so, and will need no guidance from this book.

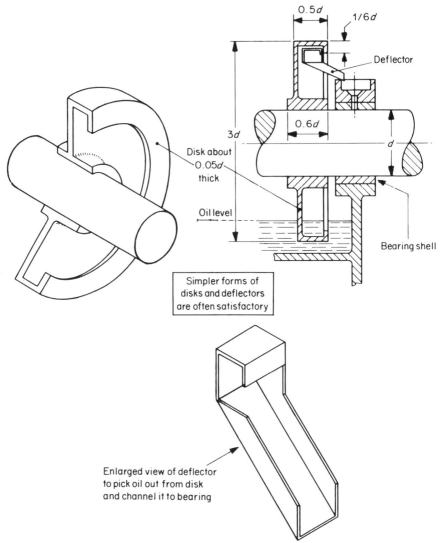

0.5*d*

1/6*d*

Deflector

3*d*

0.6*d*

d

Disk about
0.05*d*
thick

Oil level

Bearing shell

Simpler forms of
disks and deflectors
are often satisfactory

Enlarged view of deflector
to pick oil out from disk
and channel it to bearing

Figure 13.8 Scoop oiling (see paragraph 13.32)

(iv) Oil throwers and seals

13.37   Wherever possible, bearings should be totally enclosed so that ample oil may be circulated round them without spillage or loss, and the bearings themselves protected against external damp and dirt. Where a shaft finishes flush with the end of a casing, it should be protected with a drained cover, such as that illustrated in figure 13.11.

13.38   Where a shaft has to protrude beyond the casing, a decision must be taken as to whether to accept the inevitable oil leakage, or to provide some means of preventing or at least reducing it. In simple mechanisms in which the

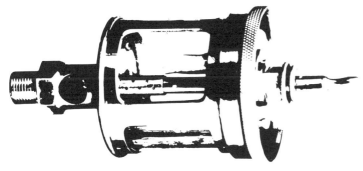

Figure 13.9 Drip-feed lubricator (see paragraph 13.34) [Reproduced by permission of J & W Kirkham Ltd]

Figure 13.10 Multiple sight-feed mechanical lubricator (see paragraph 13.35) [Reproduced by permission of J & W Kirkham Ltd]

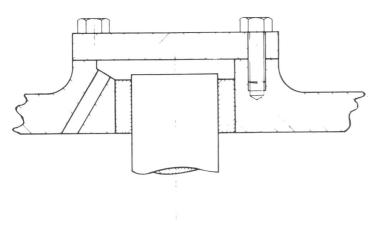

Figure 13.11 Drained cover (see paragraph 13.37)

loads, speeds and periods of usage are low (such as domestic lawn-mowers), the oil leakage is normally accepted. However, for bearings with a pressure-fed oil-supply, end leakage cannot be tolerated.

13.39   Methods of reducing leakage fall into three main groups:

(a) elastic seals;
(b) wind-back seals;
(c) oil throwers.

Elastic seals

13.40   The form shown in figure 13.12 is representative of a wide range of proprietary seals. These seals are a press fit in the housing, and are held in contact with the shaft by a garter spring which, as shown, should always be on the 'oil' side of the seal and not exposed to the outer environment. On assembly, it is usual to give these seals a light smear of grease, and the maker's instructions on this point should be observed.

13.41   Provided there is no air pressure difference across these seals, they may be regarded as totally effective. If the external conditions are severe, one of the special forms of double seal ought to be fitted.

13.42   Elastic seals are suitable for the full range of bearing sizes and speeds covered in figures 6.1 and 6.2, and for operating temperatures up to 150°C (300°F). Similar designs are available in special materials for higher temperatures.

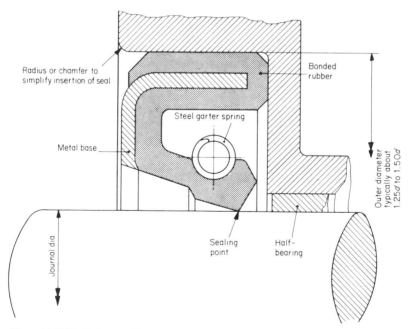

Figure 13.12 Typical rotary lip seal (see paragraph 13.40)

Wind-back seals

13.43   If a screw thread (either right-hand or left-hand as may be appropriate) is cut in either the journal or the bearing, the rotation of the shaft will produce a pumping effect that will tend to push any leaking oil back into the bearing. This wind-back screw ought to be cut in an extension of the bearing, and not in way of the loaded area. Typical dimensions for a wind-back seal are given in figure 13.13. One complete turn of thread may be adequate, but a greater

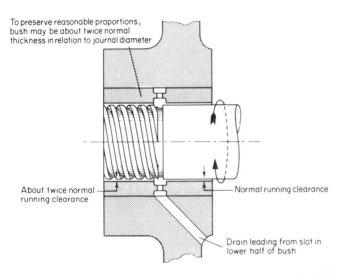

Figure 13.13 Typical dimensions of wind-back seal (see paragraph 13.43) Multi-start rectangular thread with lead of about $0.5d$ and depth $0.15 \sqrt{d}$ mm $(0.03 \sqrt{d}$ mm $(0.03 \sqrt{d}$ in). Number of starts is typically nearest whole number to $0.6 \sqrt{d}$ mm $(3 \sqrt{d}$ in). Sketch shows four starts, as for bearings of 35 to 57 mm ($1\frac{3}{8}$ to $2\frac{1}{4}$ in) diameter. Handing of thread must suit direction of shaft rotation

length is needed if the bearing is heavily flooded with oil or is liable to have oil blown out by crankcase pressure variations, as in i.c engines.

13.44   Wind-back seals are reasonably effective in retaining oil, but they do nothing to prevent the ingress of dirt – indeed, they tend to pump it in, and thus should not be used in dirty, dusty or wet situations. Some early cars had wind-back seals on the crankshaft where it emerged to carry the fan pulley, but this application was not successful. When the car was driven through a flooded section of road, the wind-back seal pumped far too much water into the crankcase.

Oil throwers

13.45   Where the external conditions are reasonably clean and dry, and where no great volume of oil is emerging from the bearing, a simple oil-throwing disc may be used instead of a seal. An arrangement of this type for a

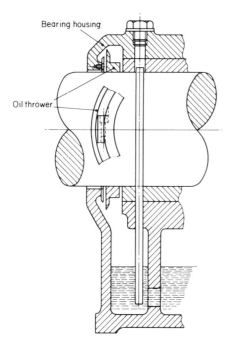

Bearing housing

Oil thrower

Figure 13.14 Simple oil thrower as applied
to ring-oiled bearing (see paragraph 13.45)

ring-oiled bearing is illustrated in figure 13.14. The oil thrower itself may be of
steel or bronze, and may be either of the slip-on type secured to the shaft by a
grub-screw, or in two halves joined by Allen screws, as in the scrap view.

13.46   An oil thrower will be ineffective unless a good, well-guarded route
is provided for the oil flung from the thrower to return to the sump without
danger of its dripping on to the shaft on the wrong side of the thrower disc. The
construction shown in figure 13.14 will achieve this, provided the inner under
surface of the top casing is smooth and painted so that the oil will run down it
steadily instead of collecting in globules.

13.47   Possibly the simplest form of oil thrower is that shown in figure
13.15, consisting merely of sharp-edged grooves cut in the shaft itself. The
drawing shows two such oil-throwing edges, but if the leakage to be dealt with
is very low, then one edge may be sufficient. The disadvantage of the
arrangement is that, quite apart from introducing undesirable stress-raisers,
the proportions shown reduce the shaft diameter under the grooves by an
amount that adds over 37 per cent to the shear stress in the shaft arising from
any given torsional load. Thus, the arrangement cannot be recommended for
applications in which the projecting shaft has to transmit any substantial
torque, although it can be useful in cases where the shaft end is left open merely
to provide a drive to a tachometer or some other small auxiliary device. When
centrifugal force at surface is less than about 8g (e.g. 50 mm shaft at 535
rev/min) very little oil will be thrown off above the shaft centreline, and the
shape of the oil-catching pocket in these circumstances is relatively unimportant. For higher speeds the pocket may be tapered as shown, so that oil
thrown off upwards will tend to drain back to the 'wet' side of the thrower.

13.48   The oil throwers and seals described are intended merely to prevent

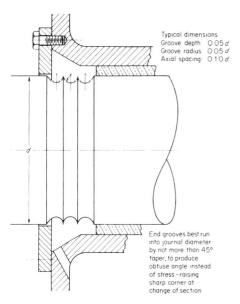

Typical dimensions
Groove depth 0.05 *d*
Groove radius 0.05 *d*
Axial spacing· 0.10 *d*

End grooves best run
into journal diameter
by not more than 45°
taper, to produce
obtuse angle instead
of stress - raising
sharp corner at
change of section

Figure 13.15 Very simple oil thrower (see paragraph 13.47). When the centrifugal force at surface is less than about 8 g (e.g. 50 mm shaft at 535 rpm), very little oil will be thrown off above the shaft centre-line, and the shape of the oil-catching pocket in these circumstances is relatively unimportant. For higher speeds, the pocket may be tapered as shown, so that oil thrown upwards will tend to drain back to the 'wet' side of the thrower.

the oil creeping along the shaft to waste. None of them is suitable for preventing leakage against any substantial pressure difference. For seals that have to hold against pressure, reference should be made to the *Design Council Guide on Seals*, No. 15, the Design Council.

(v) Oil filtration

13.49   In a bearing of, say, 50 mm (2 in) diameter running at 3000 rev/min, the normal running clearance will be of the order of 0.075 mm (0.003 in), and if the eccentricity ratio is 0.70, this will correspond to an oil-film thickness of 11.5 μm (0.00045 in). This would seem to imply that abrasive particles of less than 11.5 μm diameter could not score or damage the bearing. Unfortunately, the position is not as simple as this, because abrasive particles are rarely truly spherical, and therefore a particle that can pass through, say, a 10 μm hole in a filter, could well be greater than 11.5 μm in some other direction. Furthermore, vibration, misalignment, out-of-roundness and surface roughness, all combine to produce an oil film of non-uniform thickness, which in places will be much below its average thickness. Thus, in the 50 mm (2 in) bearing just described, abrasive particles of even 5 μm diameter could probably score and damage the rubbing surfaces.

13.50   Abrasive dirt, even of extremely fine powdery form, is therefore highly undesirable in any lubricant, and the life and reliability of plain bearings will be *much* increased by scrupulous cleanliness in the initial assembly, coupled with care in keeping the lubricant clean in service. (See also paragraph 14.05b.)

13.51   The usual proprietary full-flow filter with a disposable element, such as is fitted to car engines, is seldom claimed to remove more than about 20 per cent of 10 μm particles, and about 10 per cent of 5 μm particles.

13.52   In the face of these figures, some people thoughtlessly take the view that a filter is not worth the trouble and cost of fitting. Why, they say, expend money and effort on something that leaves 90 per cent of the dirt in the system?

13.53   To understand the fallacy of that short-sighted argument, it is useful to examine the hypothetical case of a machine, say a small machine tool, in which the sump capacity is 5 litres of oil, and the oil flow rate is 10 litre/min. For simplicity, let it be assumed that all the dirt entering the system consists of particles of 5 μm diameter, and that they are all abrasive. For round-figure calculations, let it further be assumed that the rate of dirt entry into the machine is 1 milligram per minute.

13.54   In such a machine, the whole capacity of the oil sump would be circulated through the filter twice in each minute, so that, if the filter could remove 0.5 mg of dirt on each pass, it would be 'holding its own' in respect of the dirt intake, and the dirtiness of the oil would remain constant.

13.55   If this filter were removing 10 per cent of the 5 μm particles passing through it, the required removal rate of 0.5 mg of dirt per pass would be achieved if the oil contained 5 mg of dirt per sumpful, i.e. 1 mg/litre. This would be equivalent to the level of 'dirtiness' reached after only 5 min of operation without a filter, and must thus be regarded as a very good level of cleanliness indeed.

13.56   The oil cleanliness would be maintained at this high standard until the filter became completely choked. It would, in fact, improve for a time, because a partially-choked filter has, in effect, a finer mesh than when new.

13.57   If the same hypothetical machine was run *without* a filter, the amount of dirt floating around in the oil after 100 h would be 6 g (0.2 oz). In other words, without the filter the oil would be over a thousand times dirtier.

13.58   The facts are, of course, that the rate at which dirt enters the system must not be confused with the rate at which it is presented to the filter. A filter may well remove dirt at 100 per cent of the former rate, even although it is removing dirt at only 5 per cent of the latter rate.

13.59   For any given size of dirt particle, the dirt content of the oil, in terms of the mass of dirt per unit volume of oil, is given by

$$\frac{\text{dirt intake rate (mass/unit time)}}{\text{pump capacity (volume/unit time)}} \times \frac{100}{\% \text{ particles removed by filter}}$$

This will hold with any consistent set of units, and it will be noted that the answer is independent of the sump capacity.

13.60   The same expression may be used to calculate the effect of a by-pass filter, except that the rate of oil flow through the by-pass must be inserted in place of the pump capacity. If the by-pass filter is of the centrifugal type (either motor-driven or proprietary self-acting Hero-turbine driven), the higher percentage removal of particles will tend to counterbalance the reduced filter flow rate, and in respect of particles below 3 μm will usually far outweigh it.

13.61   The ideal (but seldom used) system is to have both a full-flow paper or felt filter *and* a centrifugal by-pass. This, incidentally, is the procedure adopted by several bus companies that, in the interest of fuel economy, have changed to a less viscous grade of lubricant than that originally recommended by the engine makers. This thinner oil reduces the engine friction and markedly increases the miles per gallon of fuel – although at the risk of greater bearing and cylinder wear. Under these circumstances, with the thinner oil films arising from the less viscous oil, the cash value of oil cleanliness can become very significant.

Installation of filters

13.62   The best position for a full-flow filter is between the pump and the bearings. The filter must be provided with a spring-loaded by-pass, so arranged that oil will continue to reach the bearings even if the filter becomes choked with sludge. In addition, it is advisable to protect the oil pump by a relatively coarse filter on the suction side, the object being to keep out things such as small nuts, screws or other debris that might have accidentally dropped into the sump.

13.63   The air space above the oil in any oil tank or sump obviously needs to be ventilated to the atmosphere. In small installations, this vent may be merely a pin-hole in the centre of the filler cap, but with larger equipment, an inverted U-tube vent is often used. In very dirty or dusty conditions, it may be advisable to have some form of filter in this air vent, the appropriate filter depending on the type of airborne contaminants expected.

13.64   It is also advisable to fit a gauze filter in the filler orifice, the absolute necessity for this depending on whether, for example, the machine is to work in a clean hospital ward or on a muddy building site.

(vi) Oil cooling

13.65   It is not possible to lay down any useful rules regarding oil cooling, which is normally required only in specialised applications, where the bearing designer will have ample past experience on which to draw.

13.66   In other cases, particularly with 'standard' but somewhat highly-rated plant shipped overseas and installed in a hot climate, oil cooling can be a useful ad hoc expedient to overcome site difficulties. The amount of cooling necessary and the best way to arrange it will depend on local circumstances. It may amount to no more than diverting the delivery pipe from the oil pump on its way to the bearings, and arranging for it first to pass through a few turns of copper tube immersed in a water trough. The aim should be to keep the oil-inlet temperature to the bearings below 60°C (140°F) if at all possible.

13.67   The specific heat capacity of mineral oils varies with temperature, but for all practical purposes of oil-cooler calculations may be taken as 0.48 (thermal capacity, 2000 J/(kg °C)).

13.68   When any serious oil cooling is required, advice should be sought from one or other of the oil suppliers.

Chapter 14
Life of bearings

14.01   The life of bearings is related to

(a) number of starts and stops;
(b) duration of stationary periods;
(c) load when stationary;
(d) time between starting and application of full load;
(e) time between starting and establishment of full oil supply;
(f) time occupied in accelerating and decelerating.

14.02   Dry-rubbing bearings and oil-impregnated porous bearings have no problems associated with starting and stopping, because a low rotational speed actually improves their load-carrying capacity. In these bearings, the wear takes place slowly all the time the journal is rotating.

14.03   Hydrodynamic bearings, on the other hand, experience no wear so long as their journals are supported by the hydrodynamic film flow of their design conditions. Provided the oil is clean, wear takes place only at times of starting and stopping, when, for at least a few revolutions, the speed of rotation will be insufficient to generate the necessary hydrodynamic wedge of oil, and the bearing will run in a starved or boundary lubricated condition. The wear or damage at each start and stop will depend chiefly on the amount of dirt and debris in the oil, and on the six factors listed in paragraph 14.01.

14.04   Precise instructions cannot be given on how to cope with these influences in any given case, but two examples may be useful as broad guidelines:

(a) Large turbo-alternator bearings carry a heavy load when stationary and take a relatively long time to run up to full speed. Because of shift work, week-ends, stand-by duty, etc., they often have to stand idle for days at a time. Under these conditions, the wear would be unacceptable if the bearings were left to start unaided. Therefore, a system of high-pressure 'jacking oil' is supplied from a separate pump and led into special grooves so that the whole shaft can be 'floated' on oil for a brief period before and during each start. The normal oil pump is also separately driven and is not only brought into operation some time before each start, but at each shut-down is kept running at full output until the shaft has stopped rotating. Even with these precautions, the bearing life is seldom more than 5000 starts.

(b) Automobile engine bearings carry very little load when stationary, the average idle period is comparatively short, a reasonable engine speed is reached within a second or two of starting, and the speed drops to zero very quickly on shutting down. Full load is not applied immediately on starting (the engine turns at least half a revolution before it even fires on the first cylinder), and the full oil-supply pressure is usually established within the first few seconds. Even under these relatively favourable conditions, a bearing life of 10000 starts (say 100000 miles of 10 miles average journey length) is unlikely to be obtained without very careful oil cleaning (renewing the filter element and changing the oil every 5000 miles or, say, 150 hours running time), to get rid of the detritus that would otherwise score the journals.

14.05   The provision of separate jacking oil, or the changing of oil every 150 hours, would seldom be acceptable in ordinary industrial practice. Nevertheless, to obtain a reasonable bearing life, it is advisable to adopt as many of the following expedients as may be practicable:

(a) Make sure that the oil pump is placed below the level of oil in the sump, so that it will always be completely primed and ready to deliver a full flow immediately on starting.

(b) Pay particular attention to oil cleanliness by providing adequate filtration, either on a full-flow or by-pass system. Dirty oil is usually the principal cause of wear, and few people realize the amount of core sand, metallic swarf and ordinary dust that lies inside even the best of new machinery. With important products, some manufacturers carry out all test running with a separate filter fitted immediately before each bearing, thereby removing any dirt that may have entered the pipework during assembly. These individual filters are removed before commissioning.

(c) Arrange for the oil pump to be separately driven so that it can be started before the main machine, and kept running while the machine is slowing down.

(d) In cold situations, provide means for heating the oil sump, so that the oil pump suction will not be drawing in vain on congealed or over-viscous oil.

(e) With ring or disc lubrication, remember that no flow of oil will actually reach the bearing surfaces until one or two seconds after a start from rest. Thus, if the machine has been standing idle for many hours, much of the oil previously in the bearing may have drained away, leaving it inadequately adapted to carry full load immediately. The degree of danger associated with this will depend on how long the machine has been idle, the viscosity of oil employed, the ambient temperature, whether the bearing has been carrying a heavy load in its static condition (which may have squeezed out all the oil and established metal-to-metal contact), as well as on the time taken to reach normal speed after starting and on how many revolutions will take place before full load is applied. In applications where the summation of these effects is significant, it may be necessary to flood the bearing with oil after each lengthy stop – either by liberal application from a clean oil-can or by any other appropriate means.

(f) With drip-feed lubrication, insist on the mechanical lubricator(s) being cranked vigorously before each start to prime all the oil pipes and passages.

(g) If a long life is required from a continuously-loaded bearing subject to frequent stops and starts, then an expedient sometimes adopted is to rely on ordinary hydrodynamic lubrication during all normal operation but using, instead of a whitemetal lining, a ptfe type dry bearing lining that will cope much better with the boundary lubrication conditions at each start and stop.

(h) If the bearing load is below a certain critical value (see paragraph 4.05), the time taken to accelerate to full speed will be of no significance, but the precautions listed in (a) to (f) above will still be needed to avoid damage through starting with an unflooded bearing. (See also paragraphs 4.07 and 4.08.)

Chapter 15

Automobile engine crankshaft bearings

15.01   The main bearings and big-end bearings in motor cars are really outside the scope of this book, but it is nevertheless interesting to examine why they operate satisfactorily with diameters much smaller than those recommended for equivalent loads in figure 6.1.

15.02   The reasons for this apparent discrepancy may be summarised thus:

(a) The loads normally quoted for automobile bearings are peak loads, whereas those given in figure 6.1 are steady loads that can be carried continuously. The peak loads on automobile bearings occur briefly at intervals of every two revolutions, and thus correspond to much lower steady loads, as explained in paragraph 4.13.

(b) The main-bearing loads are not unidirectional, but are distributed around the whole circumference of the bearing, thus spreading the wear over a wide area.

(c) In many makes of engine, the main-bearing load conditions are helped by the fact that their load vectors have transiently favourable values of $(n-2N)$, see paragraph 4.21.

(d) The load on connecting rod bearings completely reverses during each cycle, thereby introducing a squeeze effect which helps to fill the clearance space with oil in precisely the right manner to assist in the carrying of a subsequent short-lived peak load.

(e) Modern car engines tend to be relatively compact, with closely spaced bearings held in extremely rigid housings machined to very close tolerances. It is usual to specify a maximum alignment error of 1 part in 100000 to 150000 between the bores of adjacent bearings, as against the 1 part in 50000 alignment tolerance built into figure 6.1. The result is that the problem of edge loading from misalignment is less severe than in most other applications.

(f) Since the top-end bearings are lubricated by splash, there are no oil grooves in the big-end bearings, the whole area of which is thus available for load carrying.

(g) Main bearings sometimes have oil grooves round only part of their circumference, thus increasing the bearing area in the more heavily loaded regions. The oil feed to the crankpin in these cases is taken into the journal through multiple radial drillings, one of which is always opposite an oil groove (see figures 13.1 and 13.2).

113

(h) Main bearings, particularly the heavily loaded centre main, are often made without any oil grooves whatever, thereby still further increasing the bearing area. In these cases the adjacent crankpins are fed from the main bearing on their other side. For comments on the effect of oil grooves, see Section 13(ii).

(i) The modern tendency is to use larger and stiffer crankshafts than formerly. It is standard practice to use some form of surface hardening on the journals, and at least one manufacturer uses crankshafts of cast-iron, which provides greater internal molecular damping than steel.

(j) The surface finishes adopted are usually of the order of 0.2–0.25 μm (8–10 μin) as against 0.4 μm (16 μin) in normal industrial practice.

(k) The bearings are invariably thin-walled.

(l) The bearing lining material is carefully chosen to ensure a high fatigue strength, and is commonly flashed with soft lead–tin or indium to assist in running in. (See paragraph 7.23.)

(m) Automobiles normally employ highly specialized lubricants carefully developed for this particular application.

(n) The oil is kept in better condition than in many other applications. The standard servicing procedures call for a complete change of oil, and oil filter element renewal, at intervals of about every 150 hours or so of operation. Attention of this kind is seldom given to normal industrial plant.

(o) Starting conditions are reasonably favourable in that even a hot engine with plenty of oil still lying around the bearings will take at least half a revolution before the first cylinder fires, and at least two-and-a-half revolutions before the last cylinder fires. There is thus a good opportunity for the oil to be spread around the whole bearing surface before any peak load is applied. Furthermore, starting always takes place with a partial throttle opening and correspondingly low peak loads. With a cold engine, where the oil may have drained to the bottom of the bearings, there is the compensation that the engine will usually turn idly through rather more revolutions before it fires.

(p) Because of the very nature of automobiles and their parking places, there is always a gap of several minutes between the time of any cold start and the time when the engine is at full throttle and top cylinder pressures.

(q) The n ormal car oil pump gives its full delivery pressure at fast idling, so that there is a full oil flow through the bearings at all times once they are warm. A relief valve copes with the surplus delivery at higher speeds.

(r) The anticipated life of a set of car bearings corresponds to perhaps 3000 to 4000 hours of operation (say, about 160000 km or 100000 miles) – and in most cases probably less than 500 hours will be at more than 75 per cent of full load. By industrial standards, these are extremely short-life conditions.

15.03 If the paragraphs (a)–(r) are studied in detail, and load-carrying capacity corrections are made in accordance with the recommendations in this book, it will be found that the design of automobile engine bearings conforms much more closely to the recommendations in figure 6.1 than might at first have been supposed.

15.04 Even so, however, they tend to be rated very nearly to the limit of their capacity and, in spite of all the accumulated knowledge on the subject, automobile crankshaft bearings are seldom satisfactory 'straight off the drawing board'. They rarely acquire their final production-line dimensions, tolerances, material specifications, oil hole and grooving arrangements, etc., without a long history of research, development and extensive road testing.

15.05   This brief review of these highly-rated bearings may be useful in suggesting ways in which readers, if in difficulty, may themselves tackle the problem of high bearing pressures in other applications.

Chapter 16

Examples of practical design problems

16.01   In the design of bearings for real applications, as opposed to academic exercises, consideration must be given to *all* the aspects that might affect bearing performance, as well as to *all* the features of the proposed design that might either improve or detract from the general viability or attractiveness of the complete machine for which the bearing is intended. This frequently means giving consideration to matters quite outside the list of normal design variables set out in figure 2.1. Some examples of this broad approach, or of the need for it, are given in the following paragraphs.

Example 1: Spindle bearings for high-precision grinding machines

16.02   For the grinding machine shown in figure 16.1a, the designer was asked to provide bearings having the absolute minimum of diametrical clearance. The reason for this request was that during normal grinding the forces A and B both acted to hold the spindle against the side of the bearing remote from the workpiece, whereas when running free there was no force A and force B then tended to tilt the grinding wheel towards the workpiece, as shown in the illustration. This meant that, in the final finishing passes across the workpiece, the grinding wheel tended to move forward in total by the amount of the bearing clearance, so that the spark-out time was unduly prolonged. Thus, for the production of accurately dimensioned diameters, it was very difficult to judge at what diameter to commence the spark-out procedure. Furthermore, at the beginning of each fresh pass, there tended to be an undesirable 'bump' when the grinding wheel in its forward free-running position encountered the workpiece and was then suddenly forced back into its retracted position.

16.03   Probably the only type of bearing to satisfy this demand would be a pressurised air bearing, which would tend to be much too expensive for small machines. The alternative approach shown in figure 16.1b was therefore evolved. In this alternative layout, the belt tension assists in holding the spindle against the back of both bearings at all times, thereby maintaining the grinding wheel in almost the same axial line when sparking out as when grinding. The difficulty of changing a one-piece belt on a pulley placed between two bearings was met by the ingenious concept of mounting the driving-end bearing on an overhung portion of the main structure, as shown.

116

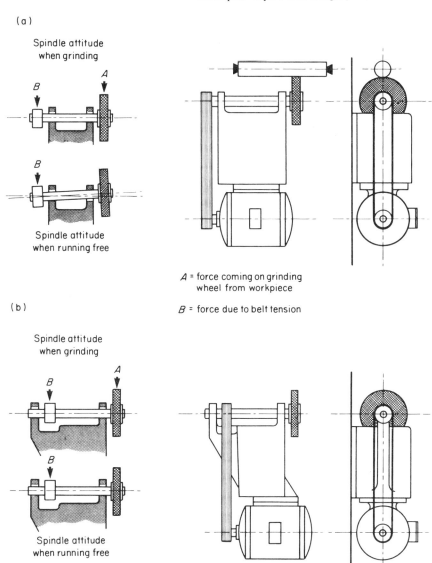

(a)

Spindle attitude
when grinding

Spindle attitude
when running free

(b)

Spindle attitude
when grinding

Spindle attitude
when running free

A = force coming on grinding
wheel from workpiece

B = force due to belt tension

Figure 16.1 Precision grinding machine spindle

16.04   Designers ought continually to be on the look-out for the possibility of obtaining benefit from indirect modifications of this kind.

Example 2: Turbo-alternator bearings

16.05   As early as 1910, the steadily increasing power and size in which turbo-alternators were being built began to create problems with the erection at site of the long lines of widely spaced bearings inherent in such machines. To

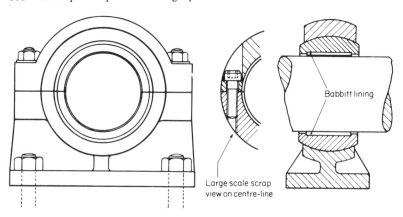

Large scale scrap
view on centre-line

Babbitt lining

Figure 16.2 Diagrammatic arrangement of self-aligning bearing (see paragraph 16.05)

provide the ideal form of support shown in figure 6.12, the bearings had not only to be located *exactly* (i.e. within tolerances equivalent to a fraction of the bearing oil-film thickness) but each had also to be mounted at the exact angle of inclination. To overcome these problems, it became standard practice for the individual shafts to be provided with couplings of the Wellman–Bibby type which would accommodate minor inter-shaft malalignments, and for each bearing to be carried in a self-aligning spherical mounting of the type shown diagrammatically in figure 16.2.

16.06   With heavy steam-turbine shafts, it is difficult to maintain a film of lubricant between the two spherical surfaces (especially as they are not in any significant relative motion) and the friction between them thus soon becomes too great for any self-aligning movement to take place during normal running. The sole purpose of the spherical construction was thus simply to allow the bearing to settle freely into its natural alignment during the initial assembly and erection stages, after which the cap was tightened down and the bearing thereby locked in its correctly-aligned position before the set was put in operation. Allen screws or the like (shown in the scrap view on figure 16.2) were provided to hold the two halves of the spherical shell together for handling purposes, but these played no part in the functioning of the assembly.

16.07   Shortly after World War 2, most turbo-alternator makers abandoned the use of flexible couplings, and relied on good foundations and accurate alignment on assembly, although still keeping the self-aligning bearing so that the work of aligning at site was concerned only with the location of the x and y-axes of the bearing seats, and did not simultaneously involve worrying about the parallelism of each bearing with its associated journal. Within the last decade, there have been further developments on turbo-alternator bearing arrangements with, in general, quite separate approaches being adopted in the USA and UK.

16.08   In the USA, there has been a move towards ignoring the natural lie of the shafts, as illustrated in figure 6.12 and, instead, mounting all bearings in a straight line and simply pulling the shafts into something near this straight line by applying brute force to the couplings between them. With this procedure,

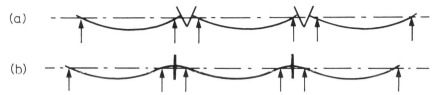

Figure 16.3 Turbine bearing arrangement (see paragraph 16.08)

the individual shafts lie somewhat as shown in figure 16.3a before they are bolted up, and as shown in figure 16.3b after they have been bolted up.

16.09 With this arrangement, the bearing alignment can be carried out with great accuracy (by optical means, for example), but at first glance this advantage would seem to be completely overshadowed by the enormous additional stresses introduced into each shaft. In fact, this is not necessarily so, as perhaps may be most readily appreciated by the analogy of the respective bending moments in the two different shaft arrangements shown in figure 16.4. In figure 16.4a, there are three separate shafts, whereas in figure 16.4b, there is a slightly longer single shaft lying on supports placed at the same spans as before. If w is the weight per unit length of each shaft and s is the bearing span, the bending moment at the centre of each of the three separate shafts will be $ws^2/8$, whereas the peak stresses in the single-shaft arrangement will all be at the

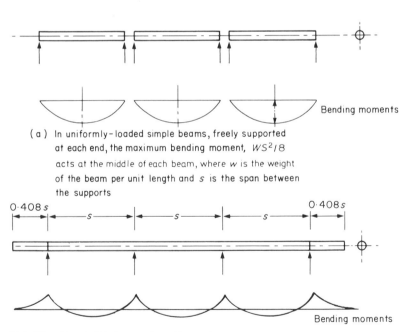

(a) In uniformly-loaded simple beams, freely supported at each end, the maximum bending moment, $WS^2/8$ acts at the middle of each beam, where w is the weight of the beam per unit length and s is the span between the supports

(b) When the shaft is in one piece, the centre spans will, in effect, be uniformly-loaded encastré beams in which the maximum bending moment, $WS^2/12$ acts at each point of support

Figure 16.4 Simple analogies to illustrate paragraph 16.09

points of support and each will amount to only $ws^2/12$, i.e. two-thirds of those in the corresponding three-shaft arrangement. This is because the centre spans of the single shaft will each, in effect, be encastré beams and the end portions will be simple cantilevers providing bending moments to balance those at the ends of the outer encastré-like parts. In the case of real turbine installations carrying loads in the form of discs, etc., the conditions will naturally not be completely analogous to those of figure 16.4b, but the same broad principles will apply and can, of course, be readily checked by calculation in any given case. The fact that USA manufacturers and users seem to encounter no problems in service with this procedure would seem to confirm that it has considerable merit. It is, of course, still necessary to use spherically-seated bearings to accommodate the changes in shaft inclination that arise during the bolting up of the couplings.

16.10   In the UK, the change of practice has centred on the virtues of spherical bearing seats compared with their disadvantages – the chief of which is their length. When spherical seats were first adopted for large turbo-alternator bearings, it was usual to make their length correspond to a b/d ratio of about 3:1. This was done to give any edge loading on the bearing ample 'leverage' to shift the spherical seat into its correct position. Later, the desire to save length resulted in the normal b/d ratio coming down to unity, and in some cases $b/d = 0.5$ was employed, with increasing reliance being placed on site erectors to see that the spherical movement had indeed settled into the correct position. The biggest step forward was, however, when it was realised that (for the reasons outlined in paragraphs 6.29 to 6.39) these shorter bearings were much less sensitive to edge loading than were the previous very long bearings, and that with short enough bearings it might be feasible to dispense with the spherical seat completely. A study of the loading conditions showed that, with reasonable alignment, a bearing length corresponding to $b/d = 0.3$ (or even less) might be adequate in many cases.

16.11   Current UK practice with some turbo-alternator makers is thus to use normal 'fixed' bearings (i.e. without spherical self-aligning seats) with $b/d = 0.3$ or even 0.25. Compared with the long-standing practice of $b/d = 1$, this saves $0.7d$ on the length of each bearing; on a large set with six bearings, this amounts to a total saving of $4.2d$. If the bearings are, say, 610 mm (24 in) in diameter, this corresponds to a saving of some 2.5 m (8 ft), not just in the length of the bearings but also in the length (and therefore the cost) of the foundations, the building, the crane rails, the pipework, etc., as well as in the cost and weight of all the turbine shaft forgings. Alignment on site is also easier since there is nothing to get misaligned in transit from the test bed to the site.

16.12   This particular application has been dealt with at some length because it is a good example of how intelligent bearing design can sometimes bring very substantial benefits in areas far removed from the bearings themselves.

Example 3: Problems with inadequate background data

16.13   The drawing reproduced in figure 16.5 is, unfortunately, typical of what is all too frequently sent to bearing manufacturers to illustrate what customers are seeking when they order standard items. In this case, two bushes

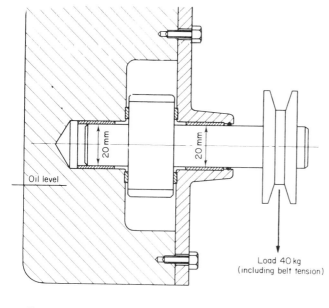

Temperature 50 °C
Speed 2000 rpm

Figure 16.5 Pulley-driven gear (see paragraph 16.13)

and two thrust washers were required and it was left to the bearing supplier to decide *which* of his various types of standard bush and/or thrust washer would be most appropriate for the duty.

16.14   The two features to note about this drawing are, first, that the mechanical layout of the bearing surroundings seriously inhibits the satisfactory performance of *any* bearing and, secondly, that it is impossible to suggest any specific improvements to the design without knowing much more about the background data than can be deduced from the drawing.

16.15   Thus, the more obvious faults are

(a) The pulley is too far overhung from the nearest bearing, thereby setting up a couple that will seriously increase the loading on both bearings.
(b) The inboard bearing is not adequately covered by the shaft, and would therefore wear with a step in it. The step would tend to carry much of any thrust in a left-hand direction and thus give rise to local over-heating, with consequent damage to the bearing lining for some distance along the bearing.
(c) The outboard bearing is held in a housing plate that is not adequately located in relation to the main housing, thereby making the mutual alignment of the two bushes far too indeterminate to achieve satisfactory results.
(d) Although the gear wheel is shown running in oil, there is no obvious route for any of this oil to reach the rubbing surfaces of the two thrust washers or of the two bushes. The amount of lubrication they are liable to receive would thus seem inadequate for them to be rated as film-lubricated

bearings, yet the pV rating is too high to permit operation with dry-bearing materials in a dry state.

(e) The possibility of oil circulating through the inboard bearing is, in any case, ruined by the blind-end configuration of this bearing. Also, detritus or other dirt, could build up in the blind end and cause trouble.

(f) The housing surrounding the outer bearing seems to be too thin to carry the stresses arising from the interference fit of the bush it contains. It would thus seem liable to become bell-mouthed, which would loosen the bush.

16.16   In order to improve the design, several other problems also need to be tackled. For example:

(a) No indication is given of the magnitude or direction of the load on the gear wheel arising from its meshing with its mating gears. This, together with uncertainty about how much of the specified 40 kg load arises from belt tension (which will pull centrally on the shaft centre-line) and how much arises from the driving effort (which will be displaced about 27 mm from the centre), makes it impossible to determine the true load on the bearings, or to discover how it will be shared between the two bushes.

(b) Although one would be tempted to recommend moving the pulley closer to the bearing, this may not be possible if the driving pulley is so placed that it cannot readily be moved to suit. In such a case, it might be necessary to move the bearing out closer to the pulley, but this would leave the gearwheel in a less well-supported position, and thereby call for either a larger shaft or some other design changes.

(c) Much will depend on the use to be made of this pulley-driven gearwheel assembly. Is it to be run continuously on full load, day and night, or to be run for only a few seconds, at a time, once or twice a day, as, for example, to close overhead blinds or shutters at night and re-open them in the morning?

16.17   An important lesson to be learnt from the foregoing is that it is seldom possible to say whether a bearing design is good or bad simply by looking at it, since so much depends on extraneous facts that would not be apparent in any normal drawing. These extraneous facts will vary from one application to another, so that a drawing representing a good design for one customer may, for another, be either quite inadequate or unnecessarily complex.

Appendix 1

Plain bearings versus rolling bearings

A1.1   Bearings are normally regarded as falling into one or other of two main categories: either rolling bearings or plain bearings. The former includes ball and roller bearings of all kinds; the latter includes bearings operating on many different principles and using a wide variety of lubricants but generally having interacting surfaces with relative motion of a sliding nature.

A1.2   The choice between plain and rolling bearings is usually made on the basis of practical, commonsense considerations rather than on the basis of a technically involved appraisal. For those seeking guidance, some of the particular advantages of each type are given below.

A1.3   Rolling bearings are usually the easier for a draughtsman to incorporate into a design, because all he need do is to select them from a catalogue which gives full details of load capacities, appropriate speeds, housing fits and tolerances, anticipated life, etc. When spares are available, rolling bearings are easier to replace because there are no wearing surfaces to be reconditioned. Although there is usually no detectable difference between the running frictional losses of the two types, rolling bearings do tend to have much less 'stiction' than most types of plain bearing when starting from rest. For many applications in which the load and speed conditions are not too severe, rolling bearings may be adapted to run in a grease-packed sealed construction that will require no further lubrication or maintenance throughout their life.

A1.4   Plain bearings, on the other hand, do not always have to be made by specialists, and in their traditional forms can often be renewed or repaired with no more than the usual workshop facilities available almost anywhere in the world, including ships' engine-rooms. Even in their more advanced forms, plain bearings are often easier to replace than rolling bearings, particularly since most types can readily be made of split construction, enabling them to be removed without withdrawing the shaft. They are also less susceptible to damage from water and dirt, can be built to carry many times higher loads than rolling bearings, and are more suitable for operation at very high speeds. Plain bearings tend to wear out more gradually and are much less prone to sudden collapse without warning. Even if they fail, it is usually possible to keep the machine running long enough to avoid any hazards of an immediate shutdown. Plain bearings are quieter and when necessary, can be designed to provide and maintain a more precise concentricity of the shaft within its housing. They do not 'brinell' when subject to vibration from an external source, and operate reliably under conditions of high cyclic loading such as occur in reciprocating machinery. In special forms arranged to use readily available pressurised gas (e.g. air) as a lubricant, they have a lower coefficient of friction (both when at rest and in motion) than any other known form of support. They also tend to be much cheaper.

Appendix 2
Glossary of bearing terms

A-bracket An outboard support for a ship's propeller shaft, consisting of two angled struts attached to the hull and converging on to a boss housing a bearing.

Abrasion, abrasive wear Wear caused by the presence of hard protuberances on the surface or by hard particles either passing between the two surfaces of a bearing or embedded in one of them. It may occur either in the dry state or in the presence of a liquid. *See also* Wear.

Abrasive erosion Erosion (q.v.) in which the relative motion of the solid particles is nearly parallel to the solid surface.

Additive A material added to a lubricant to impart new properties or to enhance existing properties. *See also* Detergent additive; Dispersant additive, Oil additive.

Adhesive force The attractive force between the adjacent surfaces in frictional contacts.

Adhesive wear Wear (q.v.) by the transference of material from one surface to another during relative motion due to a process of solid-phase welding. The process commences at the tips of the surface asperities and is usually the first stage in the seizure process.

Adsorption Atomic or molecular attachment to a solid surface, e.g. of hydrocarbons, oxides, or gases to a journal or bearing surface.

Aerostatic bearing An externally pressurised bearing (q.v.) using air as the fluid.

Aft bush The bush situated at the aft or outboard end of a sterntube, adjacent to the propeller, in a two-bearing sterngear system (q.v.).

Aftermost plummer bearing A main propulsion shaft plummer bearing (q.v.) which has the bearing lining in both top and bottom halves. So called because it is positioned aft of all other bearings and immediately forward of the sterntube (q.v.) It may be fitted to the forward end of the propeller shaft or to the aft end of the adjacent intermediate shaft.

Air bearing A bearing using air as a lubricant. *See also* Gas lubrication.

Air-cooled bearing A bearing in which the cooling medium is air.

Air seal An annular cavity, surrounding a shaft where it enters a bearing casing, into which low-pressure air is introduced to prevent oil being drawn out of the casing by air suction from the immediate environment. Used, for example, in electrical machines

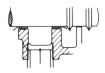

Alignment *Bearing to shaft:* the parallelism of the longitudinal axes of the shaft and the bearing bore, or of a thrust collar face to the face of the thrust bearing. *Of bearing bores:* the coincidence of the longitudinal axes of a number of bearing or housing bores with a straight line, or a predetermined curve. For example, in an engine the former will apply to the big ends and small ends, and the latter to the main bearings.

Aluminium alloys *See* Aluminium–silicon, Tin–aluminium.

Aluminium –silicon alloys Aluminium alloys containing about 8 to 16 per cent silicon, often with small additions of copper or nickel. Used for pistons (low expansion alloys); components (low density, ease of casting, some alloys used in marine applications because they have good resistance to corrosion); bearings (lined on to steel backing).

Aluminium tin *See* Tin-aluminium

Amonton's laws The two laws were propounded in 1699: (a) The friction force is proportional to the normal force. (b) The friction force is independent of the size of the contact area.

Anisotropic cracking Cracking of tin-base whitemetal linings due to thermal cycling in service, causing plastic deformation in axes of higher thermal expansion. (*Syn* Thermal ratcheting)

Annular groove A groove running in a circumferential direction. In a journal bearing, it may be in the bore or the back and is parallel to the bearing ends. It is used to feed oil to the bearing, or to connecting axial grooves, and to transfer oil between bearing and shaft. (*Syn* Circumferential groove). *See also* Groove

Anticlastic bearing A structural bearing designed to accommodate simultaneously rotation (q.v.) in two vertical planes at right angles, and translation (q.v.) in any direction in a horizontal plane (but not rotation in the horizontal plane).

Anti-seizure property The ability of a bearing material to resist seizure during momentary lubrication failure. *See also* Compatibility.

Anti-wear additive An additive used in extreme-pressure lubrication (q.v.).

Anti-wear treatments Surface chemical or thermal treatments applied to improve wear resistance.

Area of contact Between two solid surfaces, described in tribology as: (a) Apparent area (or nominal area) – area of contact defined by the macroscopic geometry of the bodies. (b) Real area – the sum of the local areas transmitting interfacial force directly between the solids.

Asperities Small-scale irregularities on a surface.

As-pressed bearing, bush A half-bearing (q.v.) or bush produced by pressing from strip or bar material where the back is formed with sufficient accuracy to avoid the need for subsequent machining. Particularly applicable to thin-walled bearings and wrapped bushes (q.v.). *See also* Precision pressed bearing.

Attitude In a bearing, the angular position of the line joining the centre of the journal to that of the bearing bore, relative to the direction of loading.

Attrition The removal of small fragments of surface material during sliding contact.

Axial clearance, total axial clearance The

operating clearance between the face of a thrust bearing and the cooperating collar face. The total axial clearance applies to double thrust bearings and to the total endwise movement allowed, usually when less than 2 mm.

Axial groove(s), grooving Groove(s) in the surface of a journal bearing parallel to the axis of the journal. *See also* Groove.

b/d ratio The ratio of the axial length (or width) of a plain bearing to its bore diameter.

Babbit metal A non-ferrous bearing alloy originated by Isaac Babbit in 1839. Currently includes several tin-base alloys containing tin, copper, antimony and lead in various proportions. It is the USA name for whitemetal (q.v.).

Back of bearing The surface on the opposite side to the bearing surface. The outside diameter of a journal bearing or bush.

Backing The supporting material in a bearing, to which the lining material is bonded. *See also* Bimetal bearing, Trimetal bearing.

Baffle A system of shaped walls and cavities, where a shaft enters a bearing casing, designed to prevent the escape of oil. It usually collaborates with flingers.

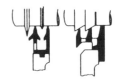

Barrelled journal A journal (q.v.) the axial contour of which is greater in diameter at the centre than at the extremities.

Barrier An electrodeposited layer of nickel or other metal superimposed upon the interlayer surface of a trimetal or overlay-plated bearing (q.v.) prior to the deposition of the overlay (q.v.) in order to reduce diffusion of tin from the overlay to the interlayer.

Barring The 'turning over' or rotation of a machine very slowly by external means for purposes of erection, maintenance, or prevention of thermal distortion by heat soak. Strictly, barring is an intermittent motion often applied by lever (bar) and ratchet, but now synonymous with turning (q.v.).

Barring gear The mechanical means provided for carrying out the operation of barring.

Base plate The structural component of a vertical thrust bearing which supports the thrust pads and forms the lower part of the oil retaining enclosure.

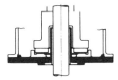

Bath-lubricated bearing A bearing whose casing contains an oil bath, with means for supplying oil therefrom to the bearing surfaces but not necessarily including a means for cooling the lubricant. *See also* Disc lubrication, Oil-ring lubrication.

Bath lubrication The lubrication of surfaces by partially or wholly submerging them in a bath of lubricant. Also used to describe a complete bearing in which the lower casing contains a bath of oil in to which a disc dips to lift oil to the bearing surfaces. *See also* Bath-lubricated bearing.

Bean hole An elongated hole delineated by two semi-circles joined by short straight lines. Often used in conjunction with a dowel to give location in one direction and freedom for translation in another direction at 90°.

Bearing A support or guide by means of which a moving part is located with respect to other parts of a mechanism. *See also* under specific bearing names, e.g. Big-end bearing.

Bearing assembly An assembly comprising at least a journal bearing and/or a thrust bearing, with bearing housing (q.v.) or casing (q.v.), but possibly including all features necessary to make a complete bearing, e.g., pedestal bearing, plummer bearing, vertical bearing, main thrust block.

Bearing bore The working surface complementary to the journal. Also used to signify internal diameter

Bearing brass An obsolete term for an insert bearing (q.v.). *See also* Brass.

Bearing bronzes The bronze alloys commonly used in bearing manufacture are gunmetal (copper, tin, zinc); leaded gunmetal (copper, tin, zinc, lead), phosphor bronze (copper, tin, phosphorus), copper lead (copper, lead, with or without tin). All these alloys may have an addition of nickel to

increase the strength and where appropriate to improve the lead distribution.

Bearing casing The main structural component which supports the bearing and which also forms the oil-retaining enclosure.

Bearing casing bottom That component of a horizontal bearing split casing which is below the shaft axis.

Bearing casing top That component of a horizontal bearing casing (q.v.) which is above the shaft axis.

Bearing clearance The space between the journal and the bearing bore, defined by the difference between their diameters or radii. In order to avoid confusion, it is preferred to use the terms diametric clearance or radial clearance (q.v.).

Bearing housing The hollow member into the bore of which a bearing liner is fitted. The housing may be made from a single block, or may be split into halves. Also a structural component supporting a bearing, or bearings, but which is itself supported inside the oil retaining casing.

Bearing length Commonly applied to the axial dimension of a journal bearing or bush. In modern theory, the term is restricted to that dimension of a bearing surface in the direction of sliding motion. *See also* Bearing width.

Bearing liner In a journal bearing, the tubular element (often in halves) the bore of which is the bearing surface. (*Syn* Insert bearing, insert liner).

Bearing lining The material forming the working surface of the bearing and which is bonded to the backing.

Bearing load (journal) The radial force imposed upon a bearing by reaction with the journal.

Bearing load (thrust) The axial force imposed upon a thrust bearing by reaction with the shaft collar.

Bearing pressure A commonly used but now non-preferred term. The preferred term is specific load (q.v.).

Bearing reaction The equal and opposite force with which the bearing opposes the load applied by the shaft. *See also* Bearing load.

Bearing shell An obsolescent term sometimes used for half-bearings (q.v.) in the rough formed state awaiting lining. Now also used for thin-walled half-bearings, but non-preferred in this usage.

Bearing system The interactive system of the bearing, the cooperating surface, and the lubricant, operating within the applied and resultant environmental conditions.

Bearing width The overall dimension of a bearing liner, measured in a direction perpendicular to the motion. (*Syn* Bearing length of a journal bearing)

Bedded arc   That part of a bearing bore in which, by design, the curvature is the same as that of the journal. It may be produced, as in cross head bearings, by deliberate machining or hand scraping, or by wear when running in (q.v.). (*Syn* Fitted arc)

Bedding in   The operation of producing conformity between a bearing surface and a shaft by hand scraping the former or by controlled wear during running at low speed.

Bedplate   The plate or girder forming a base upon which a machine is built. In engines, it often incorporates the housings for the main bearings.

Beilby layer   An amorphous layer of deformed metal and oxide particles which, according to Beilby, is formed by polishing (current thought disputes the existence of such a layer).

Big end   The larger (crankshaft) end of the connecting rod which is connected to the crankpin. (*Syn* Bottom end, Large end)

Big-end bearing   A bearing at the larger (crankshaft) end of a connecting rod in an engine. *See also* Bottom-end bearing.

Bimetal bearing   An insert bearing (q.v.) consisting of a backing, usually steel, but sometimes cast iron, bronze or brass, lined with a bearing alloy.

Black scab damage   *See* Wire wool damage

Blade rod   In a particular V-engine configuration, the narrow 'inner' rod which embraces and oscillates upon the outer surface of the fork rod bearing. *See also* Fork rod.

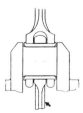

Bleed groove, hole   A small groove, or hole, leading from an oil groove at the end remote from the oil entry to increase the flow of oil for cooling purposes and/or to flush out debris or foreign matter (q.v.). Used particularly in bearings operating with thin films or small clearances. *See also* Groove.

Blind groove   A groove which terminates within the bearing surface and does not lead to the edge of the bearing.

Block half-bearing   The insert half-bearing (q.v.) housed in the cylinder block of an engine or reciprocating pump (q.v.). (*Syn* Case half-bearing)

Bolting loads   Loads imposed by bolts when tightened to withstand separating forces exerted by the interference fit (q.v.) of bearings, and by the reciprocation of moving parts.

Bond strength   The mechanical strength of the bond between the backing material and bearing lining, or between the interlayer and overlay or surface lining in overlay and trimetal bearings (q.v.).

Bore relief   An arc of reducing wall thickness towards the joint faces on half-bearings (q.v.). It is used to avoid the danger of a small step in the assembled bore due to manufacturing tolerances in the bearing wall thickness or the housing bore.

Boring allowance   The material left in the bore of a bearing or bush to permit of final boring in situ after assembly in the housing. The boring allowance should be clearly defined, either in terms of maximum permissible radial cut, or of permissible maximum diameter which is allowed at the final machining after assembly. This is particularly important with thin-walled bearings (q.v.) which have very thin linings. (*Syn* Finish allowance). *See also* Pre-finished bearing.

Boring in situ   This is the final operation of boring bearings or bushes after assembly in their housings, in order to compensate for possible misalignment, or to control clearance within closer limits than is possible with pre-finished bearings (q.v.). Care must be taken to ensure that the boring allowance (q.v.) is not exceeded.

Bossing   In marine engineering, the term given to the swelling of the hull where the shell plating encloses the propeller shafting. In a single-screw ship, the term includes the swelling of the sternframe around the sterntube or stern bearing (q.v.). In a twin-screw ship, it describes the appendage tapering from the main hull lines to the propeller, or to the point where the shaft leaves the protection of the hull.

Bottom end   The end of the connecting rod which is connected to the crankpin. (*Syn* Big end, Large end)

Bottom-end bearing   The big-end bearing (q.v.) in a vertical axis cylinder reciprocating machine. This term is usually used in marine practice with crosshead engines. (*Syn* Crankpin bearing)

Boundary lubricant   A lubricant suitable for use in boundary lubrication conditions (q.v.). Fatty acids and soaps are commonly used.

Boundary lubrication   A condition of lubrication in which the friction and wear between two surfaces in relative motion are determined by the properties of the surfaces, and by the properties of the lubricant other than bulk viscosity. It occurs where load is so high or the velocity or viscosity so low that a hydrodynamic film is not established and lubrication takes place at the molecular level on asperities (q.v.) of contact. (*Syn* Boundary lubrication). *See also* Elastohydrodynamic lubrication.

Bracket bearing   The bearing in a bracket which supports the propeller shafting of a vessel outboard from the hull or sterntube.

Brass   An alloy of copper and zinc, sometimes with additions of lead, tin, nickel, aluminium and other elements. Brass alloys are poor bearing materials but still sometimes used as a backing material for whitemetal lined bearings. The terms 'brass' and 'brasses' were used to denote the bearing liners in early engines, deriving from the use of cast brass for these parts.

Breakaway torque   The minimum torque which must be applied to overcome static friction including friction within the lubricant film, to initiate movement of the shaft or machine from rest. (*Syn* Starting torque)

Bridge gauge   A means used for checking wear-down, commonly used in marine machinery. The upper half-casing and bearing is removed to expose the shaft journal. The bridge gauge is then placed over the shaft resting on the housing joint and bolted down. Feelers are used to check the gap. Now usually replaced by wear-down gauge (q.v.).

Bridge piece   The connecting member or members between two adjacent bands of a half-bearing (q.v.) in tandem axially.

Bridge-piece bearing   A design adopted especially with thin or medium-walled half-bearings (q.v.) where the wall is too thin to accommodate a central annular groove of adequate depth for the required rate of oil flow and an additional groove is provided in the bearing housing.

Brinelling   The indentation of the surface of a solid body by repeated local impact or by static overload. *See also* False brinelling.

Broaching   A machining operation entailing the linear passage of a series of cutting edges to produce a surface of the correct form in one pass. Applied to half-bearings, it is used to finish the bores and the joint faces.

Bronze   An alloy of copper and tin, often with other elements in smaller proportions. *See also* Bearing bronzes.

Burning   In sliding contacts, the oxidation of a surface due to local heating in an oxidising atmosphere.

Burnishing   A process for polishing and/or improving the accuracy of a surface by the rubbing or sliding action of another, harder, surface causing the surface to 'flow' without cutting. Burnishing may result from running in (q.v.) at very low speed.

Burr A sharp lip of metal carried over an edge during a machining operation.

Bush A bearing member which is continuous and embraces 360° of journal circumference. Sometimes used loosely to denote a bearing liner (q.v.) but this usage should be avoided. *See also* Solid bush, wrapped bush.

Butt joint A straight axial split in a wrapped bush. *See also* Clinch joint.

Button stop A disc or 'washer', secured by a countersunk head screw in a recess in the housing joint so that it engages a half-bearing (q.v.) to secure it in position. Two stops are used to each half-bearing usually placed diagonally opposite.

Cap The removable half of a split housing retaining a bearing. *See also* Keep.

Cap half The half of a split bearing (q.v.) fitted in the housing cap.

Case half-bearing *See* Block half-bearing

Casing *See* Bearing casing, Split casing

Cast-iron process The treatment of cast-iron in order to remove surface graphite which inhibits satisfactory tinning prior to the whitemetal lining. The treatment consists essentially of immersion in an electrolytic fused salt bath, which oxidises the graphite, leaving a pure iron surface which tins readily. Surfaces so treated can be re-tinned and re-metalled after service with satisfactory results, the bond strength (q.v.) obtained being comparable with that between whitemetal and tinned low-carbon steel. *See also* Kolene process.

Cavitation erosion The impact fatigue attack of bearing surfaces caused by the formation and collapse of vapour bubbles in the lubricant film.

Centre-line height The amount by which the external peripheral length (i.e. circumference) of a half-bearing exceeds that of a true semicircle when held in contact with a housing of specified diameter. *See also* Half height, Crush, nip

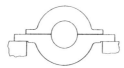

Centre-line mounted bearing A bearing supported on two plane surfaces, one at each side, which coincide with, or are close to and parallel with the shaft axis. The support surfaces are usually in a horizontal plane. The casing is usually of split construction and may contain any combination of bearing functions. The type is adopted primarily for high temperature machines to minimise movement of the shaft axis due to thermal expansion.

Centrifugal filter A filter which removes solid foreign matter from a liquid by centrifuging.

Centrifugal lining The process of lining a bearing backing by pouring molten lining material into it whilst it is rotating at a speed sufficient to throw the molten material into contact by centrifugal force.

Centrifugal loading The load vector rotating around the bore of a journal bearing in a plane normal to the journal axis, induced by out-of-balance forces imposed by rotating masses. *See also* Rotating load.

Centrifugal lubrication The supply of oil to bearing surfaces and/or a cooler by a centrifugal pump built into the bearing. Often applied to vertical bearings using radial holes in the collar.

Centrifugal separator A device similar to the centrifugal filter (q.v.) but where the centrifuging action is used to separate two immiscible liquids of different density, e.g. water and oil.

Chambered bearing An externally pressurised (q.v.) bearing with several chambers fed by individual inlet passages. Such bearings are self-centring. There is a particular type of bearing favoured by turbine makers known as a 'chambered top half'. The 'chamber' is a shallow circumferential groove having a width about $\frac{1}{4}-\frac{3}{4}$ of the bearing width or length, connecting the two axial grooves at the joint. This was adopted in the belief that it reduced power loss. It is now considered to be of doubtful value since it tends to replace shear losses by turbulent losses and moreover it may cause instability.

Chamfer A bevel or slope made by machining off the edge of a right angle, e.g. at the end of a bearing or bush.

Chatter   Elastic vibrations originating from frictional or other instability. *See also* Stick–slip.

Chatter marks   Surface imperfections resulting from chatter during machining operations.

Checking load   The load imposed upon the joint faces (q.v.) of a half-bearing (q.v.) supported in a cup when measuring its peripheral length or centre-line height (q.v.).

Churning losses   The frictional power losses caused by the churning and turbulence of lubricant between rotating and stationary parts. Particularly used in connection with tilting-pad bearings (q.v.), referring to the spaces between pads, casing and moving surfaces. It applies also to grooves and gutterways (q.v.) in journal bearings.

Circulated-oil cooling   This applies to a bath-lubricated bearing (q.v.), as distinct from forced lubrication (q.v.), where oil is circulated through the oil bath from an external system for cooling purposes only. It is used where the size of cooler needed for heat extraction precludes it being contained within the bearing casing, or where it is convenient to use a supply of cooled oil already available but it is desirable to retain the bath lubrication in case of failure in the oil supply.

Circumferential groove   *See* Annular groove

Clearance   *See* Bearing clearance, Diametric clearance, Radial clearance

Clearance ratio   The ratio of the radial clearance (q.v.) to the shaft radius.

Clinch joint   The interlocking joint at the split of a wrapped bush to keep the joint closed whilst in a free state. Also used to facilitate the grinding of the outside diameter during manufacture.

'Clip-on' flanges   Lined flanges attached to a thin-walled half-bearing by being sprung and/or peened over locating lugs, to allow assembly as a single piece into the housing.

Closure   The generic term given to the means of preventing loss of oil and/or the ingress of foreign matter at the entry of the shaft into a bearing casing (q.v.).

Coefficient of friction   The ratio of the tangential force resisting motion between two bodies in contact to the normal force pressing these bodies together.

Coining   A term used in pressing operations, whereby, after forming (e.g. into a half-bearing from flat stock) the final coining operation neutralises free stresses, and produces a dimensionally accurate and stable component. *See also* End coin.

Collar oiler   A collar on a shaft which extends into an oil reservoir and carries oil into a bearing as the shaft rotates. Wipers

are usually provided to direct the oil into the bearing. It is preferred to use the term disc lubrication (q.v.).

Compacting   The densification of porous bodies by hot or cold pressing or rolling. Applicable particularly to sintered materials. *See also* Sintering.

Compatibility   *Frictional compatibility*: a measure of the resistance against adhesive wear of a given combination of materials. Compatibility or seizure resistance is the attribute of a material, or lubricant, to work with the other components of a bearing system (q.v.), without causing damage even under unfavourable conditions. *Lubricant compatibility*: a measure of the degree to which lubricants or lubricant components can be mixed without harmful effects such as the formation of deposits. *Metallurgical compatibility*: a measure of the extent to which metals are mutually soluble in the solid state.

Conformability   That quality of a plain bearing material which allows it to adjust itself to shaft deflections and minor misalignments by deformation and/or by the wearing away of bearing material without producing operating difficulties. The term is preferred to deformability.

Conformal surfaces   Surfaces whose centres of curvature are on the same side of the interface. Non-conformal, or counter-formal, surfaces (q.v.) have centres of curvature on opposite sides of the interface, e.g. gear teeth.

Composite bearing material   A solid material composed of a continuous or particular solid lubricant phase dispersed throughout a load-bearing matrix to provide continuous replenishment of solid lubricant films as wear occurs, and effective heat transfer from the frictional surface.

Conjugate deflection   This applies where the bearing housing or support is designed so that the bearing and the cooperating surface remain substantially parallel under all loading conditions.

Connecting rod   The member connecting the piston, or the piston rod and crosshead with the crankpin on the crankshaft. *See also* Big end, Blade rod, Bottom end, Connecting rod fork, Fork rod, Large end, Little end, Master rod, Slave rod, Small end, Top end.

Connecting rod fork   The end of a connecting rod which bifurcates or forks, to terminate in two bearings spaced a distance apart on the same axis, or to carry the two ends of a pin.

Contamination   Lubricants may be contaminated by undesirable or harmful foreign matter in the following ways.

Solid particles:
(a)   built in during manufacture
(b)   ingested from atmosphere
(c)   abrasion and wear products (metallic)
(d)   from combustion processes (carbon)
(e)   entrained in make-up or replacement lubricant

Above may be referred to as dirt (q.v.).

Chemical:
(f)   from combustion processes
(g)   dilution by fuel oil
(h)   leakage of process fluids
(i)   degradation due to oxidation or high temperatures
(j)   water from condensation or leakage
(k)   anti-freeze from leakage

Contra-rotating   Two components which rotate in opposite directions about a common axis, often but not necessarily at equal speeds. Usually applies to concentric shafts, the inner rotating in the opposite direction to the outer. (*Syn* Counter rotating)

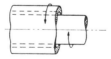

Cooler   A nest of tubes arranged in parallel through which water is circulated to extract heat from oil flowing over the tubes. The tubes are usually straight or of U-form, and are supported in baffle plates which control the flow of oil. Often the tube nest is in a separate compartment within the bearing casing, or even completely separate from the casing, with arrangements made to circulate the oil through it. Sometimes the nest may

be of circular form to fit the casing of a vertical bearing.

Cooling   The means for removing the heat generated in a bearing due to friction or to viscous shear in the lubricant. *See also* Circulated-oil, cooling, Coil, Cooler, External cooler, Fan cooling, Forced air cooling, Natural cooling, Water jacket cooling.

Cooling coil   A coil of tubing situated in an oil bath through which a liquid, usually water, is circulated to extract heat from the oil. It usually consists of one single tube but may occasionally have two tubes in parallel.

Cooperating surface   The surface of the component, e.g. journal or thrust collar, which slides over the surface of a bearing.

Copper–lead   Bearing alloys consisting primarily of copper and lead (usually about 25–35 per cent lead). Normally bonded to a steel backing by casting or sintering. Lead is present as discrete particles, the size and distribution of which depend upon the method of manufacture.

Copper–tin needles   This is copper–tin intermetallic constituent Cu_6Sn_5 which is present in tin-base alloys containing copper. This is the highest melting point constituent in the melt, and therefore determines the liquidus temperature (q.v.) of the alloy and solidifies first on cooling. This constituent, which freezes in the form of a network of needle-like crystals, helps to prevent or minimise segregation of the cuboids (q.v.) which freeze at a lower temperature. The copper tin needles are higher in density than the melt and therefore tend to segregate to the shell in centrifugal casting.

Corrosion   The chemical attack of one or more constituents of bearing material surface by acidic oil-oxidation products, water or other corrosive medium. *See also* Electrolytic corrosion, Fretting corrosion, Sacrificial corrosion, Sulphur corrosion, Tin oxide corrosion.

Corrosive wear   A wear process in which chemical or electrochemical reaction with the environment predominates.

Counter-formal surfaces   *See* Non-conformal surfaces.

Cover   A component used for closing an access opening in a bearing casing, usually occurring in a horizontal plane and fixed in position by bolts. *See also* End cover.

Crank   An arm on a shaft for communicating motion, usually of a reciprocating nature, to or from the shaft.

Crank angle The angle between the cylinder axis and the line joining the axial centres of the crankpin and the crankshaft at any given moment in the rotational cycle of the crankshaft.

Crank web(s) The member(s) which connect the main journals and crankpins in a crankshaft. On occasion, the outer surface of the crank web may also act as the main journal.

Crankpin The part of a crankshaft distant from and usually parallel with the axis, which cooperates with the connecting rod, and is connected to the main shaft by a web or webs.

Crankpin bearing *See* Big-end bearing.

Crankshaft The main shaft of an engine or reciprocating machine which carries a crank or cranks to which connecting rods (q.v.) are attached.

Crankshaft journal *See* Main journal.

Craze (crazy) cracking A network of cracks either due to rapid heating and cooling, generally on a single occasion, or to fatigue resulting from cyclic loading. *See also* Fatigue, Mosaic cracking, Thermal fatigue.

Crosshead The member fixed to the outer end of the piston rod and to which the connecting rod is coupled, which also guides the piston rod in a straight line parallel with the cylinder axis and transmits to the crosshead guide the transverse forces arising from the obliquity of the connecting rod.

Crosshead bearing The generic term for the connecting-rod bearings coupled to the crosshead. Strictly, the bearing housed in the crosshead in which the small-end pin of the connecting rod operates.

Crosshead guide The slide fixed to the engine frame which guides the crosshead in a straight line parallel with the cylinder axis and carries the transverse forces arising from the obliquity (q.v.) of the connecting rod.

Crosshead pin The journals forming part of

the crosshead on which the connecting rod small end bearings operate.

Crosshead slipper, shoe The sliding components having bearing surfaces which run in the crosshead guide (q.v.) and are attached to, or integral with, the crosshead.

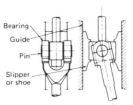

Crosswind A condition particularly applicable to thin-walled half-bearings, such that the planes embracing the ends of the bearing are not normal to the axis through the bore. This condition is caused during manufacture, and can result in bad fitting on assembly and fouling of the end faces of the bearing with the adjacent fillets. *See also* Joint face taper.

Crown A term applied particularly to half or partial bearings (q.v.) to denote the part of the bearing midway between the joint faces (q.v.).

Crowned surface Applied to tilting-pad bearings (q.v.), this indicates an operating surface which, instead of being planar, is very slightly convex. The convexity may be spheroidal or cylindrical, in the latter case the axis of curvature usually lies normal to the direction of sliding motion. Crowning may be introduced deliberately in manufacture but occurs anyway during operation due to thermal distortion and mechanical deflection.

Crush In a split journal bearing, the amount by which a bearing half extends above the horizontal split of the bore before it is assembled. The 'crush' (USA) is equal to one quarter of the 'nip' (UK). *See also* Nip.

Cuboids The name commonly given to the tin–antimony constituent, SnSb which is present in certain tin base whitemetals containing more than about 8 per cent antimony, and in certain lead-base white metals containing tin and antimony. This constituent, in the form of cubic crystals, which freezes before the final solidification of the matrix, is lower in density than the still molten metal and hence tends to segregate to the top in large statically cast bearings, or to the bore in centrifugally cast bearings, if the cooling rate is sufficiently slow. Also with very slow cooling rates, the cuboids may grow disproportionately large, resulting in plucking out or roughening of the surface during final machining. For both these reasons – segregation and crystal growth – cooling after pouring should be as fast as is practicable.

Cylinder block   The block in a reciprocating machine containing a cylinder(s) and bolted to or integral with the crankcase.

Dammed groove bearing   A plain bearing with a shallow partial circumferential groove in the unloaded half, which extends from one axial oil-feed groove in the direction of rotation, but is dammed off about 30–45° past the top of the bearing, so that oil pressure builds up due to viscous forces, and provides a stabilizing effect on shaft whirl. The bearing is suitable only for unidirectional rotation. Also known in the USA as a step bearing.

Damping   In a vibrating system, the force opposing motion and tending to cause decay of the amplitude of vibration. In a bearing, the resistance of the oil film to radial motion of the shaft.

Damping coefficient   The coefficient of the velocity-dependent terms in the equation of journal whirl motion. Normally, four coefficients are required for each bearing for the direct and the cross-coupled terms in the x and y coordinate directions.

Debris   Particles which become detached in a wear process.

Deformability   *See* Conformability.

Dendritic structure   A tree-like structure exhibited by certain metals and alloys, due to the growth of columnar metallic crystals from the surface of the mould or containing member, e.g. a liner, from which heat extraction from the molten metal occurs. In duplex alloys or mixtures such as copper–

lead and lead–bronze, this dendritic structure causes the low-melting point insoluble constituent lead to be squeezed to the extremities of the advancing columnar crystals of copper or bronze and to freeze in long stringers or particles. Contrast this with equi-axed structure (q.v.).

Detergent additive   A surface active additive which helps to keep solid particles in suspension in an oil. *See also* Additive, Dispersant additive, Oil additive.

Deva metal   A trade name for a composite bearing material which consists essentially of fine natural graphite incorporated into a matrix (q.v.) of metallic powders, such as tin bronze, lead bronze, iron or nickel, selected to provide the desired mechanical and bearing properties. Deva is the registered trade mark of Glacier GmbH, West Germany.

Diametric (diametral) clearance   The amount by which the diameter of the bearing bore exceeds the diameter of the cooperating journal. *See also* Bearing clearance.

Diesel engine types   *Low-speed* or *slow speed*: below 250 rpm rotational speed; main application is marine propulsion direct drive; two-stroke cycle. *Medium speed*: 250–1800 rev/min rotational speed; typical applications are electricity generation, rail traction, marine propulsion geared drive. *High speed*: above 1800 rev/min rotational speed; typical applications are vehicles (on and off roads), small generators, pumps.

Diffusion   The migration of metallic atoms towards a surface or interface, e.g. migration of tin in lead–tin overlays towards copper-base interlayer surface.

Dip-feed lubrication   The lubrication of parts by dipping in a lubricant or operating them partially submerged.

Direct-lined housings   Bearing housings, e.g. connecting rod big-ends (q.v.), which are lined direct with bearing metal (usually whitemetal) instead of using a separate insert bearing (q.v.).

'Directed' lubrication   An alternative method of lubricating tilting pad thrust or journal bearings (q.v.), by directing oil through jets between the pads, on to the moving surface, so avoiding the churning losses (q.v.) which occur when the bearing compartment is flooded.

Dirt A general term applied to solid particles which exist as a contaminant in the lubricant. *See also* Contamination, Foreign matter.

Disc lubrication A system of lubrication applied to horizontal bearings in which a disc attached to the shaft dips into an oil bath below the journal, collecting oil which is removed by a scraper, or by flinging action, and is channelled into the bearing. More positive than oil-ring lubrication (q.v.), but unsuitable for very high speeds owing to churning in oil bath and drop in efficiency of oil discharge due to the flinging action. The disc is usually in halves, clamped to the shaft, or the function is performed by the periphery of a thrust collar.

Dispersant additive An additive capable of dispersing cold oil sludge.

Double slope bored The term used when the bore of a bearing is machined in two parts having intersecting axes at different angles to the bearing axis. *See also* Slope bored.

Dovetail grooves *See* Keying grooves.

Dowel A cylindrical pin fitted with its axis normal to the joint faces to prevent relative displacement and to withstand shear forces. *See also* Location dowel.

Drip-feed lubrication The system of lubrication in which lubricant is supplied to the bearing surfaces in the form of drops at regular intervals. (*Syn* Drop-feed lubrication)

Dry bearing A bearing in which no fluid lubricant is deliberately introduced, e.g. certain plastic bearings, ptfe bearings (q.v.). The term is preferred to rubbing bearing.

Dry lubricant *See* Dry bearing, Solid lubricant.

Duplex alloys Bearing alloys consisting of two phases, one usually much softer than the other.

Dust seal A closure fitted to prevent the entry of dust, usually deposited from the atmosphere, into a bearing casing.

Duty parameter *See* Journal bearing dimensionless groups.

Dynamic friction The friction between two surfaces in relative motion.

Dynamic viscosity *See* Viscosity.

Eccentric A special form of crank to convert rotation to reciprocation, in which the crankpin encircles the shaft with its axis displaced from, although parallel to, the shaft axis.

Eccentric rod The rod which connects an eccentric strap to the valve spindle.

Eccentric sheave That part of an eccentric

attached to the shaft, the outer surface of which forms a journal.

Eccentric strap The bearing, or bearing housing (q.v.), which runs on an eccentric sheave.

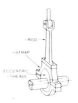

Eccentrically bored bearing A journal bearing where the axis of the bore is parallel to but not coincident with the axis of the outside surface.

Eccentricity The radial displacement of the journal centre from the centre of the bearing bore.

Eccentricity ratio The ratio of the displacement of the journal centre from bearing centre to the radial clearance. When the journal centre is coincident with the bearing centre, the eccentricity ratio is zero. When the journal centre is displaced by the radial clearance, i.e. when the journal is in contact with the bearing bore, the eccentricity ratio is unity.

Edge loading The forces imposed upon the ends or edges of bearings or bushes, due either to misalignment of the assembly or to deflection of the journal or housing under load. Edge loading can cause premature failure.

Elastic limit The maximum elastic deflection beyond which yield (permanent set) occurs.

Elastohydrodynamic lubrication A condition of lubrication in which the friction and film thickness between two bodies in relative motion are determined by the elastic properties of the bodies, in combination with the viscous properties of the lubricant at the prevailing pressure, temperature and rate of shear.

Electrical discharge attack The pitting of a bearing surface and sometimes of a journal, caused by electric discharge from one to the other through the oil film. It occurs most frequently in electrical machinery due to stray magnetic effects or faulty insulation, but can also occur in turbines and fans due to the generation of static electricity. The attack is progressive and usually results in wear (q.v.) and failure, apparently by wiping.

Electrical pitting   *See* Electrical discharge attack.

Electrolytic corrosion   The pitting of the surfaces of a bearing and/or journal in the presence of an electrolyte, e.g. sea water, due to the formation of an electrolytic cell between dissimilar metals.

Elliptical groove   A continuous groove around the bore of a bush, lying in a plane which is inclined to the axis.

Elliptical journal   A journal (q.v.) in which the cross-section deviates from a true circle in having two lobes and approximating to an ellipse.

Embeddability   Ability of a bearing material to embed harmful foreign particles and reduce their tendency to cause scoring or abrasion. Whitemetal is supreme among metallic materials in this respect.

Emboss   The formation of features such as oil grooves in a bearing by pressing in the blank before forming into a bush or a half bearing.

Emergency-run characteristic   The ability of a bearing to continue to function when there is a temporary deviation from the design conditions. Typical deviations are an increase in load, a drop in lubricating oil pressure or failure of cooling-water supply.

End cover   A component used to close one end of a bearing casing when the shaft is not required to extend through. Called top and bottom covers when applied to vertical bearings. *See also* Cover.

End coin   The axial compression of a bush in a die beyond the yield point of the material to produce dimensional accuracy. *See also* Coining.

End face   The surface, usually planar and normal to the axis, at either end of a journal half bearing or bush.

End float   The total axial movement allowed in the rotating element in a machine. (*Syn* Axial clearance (q.v.))

Engine set   *See* Set of bearings.

Entrainment velocity   The velocity of a liquid at which bubbles of gas are carried along in the stream. In a bearing oil bath, that downward velocity of liquid which equals or exceeds the rising velocity of the bubbles in still oil.

Equalising   Applied to tilting-pad thrust bearings (q.v.), this is a system of interlocking levers, or levelling plates, devised to share the thrust load equally between the thrust pads, accommodating

differences in pad thickness and changes in alignment between thrust collar and housing. Frictional effects prevent true equalisation to a greater or lesser extent. *See also* Equalising levers, Kingsbury bearing, Michell bearing.

Equalising levers   A closed circumferential chain of linked levers which support tilting thrust pads to accommodate differences in pad level due to misalignment, whilst maintaining substantially uniform loading of the pads. (*Syn* Levelling plates). *See also* Equalising.

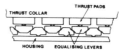

Equi-axed structure   Metallic crystals of random orientation and having approximately equal crystallographic axes. This structure is produced under certain conditions of rate and direction of cooling of the melt. Thus, in chill-cast materials such as phosphor bronze, the exterior layers may exhibit a dendritic structure (q.v.) while the core may be equi-axed. Also in sintered materials, the structure tends to be equi-axed. With this type of structure, an insoluble low-melting point constituent such as lead is normally present as random, well-distributed discrete particles. Contrast with dendritic structure (q.v.).

Erosion   The loss of material from a solid surface due to the relative motion in contact with a fluid which contains solid particles (*Syn* Erosive wear). *See also* Impingement erosion.

External-cooler cooling   Where the bearing is arranged to circulate oil through an externally situated cooler, for heat extraction.

Externally pressurised bearing   A bearing where fluid (liquid or gas) from an external supply is injected continuously between the opposing surfaces at such a pressure as to separate them completely by a fluid film. The fluid is usually injected at a number of points to provide balance and stability. Contrast with hydrodynamic lubrication (q.v.). (*Syn* Hydrostatic bearing). *See also* Aerostatic bearring.

Extreme-pressure lubricant   This type of lubricant is specially formulated for use in

extreme-pressure (ep) lubrication (q.v.). For example, e.p oils for gear teeth. They contain additives to react with the surfaces under high pressure/temperature, commonly sulphur, halogens (chlorine) or phosphorus.
Extreme-pressure lubrication A condition of lubrication in which the friction and wear between two surfaces in relative motion depends upon reaction of the lubricant with a rubbing surface at elevated temperatures.
Extrusion Plastic flow of material due to high pressure, possibly but not necessarily combined with high temperature.
Eye A ring formed on the end of a link or connecting rod to act as, or bored out to contain, a bearing.

Eyebrow A channel formed particularly in the upper half of a bearing casing (q.v.) or oil baffle (q.v.) to prevent oil dripping on to the shaft and causing leakage.

Face *See* End face, Joint face, Thrust face.
Facetting The distortion of the surface of a tin-base whitemetal bearing due to thermal cycling. *See also* Anisotropic cracking.
False brinelling A term associated with rolling contact bearings when local indentations are formed in the races due to oscillating motion of small amplitude or to static vibration, i.e. vibration imposed from an external source whilst the shaft is not rotating.
Fan cooling A type of forced air cooling (q.v.) in which the airstream is produced by a fan combined with the bearing, usually attached to the shaft.

Fatigue cracking Cracking or failure caused by repeated alternating stresses less than the ultimate stress in tension of the material. *See also* Craze cracking, mosaic cracking.
Fatigue stress A frequently repeated, usually cyclic, stress less than the ultimate stress which causes cracking or failure. (*Syn* Limit fatigue stress)
Fatigue wear Removal of particles detached by fatigue arising from high cyclic stress variations.
Features An embracing term covering all or any of those configurations of a bearing apart from the plain geometric outline, e.g. oil grooves, gutterways, holes, nick, locating pins, reliefs, etc.
Feed passages *See* Oil feed system.
Figure-eight groove An oil groove (q.v.) in the form of an 8 in the bore of a bearing, designed to spread the lubricant both axially and circumferentially on the bearing surface. Often used for reciprocating or reversing motion. *See also* Groove.

Fillets The radii at the intersection between two adjoining surfaces, commonly at right angles. Crankshaft fillets are radii between crankpins or main journals, and the adjoining webs.

Filtration Removal of foreign matter, e.g. from lubricant, by centrifuging or by passing through fine mesh or absorbent materials. *See also* Centrifugal filter.
Finishing *See* Superfinishing.
Finishing allowance *See* Boring allowance.
Finite bearing theory A solution to Reynolds equation for a journal bearing in which full account is taken of the effects of hydrodynamic pressure on flow velocity. Numerical solutions are obtained using finite element or finite difference techniques.
Firing load The loads exerted on the engine bearings as a result of the combustion of fuel between the cylinder heads and the pistons.
Fitted arc *See* Bedded arc.

Fitted bearing A bearing which has been scraped to conformity with the journal, showing no macroscopic high spots. Contrast with a pre-finished bearing (q.v.). The term is also applied to a partial bearing bore which has the same radius of curvature as the journal. *See also* Bedded arc.

Fitting test The assembly of main and big-end bearings of measured and recorded wall thickness and peripheral length (q.v.) in housings and rods, the bore diameters of which have also been measured and recorded after bolting up to a specified bolt torque before fitting the bearings. Each bearing bore is then measured after assembly, again with the specified bolt torque. The object of the test is to determine the effect of interference fit and of the housing material and design, upon the shape and size of the bearing bores. The correct wall thickness of a bearing to provide the required range of running clearance, and the optimum combination of interference fit and bolt load to produce a round bearing bore, can then be specified on the bearing drawings.

Fixed-pad (fixed-land) bearing An axial or radial-load type bearing equipped with fixed pads, the surfaces of which are contoured to promote hydrodynamic lubrication. Contrast with tilting-pad bearing (q.v.).

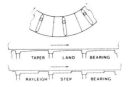

Flange A lip or face normal to the axial bore of a bearing, to locate the bearing and, in some cases, to take thrust loading. Flanges integral with the main body of the bearing are common practice with thick-walled bearings (q.v.) but separate thrust washers (q.v.) are generally used with thin-walled bearings (q.v.). The flanges are most commonly at each end of the bearing but occasionally are at the centre. *See also* Thrust bearing, Thrust face.

Flange-mounted bearing A bearing, usually horizontal, supported by a flange at one end, the face of which is normal to the shaft axis.

The flange may be of circular, semicircular or horseshoe shape. The casing may be of split or one-piece construction. Any combination of bearing functions may be contained.

Flashing A thin film, usually about 0.0025 mm (0.0001 in) thick, of pure tin or lead–tin or lead–indium alloy deposited electrolytically on the bearing surface to facilitate bedding-in. May also be applied to all surfaces of the bearing to give corrosion protection during storage.

Flexure pivot bearing A type of bearing guiding the moving parts by flexure of an elastic member or members rather than by rolling or sliding. Only limited movement is possible.

Flinger A sharp edge formed around a shaft which will shed oil by centrifugal action so limiting its travel in an axial direction. The edge may be integral with the shaft or on a component of ring form attached to the shaft. It usually collaborates with a baffle (q.v.).

Floating bush A bush which has operating clearances and bearing surfaces both on the outside and in the bore so that it is free to rotate relative to both shaft and housing. Used in oscillating applications, between concentric contra-rotating shafts, and in applications where precessing forces occur, as in epicyclic planet wheels.

Floating oil sealing ring An oil sealing ring which, instead of being fixed, is free to move transversely following any shaft movement, thereby permitting a much smaller clearance to be used. The term is preferred to floating ring seal.

Floating ring seal   *See* Floating oil sealing ring.

Flood lubrication   A system of lubrication in which the lubricant is supplied in a continuous stream. The term is usually used where the lubricant is supplied to the bearing but is not forced into it under pressure. Contrast with forced lubrication (q.v.). *See also* Circulated-oil cooling.

Fluid erosion   The wear due to the action of streaming liquid, gas or gas containing liquid droplets can be intensified by chemical action. *See also* Cavitation erosion, Erosion.

Fluid friction   The frictional resistance due to the viscous or rheological drag of fluids. (*Syn* Viscous friction)

Fluid pivot bearing   A tilting-pad bearing (q.v.) in which the discrete mechanical pivot is replaced by a hydrostatic support supplied with lubricant by bleeding from the hydrodynamic film at the bearing surface.

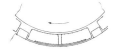

Foaming   The formation and entrainment of gas or air bubbles in the oil as a result of turbulence or churning action. The bubbles rise to the surface of the oil bath, where conditions are sufficiently still, and will coalesce and leave the liquid.

Foil bearing   A bearing made of a flexible foil held under tension against a portion of the journal periphery. The lubricant film is formed between the foil and the journal.

Fool proofing   A general term covering the various devices adopted to ensure that mating components can be assembled only in the correct way and to prevent inadvertent misassembly, e.g. an odd pitch in a bolted joint. *See also* Nick, Spoiling pin.

Footstep bearing   *See* Step bearing.

Forced-air cooling   Where the heat is removed by an airstream over a surface which is usually finned to increase the surface exposed.

Force-feed lubrication   *See* Forced lubrication.

Forced lubrication   A system of lubrication in which the lubricant is supplied to the bearing under pressure. It implies an external system of drain-tank, pump and cooler. (*Syn* Force-feed lubrication)

Foreign matter   Particles derived from machining debris, from the wear of moving parts, from the combustion of fuel, or

entrained through breathers or air filters. *See also* Contamination, Dirt.

Fork and blade bearing   A bearing for the crankpin of a particular configuration of V-engine, where each end of the bearing is held in the fork rod (q.v.), and the centre of the bearing forms a journal on which the blade rod (q.v.) oscillates.

Fork rod   In a particular V-engine configuration, the connecting rod which is bifurcated or forked at the big end to hold each end of the bearing, permitting the narrow blade rod (q.v.) to oscillate over the outer surface at the centre of the bearing.

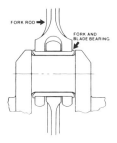

Forward bush   The bush situated at the forward or inboard end of a sterntube (q.v.).

Free spread   The difference between the outside diameter at the joint faces (q.v.) of a half-bearing in the free state, and the top limit bore diameter of the housing. Termed positive free spread when the bearing dimension is greater than the housing and negative when the bearing dimension is less than the housing.

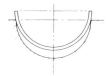

Fretting   Wear phenomenon occurring between two surfaces having oscillatory relative motion of small amplitude. *See also* Fretting corrosion.

Fretting corrosion   Corrosion which results from fretting, often characterised by the removal of particles and the subsequent formation of oxides, which are often abrasive and so increase wear. It can also involve other reaction products, not necessarily abrasive.

Friction   The resisting force tangential to the common boundary between two bodies when, under the action of an external force, one body moves or tends to move relative to the surface of the other. *See also* Amonton's laws, Coefficient of friction, Limiting state friction, Static friction.

Galling   A non-preferred term describing a severe form of scuffing associated with gross damage to the surfaces or failure.

Gallop   A term sometimes used to describe a surface which is eccentric to the axis of rotation. *See also* Swash.

Galvanic attack   *See* Electrolytic corrosion.

Gas bearing   A bearing using gas as a lubricant. *See also* Externally pressurized bearing, Gas lubrication.

Gas load   *See* Firing loads.

Gas lubrication   A system of lubrication in which the shape and relative motion of the sliding surfaces causes the formation of a gas film having sufficient pressure to separate the surfaces. *See also* Externally pressurised bearing.

Gimbal   A support, usually of ring form, arranged to pivot on two axes at right angles to each other, thus permitting angular movement in any direction.

Gland   The component which compresses the packing in a stuffing box (q.v.). The term is also used generally to describe the total system of stuffing box, packing and gland.

Grain   A generic term used to describe the crystallographic structure of a metallic material, e.g. coarse-grained, fine-grained, lamellar, dendritic, equi-axed (q.v.).

Graphite   A crystalline form of carbon having a laminar structure.

Groove   In a bearing, a channel for the distribution of lubricant. *See also* Annular groove, Bleed groove, Blind groove, Figure eight groove, Gutterway, Helical groove, Mud groove, Oil groove, Pear drop groove, Radial groove, Spreader groove, Thumb nail groove.

Gudgeon pin   A short shaft, commonly hollow, which couples a trunk piston to the connecting rod.

Gudgeon-pin bearing   The bearing or bush inserted into the small end of a connecting rod. This bearing oscillates around the gudgeon pin which is fixed in the piston. *See also* Piston-pin bearing, Small end bearing.

Guide bearing   A bearing used for positioning a slide, or for axial alignment of a long rotating shaft. The term is sometimes used to denote a journal bearing for a vertical axis shaft, deriving from the absence of any journal load and therefore its function is to guide the shaft or maintain it in position.

Guide bush   A bush used where the sliding motion is in an axial direction, e.g. reciprocating motion, as in the case of a piston rod.

Gum   In lubrication, a rubber-like sticky deposit, black or dark brown in colour, which results from the oxidation and/or polymerization of fuels and lubricating oils. Harder deposits are described as varnishes or lacquers (q.v.).

Gunmetal   Alloys of copper, tin and zinc, sometimes also with nickel and/or lead. *See also* Bearing bronzes.

Gutterway   A relatively broad axial groove to distribute oil to the bearing, usually coincident with the split lines of half-bearings (q.v.). *See also* Groove.

Half-bearing   Semi-cylindrical insert bearing having two joint faces at, or near, the centre-line. *Interchangeable half-bearings*: Half-bearings manufactured to such close limits as to be individually interchangeable so that any half-bearing may be used to replace any other made to the same design and limits. The assembled bearing may be composed of two identical half-bearings, or of two halves having differing features but each of which is individually interchangeable. *Paired half-bearings*: Bearings in which, for reasons of convenience, the two halves are manufactured together and are not individually interchangeable although they may appear alike and are usually interchangeable as a pair. *See also* Pair of bearings.

Half height   The dimension from the crown of the back to the plane embracing both joint faces of a half-bearing when held in a housing of specified diameter. *See also* Centre-line height, Crush, Nip.

Half-speed vector   A force vector which rotates at half shaft speed, a condition which theoretically renders the maintenance of hydrodynamic lubrication (q.v.) impossible. *See also* Rotating load.

Half-speed whirl   *See* Oil-film whirl.

Half thrust washer   A thrust washer (q.v.) which extends around 180° of the thrust face. Half thrust washers may be used in pairs to facilitate assembly, or may be used singly to save cost if the thrust loading permits.

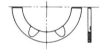

Hand pouring   *See* Static lining.

Heat dissipation   The removal of heat generated in a system, e.g. the frictional heat of oil-film shear, by conduction through lubricant or bearing housing, by radiation and convection from external surfaces, or by circulated liquid coolant, e.g. water-jacket.

Heat pipe   A tubular device for transferring heat from one end to the other by evaporation and condensation of a fluid in the sealed system.

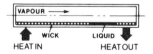

Helical groove   A groove which follows a helical path in the bore of a bearing, designed to promote oil flow in an axial direction through the clearance space with journal rotation. *See also* Groove.

Herring-bone grooves   A series of shallow grooves of V-pattern to improve the

hydrodynamic performance of a bearing, the apex of the V pointing in the direction of sliding motion of the cooperating surface. Usually applied to thrust faces operating with gas lubrication (q.v.). *See also* Groove.

Hollow (hourglass) journal   A journal (q.v.) the axial contour of which is smaller in diameter at the centre than at the extremities.

Honing   A process for improving the finish of a surface by the rubbing action of abrasive stones, usually in the presence of a liquid. This is not a process which will generate surface accuracy although it may improve it by careful attention to the stones.

Horizontal bearing   The generic term for a bearing in which the axis of the shaft lies in the horizontal plane.

Horns   Arcs adjacent to the split or joint faces of partial or half-bearings.

Housing   *See* Bearing housing.

Housing swell   Expansion of a housing bore, caused by the interference fit of the bearings. This expansion is extremely difficult to calculate owing to the complexity of many housing designs, hence the use of fitting tests (q.v.) in appropriate cases.

Hybrid bearing   A bearing operating with both hydrostatic and hydrodynamic lubrication (q.v.). It applies to a bearing operating with jacking oil and at a speed sufficient to support the load on a hydrodynamic film.

Hydrodynamic bearing   A bearing designed to operate under hydrodynamic lubrication (q.v.) conditions.

Hydrodynamic lubrication   A system of lubrication in which the shape and relative motion of the sliding surfaces causes the formation of a fluid film having sufficient pressure to separate the surfaces.

Hydrostatic bearing   *See* Externally pressurised bearing.

Hydrostatic lubrication   A system of lubrication in which the lubricant is supplied under sufficient external pressure to separate the opposing surfaces by a liquid film. *See also* Externally pressurised bearing.

Impingement erosion   Erosion (q.v.) in which the relative motion of the solid particles is nearly normal to the solid surface.

Impregnation   The filling of voids in a porous matrix (q.v.) with a selected impregnant, e.g. graphite or oil in porous sintered products, ptfe (q.v.) in a porous tin bronze lining of dry bearings (q.v.).

Inclined bearing   A bearing in which the shaft axis lies at an angle to the horizontal.

Indented bore   The provision of indentations in the bore of certain types of bearing to increase lubricant supply. Particularly applicable to oscillating conditions, and to marginally lubricated materials such as pre-lubricated bearings (q.v.).

Inertia loading   Inertia loading in engine bearings results from the acceleration forces on reciprocating and rotating components, e.g. piston and connecting rod.

Inhibitor   Any substance which slows or prevents such chemical reactions as corrosion or oxidation.

Insert bearing   A bearing member inserted into a bearing housing to support the journal. It may be a bush, half-bearing, or partial bearing (q.v.). The term is preferred to insert liner and shell bearing.

Interference   The amount by which the outside diameter of a bush or unsplit bearing in the free state exceeds the bore diameter of the housing into which it is to be inserted. *See also* Nip.

Interference fit   The excess of the diameter of an internal component over the bore of its housing, resulting in a frictional retaining force. *See also* Crush, Nip.

Interlay (interlayer)   The intermediate layer between the overlay and the backing in a multi-layer insert bearing. *See also* Overlay plated bearings, Trimetal bearing.

Inter main bearing   The main bearings situated between the front main bearing and the centre main bearing, and between the centre main bearing and the rear main bearing. The inter main bearings are normally considerably narrower than the other main bearings, in order to reduce the overall length of the crankshaft, and thus are often more heavily loaded.

Intermediate plummer bearing   A plummer bearing (q.v.) which supports the intermediate shafting in a ship and which, conventionally, has a bearing surface in the bottom half only.

Intermediate shaft(s)   The length(s) of main propulsion shafting which connect the engine coupling to the thrust shaft and propeller shaft.

Iron printing   A non-destructive test to determine the extent of ferrous contamination of a bearing surface. A filter paper soaked in potassium ferricyanide solution is pressed into contact with the previously degreased bearing surface. Ferrous particles (steel or cast-iron) embedded in the bearing surface react with the ferricyanide to form blue spots (Prussian blue) on the filter paper, which may then be washed in water to remove the ferricyanide. The blue spots remain as a permanent record.

Jacking   An arrangement by which oil is supplied to a bearing of the hydrodynamic type at a point opposite to the applied load and at a sufficiently high pressure to separate the surfaces of the bearing and the cooperating journal or thrust collar by a film of oil. It is used on heavy machinery to prevent wear during the starting and stopping cycles, or during continuous rotation at low speed on the turning gear, and to reduce friction when starting.

Jacking oil groove, pocket   The groove, or pocket, formed in the bearing surface for the introduction of jacking oil.

Joint face   The abutting faces, usually flat, of two adjoining components. In a bearing, the radial faces of a half-bearing (q.v.) or a split bush.

Joint face dowels   Joint-face location dowels are often fitted in the case of medium-walled or thicker plain bearings to locate the two halves in relation to each other, and to locate one half axially where only the other half has positive axial location. *See also* Location dowel, Medium-walled bearings.

Joint face relief   The removal of areas of the joint face to avoid contact and compressive stress where it is undesirable, e.g. in flanged bearings where the flanges are relieved from abutting at the joint.

Joint face stagger   The condition where there is a step at the joints between the bores of the two halves of a bearing assembly. Normally due to assembly error, machining error, or distortion.

Joint face swell   The swelling of the wall of a bearing near the joint faces (q.v.) due to excessive interference fit (q.v.) or to distortion during manufacture.

Joint face taper   Applicable to half-bearings (q.v.). A condition in which the joint faces are 'out of square' with the end faces of the bearing. This is a manufacturing defect and can result in the distortion of the bearing on assembly, owing to the interference fit or nip (q.v.) being greater at one end of the assembled bearings than at the other end.

PLANE CONTAINING BEARING AXIS

Journal   That part of a shaft or axle which rotates or oscillates within a bearing. A journal is part of a larger unit, e.g. crankshaft, and it is preferred that the term shaft be kept for the whole unit. *See also* Barrelled journal, Elliptical journal, Hollow journal, Lobed journal, Tapered journal, Waisted journal.

Journal bearing   A bearing in which a journal rotates, or oscillates relatively to its housing, the load acting in a plane normal to the journal axis. A full journal bearing is one which embraces 360° of circumference (excluding oil grooves). *See also* Half-bearing, Partial gearing.

Journal bearing dimensionless groups   These are dimensionless numbers used to evaluate the performance of plain journal bearings in

terms of the effective parameters. Although of similar construction, the precise form varies according to the origin and convenience of usage

Duty parameter, Δ, used by the Institution of Mechanical Engineers (UK):

$$\Delta = \frac{W}{\eta U b}\left(\frac{C_r}{i}\right)^2 = 2S_0$$

Load number, W, used by Engineering Sciences Data Unit and the Glacier Metal Co. Ltd. (UK):

$$W' = \frac{W}{\eta n d b}\left(\frac{C_d}{d}\right)^2 = 2\pi S_0$$

Sommerfield number, S_0, used in West Germany:

$$S_0 = \left(\frac{C_r}{r}\right)^2 \frac{P}{\eta \omega}$$

Sommerfeld number, $1/S$ or S, used in the USA.
Sometimes known as the Sommerfeld Reciprocal:

$$\frac{1}{S} = \frac{\eta n}{P}\left(\frac{d}{C_d}\right)^2 = 2\pi S_0$$

Any consistent set of units may be used. In the following, the quantities are expressed in SI units

C_d = diametric clearance (m)
C_r = radial clearance (m)
P = bearing specific load (N/m^2)
U = shaft peripheral velocity (m/s)
W = bearing load (N)
b = bearing width (length) (m)
d = shaft diameter (m)
n = rotational speed (rev/s)
r = shaft radius (m)
η = dynamic viscosity (Ns/m^2)
ω = rotational speed (rad/s)

Journal locus   The path described by the centre of the journal within the bearing clearance during the operating cycle of the mechanism.

Journal pad   *See* Tilting pad.

Keep   A structural component which clamps a bearing or a bearing housing to the supporting structure. It does not include the oil-retaining enclosure. It may be, but is not necessarily, similar in shape to a bearing cap (q.v.).

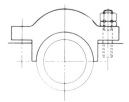

Keying grooves   Grooves cast or machined in bore of bearing housing or shell to retain lining material (usually whitemetal) mechanically in the event of poor metallurgical bonding. Not recommended for highly stressed bearings. Method is obsolescent. (*Syn* Dovetail grooves). *See also* Stress raiser.

Kinematic viscosity   *See* Viscosity.
Kingsbury bearing   A type of tilting-pad thrust bearing (q.v.) named after the inventor, Albert Kingsbury. *See also* Michell bearing, Tilting-pad bearing.
Knife-edge seal   A sharp edge cooperating with a moving surface, usually arranged in the direction of motion, with minimum clearance to minimise the passage of a fluid. The sharp edge is used to minimise the generation of heat and/or damage in the event of accidental contact with the moving surface. The inner edges of oil baffles (q.v.) are commonly made in this form.

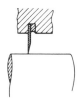

Knuckle-pin bearing   *See* Wrist-pin bearing.
'Kolene' process   A proprietary process for the treatment of cast iron to remove surface graphite. *See also* Cast iron process.

l/d ratio   *See* b/d ratio.
Lacquer   In lubrication, a deposit resulting from the oxidation and/or polymerisation of fuels and lubricants when exposed to high temperatures. Softer deposits are described as varnishes or gums (q.v.).
Lamellar   In layers. Certain materials, such as graphite and molybdenum disulphide, nave a layer-like or laminar structure which facilitates slip, so reducing friction and wear.
Laminar flow   When the lubricant in the oil film flows smoothly in streamlines through the clearance space, obeying Poiseuile's law. Under these conditions, there will be no movement of fluid particles across the thickness of the lubricant film, i.e. the film acts like a number of thin layers (laminae) that slide over each other, but do not mix. *See also* Superlaminar flow.
Land   The load-carrying parts of a bearing surface bounded by grooves or edges of the bearing. In a cylindrical-bore bearing, that part of the bearing surface between a circumferential groove and the end of the bearing, hence land width. In a thrust bearing, those parts of the surface lying between grooves, e.g. taper land thrust bearing (q.v.).
Lantern ring   A ring, commonly of H-section with perforations, used to distribute lubricating or cooling fluid to gland packing. *See also* Gland.

Lap   A rigid body, usually of metal, the surface of which is finished to the desired form for use in the process of lapping. The lap is usually, but not essentially, of softer material than the material to be lapped.
Lapping   A process for improving the finish and accuracy of a surface by the rubbing action of a lap charged with abrasive particles, usually in the presence of a liquid.
Large end   *See* Big end.
Leaded bronze   *See* Bearing bronzes.
Lemon-bore bearing   A bearing with two lobes (q.v.) or partial arcs each having a larger radius than half of the minimum bore 'diameter'. *See also* Profiled bore.

Lignum vitae   A hard, dense tropical wood used for lining the sterntube bearings of ships.

Limiting fatigue stress   *See* Fatigue stress.
Limiting static friction   The force tangential to the interface which is just sufficient to initiate relative motion between two bodies under load.
Linear bearing   A support or guide which constrains a moving part to travel in a straight line.
Lining   In a plain bearing, the material applied either direct to the housing or to the shell, to form the bearing surface which supports the journal or other moving member. *See also* Micro-lining.
Liquidus   The temperature above which an alloy is completely liquid. Conversely, the temperature at which an alloy *commences* to freeze on cooling from the liquid state.
Little end   *See* Small end.
Little-end bearing   *See* Small-end bearing.
Load-carrying capacity   The maximum load that a sliding or rolling system can support without failure, or the wear exceeding the design limits for the particular application.
Load line   The line of action of the load through the axis of the journal.

Load number   *See* Journal bearing dimensionless groups.
Lobe   The partial arc of a bearing bore with radius greater than half the minimum bore 'diameter'. *See also* Profiled bore.

Lobed bearing   A bearing with two or more lobes (q.v.). *See also* Profiled bore.
Lobed journal   A journal (q.v.) in which the cross-section deviates from a true circle, having lobes or higher segments alternating with lower segments.
Locating lug   The projection provided at the joint face of a bearing half-liner to locate it, during assembly, in the bearing housing. It is not to prevent rotation. *See also* Nick.
Locating notch   A recess in the bearing housing bore provided to receive the

locating lug of a bearing half-liner. *See also* Nick recess.
Location bearing   A bearing which locates the shaft against axial movement. The term is only used where the thrust loads are relatively light and indeterminate. Where the thrust load is greater, the term 'thrust bearing' (q.v.) is used.
Location dowel   Used to locate a bearing in the housing and to prevent rotation of thrust bearings, plain or tilting pad. *See also* Joint face dowels.
Locus   The path traced by a point moving in accordance with a fixed law. *See also* Journal locus.
Lubricant   Any substance interposed between two surfaces in relative motion for the purpose of reducing the friction and/or wear between them.
Lubrication   The reduction of frictional resistance and wear, or other forms of surface deterioration, between two load-bearing surfaces in relative motion tangentially to their interface, by the application of a lubricant between the surfaces. *See also* under specific lubrication names, e.g. Bath lubrication.
Lug   *See* Locating lug.

Main bearing   A bearing supporting the main power transmitting shaft.
Main journal   That part of a crankshaft which is concentric with the axis of rotation and which runs in a main bearing. (*Syn* Crankshaft journal)
Main propulsion shaft bearings   Bearings supporting the main propulsion shafting in vessels. *See also* Bracket bearing, Plummer bearing, Shaft carriage, Stern bearing, Sterngear, Sterntube bearing, Thrust block, Trailing bearing, Tunnel bearing, Withdrawable sterngear system.

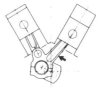

Marginal lubrication   A condition of lubrication in which the quantity of lubricant is barely sufficient. It occurs in pre-lubricated bearings (q.v.). *See also* Thin-film lubrication.
Master rod   A connecting rod to which one or more slave rods (q.v.) are connected.
Matrix   The predominant component of a material usually applied to non-homogeneous materials, e.g. whitemetals,

copper–lead, or to porous materials.

Maximum oil-film pressure   The maximum pressure built up in a hydrodynamic oil film. It depends upon oil-film thickness, clearance, and the distribution of load over the bearing surface.

Mechanical keying   *See* Keying grooves.

Medium-walled bearing   An insert bearing having a wall thickness intermediate between that of thin-walled bearing (q.v.) and thick-walled bearing (q.v.). Usually the wall thickness is 3–5 per cent of bore diameter but the contour is still dependent upon the housing. *See also* Wall.

Michell bearing   A type of tilting-pad thrust bearing (q.v.) named after the inventor, A. G. M. Michell. *See also* Kingsbury bearing, Tilting pad bearing.

Micro-lining   Very thin linings of bearings material bonded to bearing shells, usually of steel. Whitemetal linings 0.1–0.175 mm (0.004–0.007 in) thick, and copper–lead and lead bronze linings 0.125 mm (0.005 in) thick, are commonly used in thin-walled prefinished bearings (q.v.).

Microsection   A section of material, prepared by machining, filing, grinding and final polishing, to eliminate all scratches and surface blemishes, and to permit of microscopic examination of the structure of the material. Thin sections of material are usually mounted in a plastic or low melting point metal mould prior to preparation, in order to preserve edges and other features, and to produce a completely flat polished surface.

Migration   *See* Diffusion.

Milled nick   A local reduction in the bearing wall thickness, by a milling operation, in way of the nick to enable it to be formed more easily.

Minimum oil-film thickness   The thickness of the oil film at its minimum point. It depends upon the magnitude, direction and duration of applied loads at operating temperature during machine cycle.

Mirror finish   Applied to a surface which has been lapped or polished until it is free from scratches and highly reflective, as a mirror.

Misalignment   The angular displacement of a journal axis in relation to its bearing axis.

Misalignment may be built in during the erection of a machine, or may arise due to deflection or distortion of journals or machine frames caused by operating forces or thermal stresses.

Mist lubrication   Lubrication by an oil mist produced by injecting oil into a gas stream.

Monolayer   A film one molecular unit in thickness.

Monometal bearing   A bearing consisting of a single material or composite as opposed to a bimetal or trimetal bearing (q.v.). *See also* Solid insert.

Mosaic cracking   Form of fatigue cracking exhibited especially by whitemetal bearings subjected to excessive dynamic stresses.

Mud groove   An axial oil groove immediately adjacent to the joint-face or split of a half-bearing, the object of which is to promote lateral oil flow to carry foreign matter out of the clearance space. *See also* Groove.

Multi-pocket bearing   A bearing having pockets machined in the bore to provide lubricant reservoirs, or to achieve shaft stability. *See also* Lobe, Profiled bore.

Natural cooling   Where the heat is dissipated by natural radiation and convection from the exposed surface.

Neck ring   A non-ferrous (usually bronze) ring fitted at the bottom of a stuffing box (q.v.), to present a suitable wearing and non-seizing surface to the shaft. *See also* Gland.

Nick   The tongue or lug which is pressed out of the back at the joint-face, especially of a thin-walled bearing, to engage in a corresponding recess in the housing, connecting rod, or cap bore, in order to effect axial and circumferential location. It should be noted that the nick, the radial face of which bears against the radial face of the opposing housing, does not in itself prevent circumferential rotation of the bearing. It is the nip (q.v.) which prevents rotation. (*Syn* Tag, Tang)

Nick recess   The recess in the housing, rod, or cap bore into which the nick (q.v.) in the bearing engages on assembly.

Nickel barrier   *See* Barrier.

Nip   The amount by which the outer circumference or peripheral length of a pair of bearing shells exceeds the inner circumference of the housing. The 'nip' (UK) is equal to four times the 'crush' (USA). *See also* Crush (USA term).

Non-conformal surfaces   Surfaces whose centres of curvature are on opposite sides of the interface, e.g. rolling element bearings, gear teeth.

Notch   *See* Locating notch.

Oblique split   A design in which the joint faces (q.v.) of a connecting rod big end are not at right angles to the axis of the rod. This design is often adopted to permit withdrawal of the rod through the cylinder bore after removal of the big-end bearings. An oblique split housing is used in other machinery applications to rotate the bearing joint away from the load vector.

Obliquity   The acute angle between the axis of the cylinder and the axis of the connecting rod, i.e. the line joining the centres of the small and big-end bearings.

Offset halves bearing   Bearing with each half bored eccentrically in line with the split line and oil grooves and assembled so that there are continuously converging oil films in both top and bottom halves. This promotes bearing stiffness, damping and resistance to oil-film whirl (q.v.). *See also* Profiled bore.

Oil additive   A chemical added to lubricating oils to prevent or remove the agglomeration of carbonaceous products of fuel combustion, to reduce tendency to foam or to inhibit oxidation and consequent formation of corrosive oil-oxidation products which may attack bearing or journal surfaces.

Oilbaffle *See* Baffle.

Oil catcher   A trough fitted under the end of a bearing to catch oil which leaks from the closure. In vertical bearings for electrical machines, the trough surrounds the shaft and is usually combined with a safety drain (q.v.) and sometimes with a windage shield or air seal.

Oil deflector   A small spring-loaded scraper of V-form held in contact with a shaft to limit the travel of oil in an axial direction, orientated with the point of the V opposing the natural direction of travel of the oil. It is only suitable up to moderate shaft speeds.

Oil-feed system   Oil galleries, drilled passages or pipework conducting oil from supply source (e.g. pump) to bearings. (*Syn* forced lubrication system).

Oil film   In lubricated bearings, the lubricant film which separates the journal and bearing surfaces, or other components in relative motion.

Oil-film pressure *See* Maximum oil-film pressure.

Oil-film stiffness The resistance of the oil films to a displacement of the shaft.

Oil-film thickness *See* Minimum oil-film thickness.

Oil-film whirl Violent precession (q.v.) of the shaft within the clearance in a plain journal bearing, caused by oil-film instability under high speed and light load conditions. It commonly occurs with a frequency of about half the shaft speed and can be thus differentiated from a synchronous precession due to out-of-balance of the shaft, or to a rotating load.

Oil-fog lubrication *See* Mist lubrication.

Oil groove A groove or channel cut in a bearing surface to improve oil flow (or grease distribution) through the bearing. *See also* Groove.

Oil-ring lubrication In a horizontal bearing, where oil is supplied to the bearing surfaces by means of one or more loose rings encircling the shaft and the lower part of the bearing, and dipping into an oil bath. The rings are rotated by the friction of the revolving shaft. The method is not positive since ring slippage can occur. The quantity of lubricant delivered is low and reduces at high speeds.

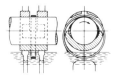

Oil seal A component of ring form, usually of elastomeric material, surrounding and in contact with a shaft where it enters a casing to prevent any passage of oil from the casing.

Oil sealing ring A fixed component of ring form surrounding, with a small clearance, a shaft where it enters a pressurised chamber in order to limit the flow of oil to a volume which can be handled by a subsequent closure, such as a system of flingers and a baffle. The clearance must be larger than the bearing clearance. The term is preferred to

restrictor ring. *See also* Floating oil sealing ring.

Oil wedge In a hydrodynamic bearing, the load-carrying capacity is obtained by the fluid being dragged through a converging tapering oil film (creating a wedge of oil). The oil pressure thus developed supports the applied load, with complete separation of the surfaces. In a journal bearing, the action is automatic. In thrust bearings, special provision must be made to enable the formation of a tapered film.

Oil whirl Instability of a rotating shaft associated with instability in the lubricant film. *See also* Oil-film whirl.

Oiler, oil cup A device for holding a small quantity of oil and delivering to a bearing, usually by gravity.

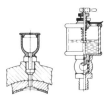

Oscillating load A load of constant or variable magnitude with cyclically changing angular direction.

Outrigger bearing A third bearing situated outside the main housing or casing of the machine, on a bracket or outrigger, to support the outer end of the shaft.

Overlay The bearing surface lining of overlay plated or trimetal bearings (q.v.), usually consisting of lead–10% tin, lead–10% tin–2% copper or lead–5% indium alloy.

Overlay thickness ranges from about 0.0012 mm (0.0005 in) to 0.1 mm (0.004 in), depending upon bearing size and imposed loadings. Overlays are usually, but not always, electrodeposited. They usually remain substantially intact for the life of the bearing.

Overlay plated bearing Three-layer bearing consisting of the backing (usually steel), the interlayer (strong bearing material, e.g. copper–lead, lead–bronze or aluminium alloy), and the bearing surface lining, usually electrodeposited, of lead alloy 0.0012–0.1 mm (0.0005–0.004 in) thick. *See also* Overlay, Trimetal bearings.

Pad A component used to support, or to press against, another part where there is sliding motion. In bearing terminology, the bearing component of a tilting-pad bearing (q.v.) (*USA syn* Shoe (q.v.))

Pad lubrication *See* Syphon lubrication.

Pad pivot That part of a tilting pad (q.v.) on which it is supported with freedom to tilt, or a component attached to a pad for the same purpose. *Line pivot*: the pivot is formed by a straight edge supported on a flat surface, with freedom to tilt in only one direction. *Point pivot*: the pivot is spherical engaging with a flat surface, giving freedom to tilt in any direction. *Spherical pivot*: the pivot is a sphere in a cup, also giving freedom to tilt in any direction. *Cylindrical pivot*: the pivot is of cylindrical form in a similar recess, tilting being in only one direction. *See also* Tilting pad bearing.

Pair of bearings Used in the car industry to mean a combination of one top-half and one bottom-half bearing. It may be a corruption of the term 'pair of half bearings'. Out of context, the term can be misleading since in other branches of engineering a pair of bearings signifies two bearings (i.e. four half-bearings) for a single shaft. *See also* Half-bearing.

Paired half-bearing *See* Half-bearing.

Partial bearing One which does not embrace 360° of journal circumference, e.g. a single half-bearing, or a 120° bearing.

Partial groove *See* Blind groove.

Partial slope bored The term used when only part of the bearing bore is machined at an angle to the axis. *See also* Slope bored.

Pear-drop groove *See* Thumb-nail groove.

Pedestal bearing A bearing supported on a column or pedestal rather than on the main body of the machine. The bearing is usually horizontal and the column or pedestal supported on, and bolted to, a flat surface, parallel to the shaft axis. The casing is nearly always of split construction and may contain any combination of bearing functions. The pedestal may be short or tall. The pedestal, or casing bottom, is usually in one piece but may be made in two or three pieces to facilitate dismantling. (*Syn* Pillow block, Plummer bearing (q.v.))

Peripheral length The length of the external circumference of a half-bearing (q.v.).

Petroff bearing An idealised condition in which a journal rotates concentrically within a cylindrical bore. A bearing operating with no load, or only a light load at high speed, approximates to this condition.

Petroff law, equation The equations linking the operating parameters of a Petroff bearing (q.v.).

force resisting motion

$$(N) = 2\pi U_\eta b\left(\frac{d}{C_d}\right)$$

resisting torque

$$(Nm) = \pi^2 n\eta bd^2\left(\frac{d}{C_d}\right)$$

power loss

$$(W) = 2\pi^3 n^2\eta bd^2\left(\frac{d}{C_d}\right)$$

where

U	= sliding velocity (m/s)
η	= dynamic viscosity (Ns/m^2)
d	= shaft diameter (m)
b	= bearing width (m)
n	= rotational speed (rev/s)
C_d	= diametric clearance (m)

The above is stated in SI units, but any consistent set of units may be used.

Phosphor bronze *See* Bearing bronzes.

Pick-up Transfer of material from one surface to the other due to strong adhesive forces in the sliding or rolling interface.

Pillow block The term (now obsolete) applied historically to a pedestal bearing (q.v.).

Pin *See* Dowel.

Piston-pin bearing The bearing or bush inserted in the piston to transmit the force between the piston and the connecting rod. The piston-pin is fixed in the small end of the connecting rod and oscillates within the piston-pin bearing. *See also* Gudgeon-pin bearing, Small-end bearing.

Pitting Removal or displacement of material resulting in the formation of surface cavities.

Pivot   That part of a spindle which runs in a bearing. The term is used mainly in instrument work or horology and usually implies a combination of journal with axial location in one direction. *See also* Pad pivot.

Pivoted-pad   *See* Tilting pad, Tilting-pad bearing.

Plain bearing   Any simple sliding type bearing, as distinct from pad or rolling type bearings.

Plain journal bearing   A plain bearing in which the relatively sliding surfaces are cylindrical and in which there is relative angular motion. One surface is usually stationary, and the force acts perpendicularly to the axis of rotation. Sometimes loosely used to denote a bearing with a plain back as distinct from one with flanges.

Plain thrust bearing   A plain bearing of the axial load type, with or without grooves. Usually a bearing where the surface lies in one plane. *See also* Fixed-pad bearing, Taper land thrust-bearing, Thrust-washer.

Planishing   A process for polishing and/or improving the accuracy of a surface by the pressure of a hard tool applied in a normal direction and without sliding or cutting.

Plastics bearing   A bearing constructed from polymeric material, or having a polymer lining.

Plastics bearing materials   Synthetic plastic (non-metallic) materials e.g. phenolic and urea formaldehydes, nylons, polymers, ptfe, which function as bearing materials with or without lubricant.

Plastic flow   Irreversible deformation under mechanical stress.

Ploughing   The plastic deformation of the surface of the softer component of a friction pair. Usually caused by particles of foreign matter dragged through the oil film by the moving surface.

Plummer bearing   A bearing primarily journal, whose function is to support a horizontally disposed power-transmitting line shaft. In marine engineering, the bearings which support the main propulsion shafting. (*Syn* Shaft carriage, Tunnel bearing – obsolescent terms). *See also* Aftermost plummer bearing, Intermediate plummer bearing.

Plummer block   An obsolescent term for plummer bearing (q.v.) or pedestal bearing (q.v.).

Poker gauge   A gauge inserted through a casing or housing adjacent to a bearing to measure shaft position, when stationary,

using a feeler gauge in a gap. *See also* Wear-down gauge.

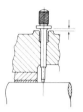

Polar load diagram   A curve plotting the direction and magnitude of the resultant loading upon a crankpin or journal surface relative to a given axis, e.g. connecting rod or journal axis.

Polishing   A process for producing a high degree of surface finish by rubbing with a soft pad charged with fine abrasive particles.

Porous bearing   A bearing made from porous material such as compressed and sintered metal powder. The pores may act as reservoirs or passages for supplying fluid lubricant or the bearing may be impregnated with solid lubricant. *See also* Impregnation, Ptfe.

Pour point   The lowest temperature at which a lubricant can be observed to flow under specified conditions.

Pouring temperature   The temperature at which a bearing material is cast or poured into the housing, fixture or shell which is to be lined. The appropriate pouring temperature is determined by the liquidus temperature (q.v.) of the bearing metal, and the size and thickness of the lining.

Power loss   In a bearing, the power absorbed in overcoming the resistance of the bearing to motion, either from dry friction or from fluid shear, and which reappears as heat which must be dissipated.

Precession   Orbital movement of the journal centre within the bearing clearance.

Precision plating   An electrodeposited overlay of sufficient accuracy to avoid the need for subsequent machining. *See also* Overlay plated bearing.

Precision pressed bearing   A half-bearing (q.v.) produced by pressing from strip material where the back, the joint faces and sometimes the end faces also, are formed with sufficient accuracy to avoid the need for subsequent machining. *See also* As-pressed bearing.

Pre-finished bearing   A bearing which is precision machined ready for assembly, and which requires no further machining or scraping after assembly. The wall-thickness and peripheral length of pre-finished bearings are so controlled in manufacture that the correct clearance and interference fit are provided upon assembly. *See also* Boring allowance, Semi-finished bearing, Undersize bearing.

Pre-lubricated bearing   A bearing which will operate for long periods after initial lubrication on assembly. Certain plastics bearing materials (q.v.) such as acetal resin copolymers require only a trace of oil or grease for satisfactory performance.

Preload   In a bearing, a condition where the cooperating surface is held in contact with the bearing by a predetermined force independently of, and in addition to, the load which the bearing is to support. It is applied in cases where variation in the position of the rotating member caused by varying operating conditions is to be minimised, e.g. in precision spindles. The term is sometimes used to describe a particular geometric condition in a tilting-pad journal bearing (q.v.), which is better described as preset (q.v.).

Preset   In a tilting-pad journal bearing (q.v.), this term is applied when the radius of curvature of the pad surface is greater than the sum of the shaft radius and radial bearing clearance. Sometimes, and particularly in the USA, this is termed preload (q.v.).

Pressed bush   *See* Wrapped bush

Pressure bonding   A method of bonding materials together, e.g. aluminium–tin alloys to steel, by hot or cold rolling with large extensions of the surfaces, resulting in the removal or disintegration of surface oxide films, and the direct alloying or welding together of the opposing materials. (*Syn* Roll bonding)

Pressurised gas lubrication   A system of lubrication in which a gaseous lubricant is supplied under sufficient external pressure to separate the opposing surfaces by a gas film. *See also* Externally pressurised bearing.

Profiled bore   A bearing bore shape which is not truly cylindrical but has been machined in such a manner as to produce a shape or profile which will promote hydrodynamic lubrication (q.v.) with minimum clearance on a selected arc or radius. *See also* Lemon bore bearing, Offset halves bearing. Lobed bearing.

Profiled surface bearing   A thrust bearing which has the face formed so as to promote hydrodynamic lubrication. *See also* Fixed-pad bearing.

Projected area, surface   The area of the bearing surface as projected on to a diametric plane normal to the load line. It is used to calculate the specific load on a

journal bearing. For a cylindrical bearing without grooves, the projected area is the bore diameter × length.

Propeller shaft   The shaft to which the propeller is attached. (*Syn* Tailshaft – obsolescent term).

Propeller shaft bearing   *See* Stern bearing.

Ptfe   Polytetrafluoroethylene, a synthetic plastic material possessing remarkable frictional properties in the dry state, e.g. static friction = kinetic friction = no stick–slip (q.v.).

Puddling   A method of assisting the escape of entrapped air, and of preventing the formation of contraction cavities, during the solidification of whitemetal linings in statically cast bearings.

pV **factor**   The product of the bearing specific load and the surface rubbing velocity. The accepted units are $N/mm^2 \times m/s$ or $lbf/in^2 \times ft/min$.

pV **limit**   The maximum permissible pV factor (q.v.) for a bearing.

Quadrimetal bearing   A bearing consisting of four layers: a backing (usually steel) a strong bearing alloy, a softer bearing alloy and a soft overlay deposited electrolytically. An example is steel/tin–aluminium/whitemetal/lead–tin overlay.

Radial clearance   One-half of the diametric clearance (q.v.).

Radial grooves   Grooves (q.v.) disposed radially across the bearing face of a thrust washer (q.v.) to distribute lubricant.

Radiation   *Nuclear*: particles and other radiation emitted by atomic nuclei which may affect lubricants and certain bearing materials in nuclear power installations. *Thermal*: the dissipation of heat from a bearing by the electromagnetic waves excited by its temperature difference above the surroundings.

Rayleigh (step) bearing   A multi-land thrust bearing (q.v.) where each land has a shallow step, resulting in a film of two discrete thicknesses. The fluid pressure caused by the constriction at the step is high enough to support the applied load. *See also* Fixed-pad bearing.

Relief   *See* Joint face relief.

Remetalling   The relining of a bearing shell or insert with new bearing material, after removal of the original lining which has been damaged in service. The relining of thin-walled bearings (q.v.) is not recommended owing to danger of distortion and loss of peripheral length of backing. The relining of thick-walled bearings is acceptable under carefully controlled conditions.

Replacement bearings Bearing liners or bushes to replace those already fitted or intended for use in machinery that has become worn or damaged in service.

Resizable bearing *See* Semi-finished bearing.

Resonance changer A device incorporated in the thrust block to change the natural frequency of axial vibration of the shafting and thrust system. An axial de-tuner. The axial stiffness of the thrust bearing is reduced by introducing an hydraulic support.

Restrictor A device fitted into a fluid flow path to cause a pressure drop or to limit the flow. Usually it consists of an orifice or series of orifices but sometimes a capillary is used.

Restrictor ring *See* Oil sealing ring

Reticular Having a network-like structure; formed with interstices. *See also* Tin–aluminium.

Rigid-walled bearing An insert bearing (q.v.) having a wall thickness greater than about 20 per cent of bearing bore diameter. *See also* Wall.

Ring-oiled bearing A bearing carrying a ring of larger diameter which can rotate with the journal and dips into a lubricant reservoir beneath. *See also* Oil-ring lubrication.

Roll bonding *See* Pressure bonding.

Rolled bush *See* Wrapped bush.

Rolling contact bearing A bearing in which the relatively moving parts are separated by elements which roll. Also called rolling element bearing, roller bearing, ball bearing.

Rotary lining *See* Centrifugal lining.

Rotating load This occurs where the load vector rotates relative to the bearing and/or the journal. Such rotating loads are the result of out-of-balance rotating masses (crankshaft unbalance), reciprocating masses (connecting rod, piston) and superimposed forces (gas pressure). The load vector may vary considerably, as in an ic engine (*see* Polar load diagram), or may be of constant magnitude rotating synchronously with the shaft (as with pure out-of-balance). *See also* Centrifugal loading.

Rotation In a structural bearing (q.v.), the rotational movement about an axis or axes transverse to and/or parallel with the plane of contact between the bearing surfaces.

Rubbing bearing *See* Dry bearing.

Rumble The name given to main bearing noise, sometimes experienced in high speed ic engines, e.g. car engines. It is attributed to movement of the crankshaft in the bearings

caused by combustion pressure.

Run Sometimes used in place of the term wiping or smearing (q.v.) when a bearing has overheated and smearing of the surface has occurred.

Running in The process by which machine parts improve in conformability (q.v.), surface topography and frictional compatibility during the initial stage of use.

Sacrificial corrosion Electrolytic corrosion (q.v.) of the bearing material selected to become anodic to the more expensive journal in the presence of an electrolyte.

Safety drain An additional baffle (q.v.) at the end of a bearing with a separate drain having an open, visible discharge to act as an extra safeguard to prevent oil entering the adjacent machine. Applied particularly to electrical machines to prevent oil entering the windings.

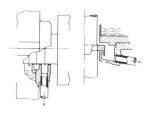

Save-all A channel or trough formed around the edge of a bedplate to prevent oil, water, etc. running over the surrounding floor. Sometimes fitted at the ends of a bearing casing underneath the shaft to catch any oil which may leak from the closure. *See also* Oil catcher, Safety drain.

Saw-cut nick A nick (q.v.) which has a saw cut at each side to enable it to be formed without the need to shear the metal. Usually adopted in bearings with thicker walls, sometimes in addition to milling. *See also* Milled nick.

Scabbing In wear, a loosely-used term referring to the formation of bulges in the surface.

Scoop lubrication A bearing in which oil is delivered by a scoop collaborating with an internal channel formed in the periphery of a ring attached to the shaft. During normal operation, lubricant passing through the bearing is directed into the channel, so that the ring does not dip into the oil bath, thus avoiding churning (q.v.). The lubricant is held in the channel by centrifugal force until removed by the scoop. The device may be used for either horizontal (more usual) or vertical bearings, and efficiency is high since delivery is independent of speed.

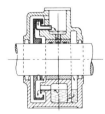

Scoring   The formation of severe scratches in the direction of sliding due to abrasion by solid particles. *See also* Scuffing.

Scuffing   Localised damage caused by the occurrence of solid phase welding between sliding surfaces, without local surface melting. In UK, scuffing implies local solid-phase welding only. In USA, scuffing usually includes abrasive effects. *See also* Scoring.

Seat   The surface of the supporting structure which is in contact with the back of a bearing. *See also* Spherical seating.

Segregation   The separation of the constituents of a bearing material caused by gravitational or centrifugal forces, e.g. lead in copper–lead bearings, copper–tin or tin–antimony compounds in whitemetal bearings.

Seizure   The stopping of relative motion as the result of interfacial friction. It may be accompanied by gross surface welding.

Self-aligning bearing   A bearing supported in such a manner that it is free to align itself to the shaft, or thrust collar, and to follow angular displacements of the axis of the shaft which may occur during operation. *See also* Gimbal, Spherical bearing, Spherical seating.

Self-contained bearing   A bearing whose casing contains the lubricant, with a means for supplying it to the bearing surfaces, cooling it and preventing loss through leakage.

Self-lubricating bearing   A bearing independent of external lubrication. Such a bearing may (a) be sealed for life after packing with grease; (b) be made of self-lubricating material (q.v.); or (c) contain an oil bath and the means for supplying the oil to the bearing surfaces. *See also* Lubrication.

Self-lubricating material   Any solid material which shows low friction without application of a lubricant, e.g. graphite, molybdenum disulphide, ptfe. The term is also applied to porous bearings (q.v.). *See also* Solid lubricant.

Semi-finished bearing   A bearing which has been finish machined except in the bore, to permit of final boring on assembly. To improve availability, replacement bearings may be stocked in this condition and bored to their finished size when the shaft diameter is known. (*Syn* Resizable bearing). *See also* Boring allowance, Pre-finished bearing, Undersize bearing.

Serrated joint   The joint faces (q.v.) of a connecting rod and cap (q.v.) which have axial serrations machined in them, which mate with corresponding serrations in the joint faces of the opposing part upon assembly. This design is often employed in V-engine connecting rods in which the split of the rod is not at right angles to the rod axis, in order to prevent or reduce distortion and/or displacement of the rod and cap bores under load. *See also* Oblique split.

Set of bearings   A complete outfit of bearings, or spare bearings, for a machine. It is always necessary to state precisely what a particular set comprises.

Shaft carriage   *See* Plummer bearing.

Shaft deflection   The elastic bending of a shaft or journal caused by applied forces.

Shaft locus   The movement of the geometric centre of a shaft or journal within the bearing clearance space, usually in a cyclical path.

Shell bearing   A non-preferred term sometimes used to denote an insert bearing (q.v.).

Shoe, slipper   A slider bearing having a plane surface. The name used by Kingsbury for a tilting thrust pad (q.v.).

Short bearing theory   An approximate, analytical solution to Reynolds equation for a journal bearing which neglects the effect of hydrodynamic pressure on circumferential oil flow.

Side foot mounted bearing   A bearing similar to the centre-line mounted type (q.v.)

but where the support surfaces are in a plane at an appreciable distance from the shaft axis (commonly this would be about $\frac{1}{2}\times$ shaft diameter). The casing is usually of split construction, and may contain any combination of bearing functions.

Sintering Heating powdered materials (usually metals or metal ores) in a reducing atmosphere to a temperature below the melting point of main constituent, when agglomeration of particles occurs. *See also* Impregnation, Porous bearing.

Skeg A strut attached to the hull of a ship to support an outboard propeller shaft.

Slave rod A member connecting a piston or crosshead to another connecting rod, the master rod (q.v.), instead of directly to the crankpin.

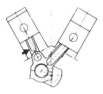

Slope bored The term applied when the bore of a bearing is machined to an offset and/or at an angle to the axis of the bearing outside diameter so as to bring the bearing bore into alignment with the shaft journal. It usually refers to ship's sterntube bearings. *See also* Double slope bored, Partial slope bored.

Small end The end of the connecting rod (q.v.) which is connected to the piston. (*Syn* Little end, Top end)

Small-end bearing The bearing at the smaller (piston) end of a connecting rod in a reciprocating machine. (*Syn* Gudgeon-pin bearing, Little-end bearing, Piston-pin bearing, Top-end bearing)

Smearing The removal of material from a surface, usually a bearing surface, and its redeposition as a thin layer on one or both of the cooperating surfaces. (*Syn* Wiping)

Solid bush Usually applied to a bush made of a single material.

Solid insert An insert bearing (q.v.) or bush consisting of a single material or composite as opposed to a bimetal or trimetal insert.

Solid lubricant Any solid used as a powder or a thin film on a surface to provide protection from damage during relative movement, and to reduce friction and wear. Many solid lubricants have a lamellar structure (q.v.), e.g. graphite, molybdenum disulphide.

Solidus The temperature at which, on cooling from the liquid state, an alloy becomes completely solid. Conversely, the temperature at which a solid alloy *commences* to melt upon heating. *See also* Liquidus.

Sommerfeld number *See* Journal bearing dimensionless groups.

Spalling The separation of particles from a surface in the form of flakes. Spalling is usually a result of sub-surface fatigue and is commonly associated with rolling element bearings and with gear teeth. *See also* Wear.

Specific load The load per unit area on a bearing surface, or projected surface (q.v.), in the case of a journal bearing, expressed in units of MN/m^2 or MPa, kp/cm^2, lbf/in^2 or psi. The term is preferred to bearing pressure.

Spherical bearing A bearing in which the two relatively sliding surfaces are of spherical form.

Spherical seating A bearing-supporting surface of spherical form to provide a self-aligning feature. This term is usually applied to the bore of the housing.

Spherically-seated bearing A bearing which has a back of spherical form supported in a similar housing to permit the bearing to align itself, or to be aligned, to the shaft. *See also* Self-aligning bearing, Spherical bearing.

Spindle Synonymous with shaft, this term is confined to specialist application, e.g. lathe headstock spindle, and usually to small sizes, as in horology and instrument mechanisms.

Spiral groove An oil groove (q.v.) which follows a spiral path in a thrust face. Sometimes used incorrectly to denote a helical groove (q.v.) in the bore of a bush.

Splash lubrication A system of lubrication

in which the lubricant is splashed on to the moving parts.

Split bearing, bush   A bearing or bush in two pieces, usually halves, for purposes of ease of assembly. *See also* Half-bearing.

Split casing   The term is confined to casings in two parts where the plane of the joint, or 'split' is coincident with, or parallel to, the shaft axis.

Spoiling pin, PEG   A small pin or dowel inserted in one of two mating components to prevent their being incorrectly assembled.

Sponge effect   *See* Squeeze effect 1.

Spreader groove   An axial oil groove (q.v.) designed to spread the lubricant across the width of the bearing. *See also* Groove, Mud groove.

Squeeze effect   This term has two preferred meanings: (a) The production of lubricant from a porous retainer by the application of pressure. (b) The persistence of a film of fluid between two surfaces which approach each other in the direction of their common normal.

Squeeze film   A load-supporting fluid film between two surfaces approaching each other in the direction of their common normal. It is able to resist very high specific loads without rupture. *See also* Squeeze effect.

Starting torque   *See* Breakaway torque.

Static friction   The tangential resistance to the relative motion between two bodies under load. *See also* Friction, Limiting static friction.

Static Lining   The lining of bearing shells or housings by pouring bearing metal into static jigs or fixtures. Contrast with centrifugal lining (q.v.).

Static vibration   A situation where a machine which is itself stationary is subjected to vibration imposed from an external source, commonly through the supporting structure, e.g. a diesel generating set on stand-by alongside another one in operation.

Stave   One of a number of similar sectors which form the lining of a sterntube bearing (q.v.). The staves may be of *Lignum Vitae* (q.v.), rubber or reinforced plastics.

Steel-backed bearing   An insert bearing (q.v.) having a steel backing lined with appropriate bearing material. *See also* Bimetal bearing, Overlay plated bearings, Trimetal bearing.

Step bearing   A plane surface bearing which supports the lower end of a vertical shaft. In USA, applied to dammed groove bearing (q.v.). (*Syn* Footstep bearing).

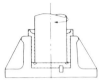

Stern bearing   The bearing which supports the propeller at the end of the propulsion shafting. (*Syn* Propeller shaft bearing).

Sterngear   The generic term for the equipment associated with the support and sealing of the propulsion shafting where it passes through the hull to the propeller. *See also* Aft bush, Forward bush, Stern bearing, Sterntube bearing, Withdrawable sterngear system.

Sterntube   The tube, through the sternframe or hull of a vessel, through which the propeller shaft passes.

Sterntube bearing, bush   The bearing, or bearings, situated in the sterntube (q.v.) to support the propeller shafting where it passes through the hull of a vessel. Usually, but not always, the outer bearing is adjacent to, and supports, the propeller.

Stick–slip   A relaxation oscillation usually associated with a decrease in the coefficient of friction (q.v.) as the relative velocity increases. It was originally associated with the formation and destruction of interfacial junctions, on a microscopic scale – often a basic cause. The period depends on the velocity and on the elastic characteristics of the system. Stick–slip will not occur if the static friction is equal to or less than the dynamic friction.

Stiffness   *See* Oil-film stiffness.

Stiffness coefficients   The coefficient of the displacement-dependent terms in the equation of journal whirl motion. Normally, four coefficients are required for each bearing, for the direct and cross-coupled terms in the x and y-directions.

Stress raiser   Any sharp corner or sudden discontinuity, e.g. a score mark or scratch, which can cause stress concentration in the affected area, and may lead to premature failure under load. Fillets and undercuts (q.v.) are provided to prevent such stress concentration, but poor machining or accidental surface damage can produce stress raisers.

Stribeck curve A graph showing the relationship between the coefficient of friction (q.v.) and the dimensionless number $\eta n/p$ where η is the dynamic viscosity, n the rpm and p the specific load.

Structural bearing A bearing designed to accommodate movement, due to thermal or stress changes, between structural members such as bridge piers and decks, building and heavy machinery supports. *See also* Rotation, Translation.

Stuffing box An annular cavity filled with packing to make a pressure-tight seal surrounding a shaft or spindle. The packing is compressed in the stuffing box by a gland (q.v.).

Sulphur corrosion The attack by sulphur compounds from oil additives or fuel combustion products upon copper-base alloys, especially phosphor bronzes, at high temperatures.

Superfinishing A process for producing a very high degree of surface finish by the rubbing action of a fine abrasive stone oscillated at high speed and traversed over the surface.

Superlaminar flow Superlaminar flow conditions pertain when there is movement of fluid particles across the thickness of the bearing oil film (encompasses vortex and fully turbulent regimes). *See also* Laminar flow.

Surface finish The topography of a surface expressed in units normal to the surface. It is usually measured along a line normal to the direction of finishing. The term is preferred to surface roughness. *See also* Surface topography.

Surface roughness *See* Surface finish.

Surface topography The geometrical form of the finish of a solid surface. It is usually viewed as a plane section, normal to the surface and to the direction of finishing. *See also* Surface finish.

Suspended crankshaft *See* Underslung crankshaft.

Swash Occurs where a component has its axis at an angle to the axis of rotation (*Syn* Gallop, Wobble).

Swashplate A component with a planar surface attached to a shaft so that the

surface is not normal to the axis of rotation.

Swell *See* Housing swell, Joint face swell.

Syphon lubrication The supply of oil to a bearing from a reservoir by the capillary action of a wick drawing the oil over a lip and thence by gravity to the bearing. (*Syn* Wick lubrication)

Tag *See* Nick.

Tail shaft *See* Propeller shaft.

Tang *See* Nick.

Taper land thrust bearing A thrust bearing (q.v.) having multiple fixed lands (q.v.), the surfaces of which are made to a converging taper with respect to the moving collar. Each land usually has part of its surface parallel to the collar and the remainder formed to a taper. *See also* Fixed-pad bearing.

Tapered (conical) journal A journal (q.v.) which is greater in diameter at one end than at the other.

Thermal cycling Cyclic variations of temperature in service, which may lead to fatigue failure or to anisotropic cracking (q.v.).

Thermal fatigue Cracking caused by repeated thermal cycling through an excessive range of temperature.

Thermal ratcheting *See* Anisotropic cracking.

Thermal taper Where a load-supporting oil-film wedge is produced by thermal distortion of the bearing surface caused by the heat generated within the oil film. *See also* Oil wedge, Thermal wedge.

Thermal wedge The increase in pressure due to the thermal expansion of the lubricant, e.g. in a parallel thrust bearing. *See also* Oil wedge, Thermal taper.

Thermoplastic materials Synthetic plastic materials, e.g. Nylon, ptfe, which undergo reversible changes with change of temperature up to a defined maximum temperature.

Thermosetting resins Synthetic plastic materials, e.g. phenolic and urea formaldehydes, epoxy, which are cured (i.e. solidified) by non-reversible changes at elevated temperatures during manufacture.

Thick-film lubrication A condition of lubrication in which the film thickness of the lubricant is appreciably greater than that required to cover the surface asperities (q.v.) when subjected to the operating load, so that the effect of the surface asperities is not noticeable.

Thick-walled bearing An insert bearing (q.v.) having a wall thickness (usually about 10–20 per cent of bearing bore diameter) such that the bore contour is substantially independent of the contour of the bearing housing (q.v.). *See also* Wall.

Thin-film lubrication *See* Boundary lubrication.

Thin-walled bearing An insert bearing (q.v.) having a wall thickness (usually less than about 3 per cent of bearing bore diameter) such that the bore contour is determined by the contour of bearing housing (q.v.). *See also* Wall.

Thrust bearing A bearing in which the load acts in the direction of the axis of rotation.

Thrust block The double thrust bearing which transmits the propulsive thrust of the propeller to the hull structure.

Thrust collar A thick disc or collar integral with, or attached concentrically to, a shaft and having an operating surface, usually planar, normal to the axis for transmitting thrust force to a thrust bearing (q.v.).

Thrust face The operating surface of a thrust bearing (q.v.), either as part of an independent unit or integral with a journal bearing or bush.

Thrust load A force acting parallel to the shaft axis but not necessarily coincident with it.

Thrust pad *See* Pad.

Thrust shaft The shaft in a ship's main propulsion shafting, which has a thrust collar (q.v.) and runs in the thrust block (q.v.).

Thrust washer A plain or contoured bearing member of washer form designed to sustain axial loading. Sometimes termed a plain thrust bearing (q.v.). It may be in halves or in one piece.

Thumb-nail groove A wide, shallow groove provided on a thrust face (q.v.), leading from the inside diameter and finishing short of the outside diameter. (*Syn* Pear-drop groove). *See also* Groove.

Tilting pad That component in a tilting or pivoted-pad bearing (q.v.) which carries the bearing surface.

Tilting-pad bearing A bearing in which the surface is divided into two or more sectors or pads (shoes in USA), which are individually supported and free to take up a position at a small angle to the opposing surface, according to the hydrodynamic pressure distribution over the surface. The bearing may be an axial thrust or radial journal load type. *See also* Pad.

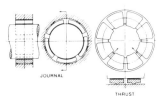

JOURNAL

THRUST

Tin–aluminium Bearing alloys of aluminium with tin (5–40 per cent) and usually with small additions of copper and/or nickel. With tin contents below about 10 per cent, the tin is present as discrete particles. When the tin content is above about 10 per cent, suitable cold working and heat treatment produce a reticular structure (q.v.) which is in the form of two interlocking three-dimensional networks. This structure closely resembles that of a sponge where the skeleton is replaced by tin and the intersecting passages are filled with aluminium.

Tin flash A thin layer of tin applied to bearings as a finishing layer.

Tin oxide corrosion The attack on the tin constituent in tin-base whitemetal bearings resulting in the formation of a hard, dark grey/black oxide on the bearing surface. The oxide is predominantly SnO_2 with some SnO. The attack is confined to the tin-base matrix and does not extend to the intermetallic copper–tin and tin–antimony compounds. It is caused by the presence of water, particularly sea water, in the lubricating oil and hence occurs more frequently in marine machinery.

Tinning The application of a coating of tin or of a tin–lead alloy to a surface, prior to lining with the appropriate whitemetal. *See also* Cast iron-process.

Top end The term used in marine engineering for the end of the connecting rod which is connected to the piston or crosshead. (*Syn* Little end, Small end)

Top-end bearing The term used in marine engineering for the upper bearing of connecting rod (*Syn* Crosshead bearing,

Gudgeon-pin bearing, Little end bearing, Small-end bearing)

Trailing bearing A main propulsion shafting bearing equipped to take the thrust from a freely rotating (trailing) propeller when the shafting is uncoupled aft of the main thrust block (q.v.). Used on multi-screw ships to reduce drag if one set of machinery is immobilised, as in warships prone to combat damage. Usually combined with the aftermost plummer bearing (q.v.).

Translation In a structural bearing (q.v.), the relative movement between the bearing surfaces in the plane of contact.

Tribology The science and technology of interacting surfaces in relative motion, and of the practices related thereto. It includes the study of wear by cavitation erosion (q.v.). *See also* Smearing, Wear.

Trimetal bearing A bearing consisting of three layers: a backing (usually of steel), an interlayer (of a strong bearing alloy, copper-lead or lead bronze), and a surface of cast whitemetal (tin or lead base) usually 0.005–0.04 inch (0.012–1.0 mm) thick. Continental European and USA practice includes overlay plated bearings (q.v.) in this term.

Trunk piston A piston which is attached directly to the connecting rod by a gudgeon pin (q.v.).

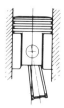

Trunnion bearing A bearing used as a pivot to swivel or turn an assembly. The term is also applied to the bearings which support the ends of ball pulverising mills, and to the bearings carrying the intermediate supporting rollers of rotating cement kilns.

Tube shaft A shaft in a ship's main propulsion shafting which passes through a sterntube (q.v.) in the hull but does not itself carry the propeller.

Tunnel bearing *See* Plummer bearing.

Tunnel shaft An obsolescent term for a ship's main propulsion shafting. The term originated with machinery amidships when a tunnel was provided through the after holds to take the shafting and also provide access to the sterntube (q.v.) and gland (q.v.).

Turbulent flow *See* Superlaminar flow.

Turning The continuous very slow rotation or 'turning over' of a machine, usually a

prime mover, by external means for purposes of erection, maintenance or prevention of thermal distortion by heat soak. *See also* Barring.

Turning gear The mechanical means provided for carrying out the operation of turning (q.v.). *See also* Barring gear.

Ultrasonic test A test of bond continuity between lining and backing (q.v.) by ultrasonic vibration emitted by a probe held against the lined surface. The echo pattern is observed on a cathode-ray tube. If the bond is continuous, the echo occurs from the back of the shell, but if discontinuous, from the interface between the lining and the shell.

Undercut A groove or channel machined into a surface at its junction with another surface at right angles, i.e. the reverse of a fillet (q.v.). The object of an undercut is (a) to eliminate stress raisers (q.v.) in highly-stressed components such as crankshafts; (b) to increase the effective axial length of a bearing, e.g. a big-end bearing, by recessing the fillet into the crankweb.

Undersize bearing A bearing finished to a bore diameter less than that of a standard bearing, in order to accommodate journals or crankpins which have suffered wear in service, and/or have been machined undersize. Such bearings may be pre-finished (q.v.) or semi-finished (q.v.).

Underslung crankshaft A crankshaft which is supported beneath the crankcase frame by bearing caps, fitted with appropriate bearings, bolted to the underside of the frame. *See also* Bedplate.

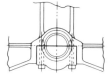

Unidirectional load An applied load or force in a constant direction in relation to the bearing concerned. Thus, the loading on rotating machinery bearings, such as turbine rotor or electrical generator journal bearings, is normally unidirectional though it may vary in magnitude.

Varnish   In lubrication, a deposit resulting from the oxidation and/or polymerisation of fuels, lubricating oils, or organic constituents of bearing materials. Harder deposits are described as lacquers, softer deposits as gums (q.v.).

Vertical bearing   A bearing in which the shaft axis is vertical.

Viscosity   That bulk property of a fluid, semi-fluid or semi-solid substance which causes it to resist flow. There are two types of viscosity measurement:

(a) *Dynamic viscosity* (η): This used to be termed the absolute viscosity but the name is now non-preferred. It is defined by the equation

$$\eta = \frac{\text{shear stress}}{\text{rate of shear}}$$

The units are N.s/m^2 or Pa s (SI): lbf s/in^2 = Reynold (Imperial); dyne s/cm^2 = poise (cgs). The commonly used unit is the centipoise (cP), where 1 cP = 0.001 Pa s.
(b) Kinematic viscosity (v): This is the ratio dynamic viscosity/density at a specified temperature and pressure. The units are m^2/s (SI); in^2/s (Imperial); cm^2/s = Stoke (cgs). The commonly used unit is the centistoke (cS), where 1 cS = 10^{-6} m^2/s.

Commercial viscosities are kinematic and usually measured by the time taken in seconds for a given volume of oil to flow through an orifice of specified size and shape: Redwood sec (UK); Engler number (Continental Europe); Saybolt Universal seconds (USA).

Viscosity index   A commonly used measure of a fluid's change of viscosity with temperature. The higher the viscosity index, the smaller the relative change in viscosity with temperature.

Viscosity pump   A device consisting of a duct, one side of which slides longitudinally relative to the other, so that a fluid, by virtue of its viscosity, is delivered against a back pressure. For practical convenience, it takes the form of a shallow circumferential groove in a stationary component surrounding a revolving cylindrical surface. It may be a separate component or may be combined with a plain journal bearing. The fluid flow increases with speed. The pressure generated increases with viscosity and speed. The device is simple and reliable although the mechanical efficiency is very low.

Viscosity pump lubrication   The supply of lubricant to bearing surfaces by a viscosity pump (q.v.).

Viscous friction   *See* Fluid friction.

Waisted journal   A journal (q.v.) whose axial contour is smaller in diameter at the centre than at the extremities.

Wall, walled   Refers to the thickness of an insert bearing (q.v.) between its bore and its back. *See also* Medium-walled bearing, Rigid-walled bearing, Thick-walled bearing, Thin-walled bearing.

Wall thickness   The total radial thickness of a bearing or bush measured from its bore to its outside diameter.

Washer   *See* Thrust washer.

Water-cooled bearing   Where the cooling medium is water, circulating through a water-jacket, coil or cooler (q.v.). *See also* Cooling.

Water-jacket cooling   Where the casing is arranged with a separate compartment adjacent to the oil bath through which water is circulated to extract heat from the oil.

Wear   The progressive loss of substance from the operating surface of a body occurring as a result of relative motion at the surface. Wear is usually detrimental, but in a mild form may be beneficial, e.g. during running in (q.v.). *See also* Abrasion, Adhesive wear, Cavitation erosion, Erosion, Fluid erosion, Fretting, Galling, Ploughing, Scoring, Scuffing, Seizure, Smearing, Spalling.

Wear-down gauge   A gauge inserted through a casing or housing adjacent to a bearing to measure the shaft position, when stationary, to determine if it has changed over a period due to wear in the bearing. The gauge may be of micrometer type, or a simple poker gauge (q.v.) may be used.

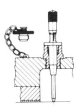

Wear rate   The rate at which wear (q.v.) occurs. It may be measured in terms of the quantity of material removed in unit time, in unit distance of sliding, in an arbitrary number of revolutions or oscillations or as a change in dimensions. It is important that the conditions and units should be precisely defined.

Weatherproof closure Fitted to bearings which operate in the open, to exclude rain, snow and airborne dust, often in areas of heavy dust deposition.

Wedge *See* Oil wedge.

Whirl *See* Oil-film whirl.

Whitemetal Alloys of tin or lead with copper and antimony and sometimes with additions of other elements, e.g. nickel and cadmium, to improve their strength and structure for use as bearing material. (*US syn* Babbitt metal)

Wick lubrication *See* Syphon lubrication.

Windage shield A wall, usually of flat sheet material, through which the shaft passes with a small clearance, interposed between a bearing casing and a machine to prevent windage from the machine drawing oil, or oil mist, out of the bearing.

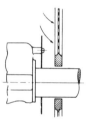

Wind-back seal An oil sealing ring (q.v.), fixed or more usually floating, having in its bore a number of helical grooves which cause entrained fluid to travel in an axial direction by viscous action. It is used to prevent the leakage of oil from a bearing, and axiomatically can only be used with unidirectional shaft rotation.

Wiping *See* Smearing.

Wire-wool damage The formation of hard black scab on a whitemetal bearing surface, causing machining damage to the journal, which can be very severe, producing long fibres, like steel wool from the journal. The cause is obscure, but is thought to be triggered by foreign particles bridging the oil clearance. The phenomenon is self-propagating once started. It occurs most frequently with chromium steel (3–Cr% $\frac{1}{2}$% Ni) shafts, due to the formation of iron carbides from shaft material and oil at high temperature. But it can occur with a wide variety of steels and with aluminium bronze journals and whitemetal.

Withdrawable sterngear system An arrangement which allows the stern bearing and seals to be withdrawn inboard for inspection and maintenance, avoiding the need to drydock or tip the vessel. *See also* Sterngear.

Wrapped bush 360° bearing member with axial or oblique split, made by pressing or rolling from a flat blank.

Wrist-pin bearing A bearing supporting a shaft or member which oscillates. (*Syn* Knuckle-pin bearing)

Index

The Classical Library of Chinese Literature and Thought makes available in unabridged, bilingual editions great works of literature, history, and philosophy from China. For each volume, leading scholars in China and from around the world have collaborated to produce an authoritative text that will be the standard for years to come.

Part of The Culture and Civilization of China series, The Classical Library of Chinese Literature and Thought is jointly published by Yale University Press and the Foreign Languages Press.

Balanced Discourses

中論

A Bilingual Edition

Xu Gan

English translation by John Makeham
Introductions by Dang Shengyuan and John Makeham

Yale University Press
New Haven and London

Foreign Languages Press
Beijing

The Chinese text comes from the Chinese Ancient Texts Center.

CHANT is a computerized database of the entire body of Han and pre-Han traditional Chinese texts. Located at the Chinese University of Hong Kong, it is designed to provide the international community of scholars with ready access to all extant ancient Chinese texts of the period, thus facilitating a wide range of research, whether in Chinese literature, history, philosophy, linguistics, or lexicography. For the database, the best early editions have been chosen, mainly from the Sibu Congkan 四部叢刊 collection. Punctuation and textual notes are then added, and emendations are clearly marked in such a manner as to permit easy recovery of the original text by the reader.

Professor D. C. Lau 劉殿爵 and Dr. F. C. Chen 陳方正 are chief editors, and Professor Ho Che Wah 何志華 is the executive editor of the CHANT database. Additional information about CHANT is available at its Website (www.chant.org).

Designed by Sonia Shannon
Set in Adobe Garamond by Asco Typesetters, Hong Kong
Printed in the United States of America by R. R. Donnelley & Sons, Crawfordsville, Indiana

Library of Congress Cataloging-in-Publication Data appear at the end of the book.

A catalogue record for this book is available from the British Library.

The paper in this book meets the guidelines for permanence and durability of the Committee on Production Guidelines for Book Longevity of the Council on Library Resources.

10 9 8 7 6 5 4 3 2 1

Publication of this book was made possible by the generous support of John and Cynthia Reed.

Yale University Press gratefully acknowledges the financial support given to The Culture and Civilization of China series by The Starr Foundation.

The Classical Library of Chinese Literature and Thought

Contents

Acknowledgments

The work of translating and annotating Xu Gan's *Zhonglun* has occupied me, on and off, for many years. It is satisfying finally to be able to bring the task to a conclusion. It is also sobering to realize that Chinese text and facing English translation are now irretrievably open to public scrutiny. Many teachers and colleagues have generously guided me and given me invaluable feedback over the years. Rafe de Crespigny, K. H. J. Gardiner, Pierre Ryckmans, Chi-yun Chen, and Roger Ames read early versions of the translation and provided criticism and advice for improvement. Yoav Ariel generously sent me material on *Zhonglun* that he had been collecting in preparation for his own translation. The anonymous readers for Yale University Press, at various stages in the manuscript's evaluation, offered encouragement and corrections. Ning Chen kindly brought to my attention a number of inaccuracies in the translation. My manuscript editor at Yale University Press, Dan Heaton, has saved me from many infelicities of style and grammar. Not only has he professionally edited the English text, but he has also had to deal with a host of problems presented by a bilingual edition. Jim Peck, executive director of The Culture and Civilization of China series, has overseen publication at all stages and arranged for Professor Dang to write the introduction. My biggest debt is to Michael Nylan and Andrew Plaks for their painstaking corrections and detailed suggestions for revision. Without their constructive critical input I would have even greater cause to be walking in "fear and trembling."

Translator's Note on the Chinese Text

The Chinese text accompanying the translation is based on the text in Lau and Chen, *Concordance to Zhonglun*, or CHANT text. Punctuation and paragraph breaks at times differ from those adopted in the CHANT text, reflecting my interpretation of the Chinese text.

Round brackets ( ) signify deletions; square brackets 〔 〕 signify additions. This device is also used for emendations. An emendation of character 甲 to character 乙 is indicated by (甲)〔乙〕. For example:

(倦)〔倚〕[b] 立而思遠

Notes to the Chinese text give details of the authority for deletions, additions, and emendations. In all cases, the CHANT readings are reproduced in the citation text. I have noted all instances in which the readings I follow differ from those adopted in the CHANT text. In the above example, I have not followed the CHANT emendation, preferring the original reading. This is indicated in the corresponding note.

A footnote callout appended to a character or passage of text that is not marked by brackets indicates that the reading I have followed is not one recorded in the CHANT text. For example:

志者、學之師[d] 也

In this example, as is indicated in the corresponding note, I have followed the alternative reading of 帥. In all cases where I have followed an alternative reading or emendation, the translation is based on that alternative reading or emendation. The corresponding note gives details and the authority for the preferred alternative reading or emendation.

Introduction by Dang Shengyuan

Zhonglun (*Balanced Discourses*; hereafter *Discourses*) was written by Xu Gan (170–217) at the end of the Eastern Han (25–
220) dynasty. *Discourses* is a work of political commentary, the
aim of which is "to expound normative principles, trace their
sources in the classical teachings, and attribute these principles
to the way of the sages and worthies."[1] Xu is thus classified
under Confucian thinkers in the "philosophers" division of all
bibliographical lists of the standard histories, except for that
of *Songshi*, where he is listed among miscellaneous writers.

Xu Gan (style Weichang) was a native of Ju prefecture in
Beihai county (the eastern part of the today's Lechang county
in Shandong). As one of the Seven Masters of the Jianan reign
period (196–220) he was a celebrated literary figure and thinker
during the period of transition from the Han to the Wei. A
gifted youth, he developed a deep love of learning. He was
painstaking in his studies and skilled at reflective thought.
While still very young, he read a wide range of books and was
exceptionally familiar with the Confucian classics. He was able
to recite at length and had a natural talent for writing. He also
had exceptional literary talents.

After the rebel general Dong Zhuo (d. 192) staged his rebellion in 190, Xu Gan left his residence in Linzi and went
into hiding on the Jiaodong peninsula in Shandong.[2] Later he
returned to Linzi and resumed his life there in retirement and
isolation. He continued to study and reflect, frequently declining requests from commandery and prefectural officials to
take up office.[3] In the Jianan reign period he served under the
famous general Cao Cao (155–220) in positions on the staff of
minister of works and was later appointed as instructor to the
Leader of Court Gentlemen for Miscellaneous Purposes.[4] In
214 he was appointed as instructor of Linzi District. After this
he retired, pleading illness. When later he was offered the post
of magistrate of Shang'ai District, he again declined on the

grounds of ill health. Before this, he had served in several of Cao Cao's military campaigns and had frequented the sporting and banquet activities at Ye (alt. Yezhong), the future Wei capital. Moreover, he composed poetry and rhyme-prose to commemorate these experiences. Together with fellow "Jianan Masters" Ying Yang, Chen Lin, Liu Zhen, and Wang Can, he died in the pestilence of 217.

In the last years of the Eastern Han, according to the author of the preface to *Discourses*, "the canons of the state had broken down and fallen into disuse. The younger male members of families who held official positions formed cliques that were aligned to powerful families. They formed social connections with these families and supported them so that they might make a name for themselves, and they vied to better one another in the attainment of ranks and titles." On these grounds, Xu Gan "cut off his contacts with others to guard his own integrity, having nothing to do with them, and derived his pleasures solely from the Six Classics." When he retired from office after serving Cao Cao, he became even more committed to "hiding himself away in a back street, nurturing his true aim, and keeping his genuineness intact. Leading a plain, simple existence, he initiated no unnecessary action…. He nourished his floodlike vital energy and practiced the arts of longevity."[5] He also began to write *Discourses*.

Because Xu Gan neither attached importance to office and emoluments nor indulged in worldly honors, devoting himself instead to the pleasures of reading and writing, he "commanded a reputation in Qingzhou" making him quite an influential figure amongst scholar-literati.[6] In the year following Xu Gan's death, 218, Cao Pi wrote the following evaluation of him: "When one takes a close look at ancient and modern men of letters, most of them did not closely monitor the details of their own behavior. Few of them would have been able to stand tall on the strength of their moral integrity. Weichang was unique in possessing the qualities of cultural refinement (*wen*) and unadorned simplicity (*zhi*). He was indifferent to worldly success and had few desires. He had the same sense of commitment as Xu You.[7] Indeed he can be called a well-balanced gentleman. He wrote *Discourses* in twenty chapters, establishing a school of thought in its own right. The form and content of his literary style are classical and elegant—well worth passing on to posterity. This gentleman will endure!"[8] Moreover, in his commentary to *Sanguozhi*, Pei Songzhi (372–451) cites a similar evaluation from *Xianxian xingzhuang* (Accounts of the Deeds of Former Worthies): "(Xu) Gan's embodiment of the way was pure and mysterious (*qing xuan*), and his cultivation of the six types of virtuous conduct was complete. Discerning in his understanding and broadly informed in his learning, he was an accomplished essay writer. He neither attached importance to office and emoluments nor indulged in worldly honors."[9]

The unsigned preface to *Discourses* relates that Xu Gan's "natural inclination was such that he constantly wanted to reduce that of which the age had a surplus and increase that in which the ordinary people of the day were deficient. He saw men of letters follow one another in the contemporary fad of writing *belles lettres*, but there was never one among them who elucidated the fundamental import of the classics to disseminate the teachings of the way; or who sought the sages' point of balance to dispel the confusion of popular contemporary *mores*. For this reason, he abandoned the literary genres of the ode (*shi*), rhyme-prose (*fu*), eulogy (*song*), inscription (*ming*), and encomium (*zan*), and wrote the book *Balanced Discourses* in twenty chapters." He thus decided no longer to practice writing in the fashionable literary genres but to live in seclusion, far removed from the hubbub of the mundane world, to devote his energies to writing *Discourses* so that he might "seek the sages' point of balance."[10] *Discourses* is the sole extant philosophical book written by any of the Seven Masters of the Jianan period.

Discourses is divided into two fascicles (*juan*) of ten chapters (*pian*) each (or twelve chapters in the second fascicle if we include the two incomplete reconstructed chapters, "Reinstitute the Three-Year Mourning Period" and "Regulate the Allotment of Corvée Laborers"). Generally speaking the first fascicle is concerned with moral cultivation and the second with statecraft and strategies for achieving political order. The form and content of the book's literary style are classical and elegant. The book frequently takes the teachings of the former kings, Confucius, and Mencius as models and gives expression to the precepts of "inner sageliness and outer kingliness."[11] The book may indeed be described as the successful realization of Xu Gan's aim of "elucidating the fundamental import of the classics to disseminate the teachings of the way, as he sought the sages' point of balance to dispel the confusion of popular contemporary *mores*."[12]

In titling the book *Balanced Discourses*, Xu Gan gave expression to the emphasis that he placed on the concept of balance in Confucian thought and values. There are many passages in which he elaborates on the meaning of balance. In "Rewards and Punishments," for example, he writes: "The former kings sought to clarify the bearing of individual circumstances, and to be balanced when they weighed up their judgments, thereby always maintaining a sense of appropriate measure." This shows that balance means "maintaining a sense of appropriate measure." In "Valuing Words," Xu Gan selects the two stories of Cangwu Bing and Wei Sheng to illustrate this: "In the past, Cangwu Bing took a wife, but because she was beautiful, he gave her to his elder brother. It would have been better not to have deferred to him at all than to have been deferential in this manner. Wei Sheng arranged to meet his wife at the edge of a river. When the water suddenly rose, he would not leave and so drowned. It would have been better not to have kept his word to her at all than to have done so in this manner." For Xu Gan the mistake these men made was to carry matters to an extreme, losing proper measure, thus resulting in them abandoning the principle of balance. For Xun Gan balance was both a conceptual value and a methodological principle to be employed in the process of thinking and recognition. This concept was clearly influenced by such ideas as *appropriateness* (*shi yi*) and *being in due measure* (*you du*), which are featured in *Huangdi neijing* and *Huainanzi*. The concept also contains rudimentary dialectical elements.

Against a background of political corruption and moral depravity at the end of the Eastern Han period, Xu Gan wrote *Discourses* to remedy the ills of his day. The various arguments and opinions raised in the book are all directed at contemporary social realities. Their political goal was to achieve order and stability. The desire expressed in the book to reestablish Confucian political ideals and ethical models is similarly motivated by the practical need to remove corrupt political practices and to remedy the degenerate state of social mores. For example, land annexation and slave ownership were two acute social problems in the Eastern Han period. They had set off a series of political crises and were also the fundamental cause of social instability. Addressing these issues, Xu Gan called for restrictions on the land annexation and excessive slave ownership practices of high-ranking and "noble" families. He argued that annexation and theft of land was responsible for the inequitable distribution of wealth in society and the impoverished state of many scholar-literati. His proposal to limit the number of slaves that "nobles" could own was directed at enriching the state and alleviating social disparities so as to effect political order throughout the realm. His proposal is furthermore related to his insistence that "the people are of foremost importance (*ren wei gui*)," a central tenet of the theory of "the people as the foundation of the nation (*minben*)." Having expressed the view that "of all the creatures in the world that live by breathing, none has greater awareness than man," Xu Gan believed that slaves, born as humans, must therefore be extended the principle of "spreading loving concern (*bo ai*)."[13] Thus in "Regulate the Allotment of Corvée Laborers" we read: "Although slaves are base in rank, they nevertheless possess the five constant virtues. Originally they were the good people of emperors and kings, but they have been entered into the household registers of small men as their personal slaves. Forlorn and destitute, they have lost their homes and yet have no one to tell of their plight. Have they not been wronged?"[14]

In order to bring a chaotic world to order and to remedy social crisis, Xu Gan advocated that government encourage learning, hold ritual in esteem, and be strict and impartial in meting out rewards and punishments. He maintained that if political order were to be effected throughout the realm, then the rites and laws of the sage kings must be followed, and that according to these rites and laws "the noble held positions that were always honored, while the lowly held positions that were hierarchically ranked. The gentleman and the small man each served in a different office." Only by upholding these distinctions could there be "no transgression by those below encroaching upon the authority of those above them, and so corvée labor and the strength derived from wealth were able to be supplied in sufficiency."[15] This sort of mindset, reflecting a staunch belief in Confucian political principles and social ideals, often was the guiding ideology when Xu Gan sought to analyze and solve social problems.

Xu Gan attached great importance to the role of rewards and punishments in governing a state, even elevating them to the lofty status of "fundamental tenets of government." Thus in "Rewards and Punishments" we read: "If rewards and punishments are not made clear, this will not only affect how well ordered the people are, it could even lead to the destruction of the state and the death of the ruler. Can one afford not to be careful?" Although it is natural for people to fear punishments and to take pleasure in rewards, rewards and punishments must be judiciously applied. In the same chapter Xu Gan states: "Nor can rewards and punishments be too light or too heavy. If rewards are too light, then the people will not be encouraged to do good; while, if punishments are too light, then the people will not be afraid. If rewards are too heavy, then people will benefit gratuitously; while, if punishments are too heavy, then people will be forced into a hopeless situation." Similarly, rewards and punishments should be neither too numerous nor too few because "if they are too numerous, then many people will receive them, while if too few, then many people will be left out." In neither case would their purpose be served. The correct way to apply them is "to be balanced [in weighing up] judgments, thereby always main-taining a sense of appropriate measure." This is a further example of Xu's advocacy of the principle of "the way of balance."

The Eastern Han was the beginning of a period of extreme darkness and corruption, leading some scholar-literati to regard politics with disdain. Some individuals came to adopt pessimistic and nihilistic attitudes toward society and history, developing decadent and hedonistic outlooks on life, such that they abandoned their individual social duties and also their responsibility as scholar-literati to provide social criticism. Xu Gan, however, was different. In addition to addressing contemporary social realities by advancing his own political propositions, he also placed emphasis on the need for scholar-literati to fire their ambitions and to have the courage to forge ahead. A concentrated expression of this thinking is the chapter "Titles and Emoluments" in which he proposes that individual scholar-literati should actively seek titles and emoluments on the strength of their virtue and meritorious service. Thus the philosophical concepts and political proposals that Xu Gan elucidates in *Discourses* are in fact representative of the critical social consciousness of a group of positive-minded scholar-literati whose family backgrounds were of the commoner class. These concepts and proposals also represent the political demands made by this group for social and political standing. Because the hereditary elite families monopolized the avenues to office, commoners who sought to gain office and official promotion faced an arduous task. Accordingly, some of the anguish that welled from the deep recesses of Xu Gan's mind is apparent in *Discourses*. In "Titles and Emoluments" he writes:

> Hence a good farmer does not worry that the borders of his fields are not maintained, but rather that the winds and rains will not be in proper measure. Similarly, the gentleman does not worry that the way and its power

are not established, but rather that he will not encounter the right times. The *Book of Odes* says:

> I yoke my four stallions,
> My four stallions stretch their necks.
> I gaze to the four quarters,
> Frustrated that I have nowhere to drive.

He is hurt because he has not encountered the way. Yet, is this so of one age only? Is it so of one age only?

The emotional frustration he feels at not being able to encounter the right times prompts Xu Gan to express disorientation and grief. This tragic outlook on life results from the frustration that scholar-literati from commoner origins with inadequate social resources met in their plans to seek titles and emoluments and to render meritorious service. In the same chapter Xu Gan maintains that in matters of titles and emoluments one should adopt the attitude that "there is a way to be followed in seeking them and their attainment is also determined by conditions which lie beyond human control." He continues: "Shun, Yu, and Confucius may all be said to have had a way to seek them. Shun and Yu's successful attainment of them and Confucius's failure to have done so may be said to have been determined by conditions lying beyond human control. This is so not only for sages; the same applies to worthies." Here he admits the existence of a destiny which controls humans but which itself cannot be controlled, rendering the human condition utterly helpless. In "Examining Falsity" we read: "The gentleman is capable of perfecting his heart and mind. When his heart and mind are perfected, then internally he becomes settled. When internally he is settled, then things cannot upset him. When things cannot upset him, then he takes singular pleasure in his way." This passage may explain why Xu Gan "had the same sense of commitment as Xu You," "nurtured his true aim and kept his genuineness intact," "was indifferent to worldly success and had few desires," "embodied the way in a pure and mysterious manner," and cultivated "the arts of longevity." This passage also reflects the influence of Lao-Zhuang thought on Xu Gan's outlook on life, revealing that in some respects there was not complete consistency between his political and moral principles and his outlook on life.

Because Xu Gan sought "balance" and "appropriateness," his arguments in *Discourses* are, as one might expect, unembellished and compact. This is, of course, one of the strengths of the book. Excessive compromise, however, can influence the incisiveness of thought and theoretical critique. Compared with *Changyan*, the work of Xu Gan's contemporary Zhongchang Tong (179–220), *Discourses* is superior in terms of the form and content of his literary style, but in terms of critical acuity it is less incisive. Precisely because of this, *Discourses* has been insufficiently regarded in research on the history of traditional Chinese thought. In many works on the history of Chinese philosophy or the history of Chinese thought, it is not even mentioned. In fact, *Discourses* boasts some unique intellectual qualities and in its approach to many philosophical problems it makes some definite advances. But it is of unique value to the study of the transition in thought between the Han and Wei periods. Below I will separately examine Xu Gan's views on the way of heaven (*tiandao*), name and actuality (*mingshi*), human talent (*rencai*), and learning (*xuexi*), to discuss some of my observations concerning Xu Gan's philosophical thought as expressed in *Discourses*.

The way of heaven. Xu Gan's views on the way of heaven were clearly influenced by Wang Chong (27–c. 100), who perceived the operations of the way of heaven to be "so of itself" (*ziran*). Like Wang Chong, Xu Gan maintained that "the way of heaven is distant, vast, dim, and obscure." He was mystified by the forces of nature and society which overrode the power of individual humans. He similarly agreed that heaven has no design, no will, and certainly no control over human affairs. For Xu Gan, in both the natural world and in human society there exist a "constant way" and "departures from regularities" (*bian shu*). Because the pattern of things is a combination of the constant way and departures from regularities, it would be mistaken to deny the existence of either. For example, there is a relation between the length of a person's human life and his character, psychological makeup, and cultivated temperament. Confucius's saying that "the humane are long-lived" may be taken to be a "constant way." Its validity should not be denied simply because of such individual departures from regularities as the early death of his disciple Yan Yuan, despite his exemplary quality of humaneness. Xu Gan asserted that good fortune and misfortune are not meted out by heaven but are the consequences of human behavior. For example, Bigan and Wu Zixu resolutely adhered to their beliefs out of a sense of moral principle. They acted contrary to the will of their rulers. Their bitter remonstrances with their rulers led to their being charged and then killed. This is what is meant by "knowing that their deaths were inevitable yet they were still happy to do what they did. What fault rested with heaven?"[16] For Xu Gan, even if heaven wanted to bestow good fortune on people, there would be no way for it to realize this aim. Given that people are not critical of the natural world when it comes to the appearance of irregular phenomena, then they should respond similarly to the random and blind manner in which misfortune and good fortune appear in the human world. Therefore, Xu Gan pointed out, when one investigates a matter or analyzes a problem, one should not be confused by randomness or blindness. Rather, one must investigate matters of the same category under normal, appropriate circumstances and establish one's argument on the basis of the principle which pertains to those matters. Another example of the dialectic pattern of existence concerns "ordering the realm." Under normal circumstances, when the constant way (*chang dao*) prevails, "practitioners of good reap good fortune, while wrongdoers meet with misfortune." Under abnormal circumstances, however, when departures from regularities (*bian shu*) prevail, "practitioners of good do not reap good fortune, while wrongdoers do not meet with misfortune."[17] That there are abnormal circumstances is no reason to doubt the constant way.

Name and actuality. *Ming* (name) refers to terms, reputation, and concepts; *shi* (actuality) refers to facts, to what is real. Their relation has been a traditional problem in the history of Chinese thought. As long ago as the pre-Qin period it had been discussed by each school and subschool. In the period of transition between the Han and Wei, under the influence of Wang Chong's critiques of irrational beliefs, a trend had developed in scholarship "to value proof" and "to attach importance to reflective examination." Consequently, the old problem of the relation between name and actuality once again became an issue of general concern to contemporary thinkers. Such writings as Zhongchang Tong's *Changyan*, Xu Gan's *Discourses*, and Liu Shao's (c. 180–c. 245) *Renwuzhi* discussed it as an important problem. In their discussions of the name-actuality relation, pre-Qin School of Names thinkers still had not developed their articulation and understanding of the relation beyond that of ordinary concepts. They had certainly not begun to connect it to real political issues in society. Discussions of name and actuality in the Han-Wei transition concerned real issues that confronted society at the end of the Han dynasty. These discussions were directed at the various forms of political abuse in selecting men for office, when names no longer matched actualities. This had been brought about by the mindset of the Han dynasty "teaching of names."[18] Consequently, discussions of name and actuality are characterized by a blatant ideological critique. In other words, contemporary discussion of name and actuality was not merely, or even principally, a question concerning logical categories but rather was a question concerning political and ethical categories. It is evident that such thinkers as Zhongchang Tong, Xu Gan, and Liu Shao were not particularly interested in the writings of Hui Shi (fourth century BCE) or Gongsun Long (third century BCE). Instead, they were more receptive to the intellectual tradition of "verifying what is actually the case" (*yan shi*) associated with the teachings of such thinkers as Wang Chong, Wang Fu (c. 90–165), and Xun Yue (148–209). Taking the discussion of the name-actuality relation as their point of departure, they developed trenchant critiques of a political mentality in the late Han which had encouraged undue importance to be placed on checking nominal conformity to certain moral categories while overlooking the underlying reality of an individual's behavior. The prescription of certain moral categories (*ming*) as the basis of moral teachings (*jiao*) is what is known as "the teaching of names."

Xu Gan's views on name and actuality are given prominence in *Discourses*. "Examining Falsity," "Valuing Proof," "A Rebuke of Social Connections," and "An Examination of Disputation" all introduce arguments relating to the name and actuality question. In "Examining Falsity" he points out that an outstanding problem concerning social and political corruption was the widespread practice of seeking reputation (*ming*). The hypocritical and ostentatious practice of those who vied for reputation and cheated one another was such that for a typical one, "If he finds himself in a situation where he can earn himself a reputation without necessarily securing an actual achievement, then he will stay put. If, however, he finds himself in a situation where he can secure an actual achievement without necessarily earning himself a reputation, then he will leave." Even within the family, such practices had become common: "fathers stealing their sons' names, elder brothers stealing their younger brothers' reputations, family members deceiving one another, and friends cheating one another." This had become a common social phenomenon that, once manifest, made it difficult to avoid "truth and falsehood [appearing] as their opposites, and right and wrong [changing] places." Xu Gan's thesis on the name-actuality relation— "when an actuality has been established, its name follows after it"—was made in response to such vulgar practices as holding empty reputations in high esteem and being contemptuous of attending to real tasks. Such behavior might well be described as a national calamity. He writes:

A name is that which is used to name an actuality. When an actuality has been established, its name follows after it; it is not the case that a name is established and then its actuality follows after it. Thus if a long shape is established, then it will be named "long," and if a short shape is established, then it will be named "short." It is not the case that the names "long" and "short" are first established and then the long and short shapes follow after them. That which Confucius valued were those "names" which truly name actualities. In so valuing names, he thereby valued actualities. Names are tied to actualities just as plants are tied to the seasons. In spring, plants blossom into flower; in summer, they are covered in leaves; in autumn, their foliage withers and falls; and in winter, they bear fruit. This is not acting for a specific purpose and yet the plant completes itself. If a plant is forced, its natural tendencies will be harmed. It is the same with names. Hence false name makers are all those who would harm names. People are aware only of the good that is done by names and are ignorant of the bad that is done by false name makers. They are greatly confused!

In this passage Xu Gan describes with utter clarity both the relation between names and actualities and the harm caused by false name making and the abandonment of actualities. He maintains that the names of concepts reflect the real existence of objective things. Accordingly, things come first and their names come after; when an actuality is established, its name follows after it. The sequence of this order cannot be reversed. With this view of the relation between names and actualities, Xu Gan advanced a trenchant critique of the various manifestations in contemporary society of false name making and the abandonment of actualities. He writes: "Because fame seekers cause truth and falsehood to appear as their opposites, and right and wrong to change places, this influences the people. This is a great calamity for the state." Thus he proposes "the examination of falsity" to oppose "fame seeking" and prescribed "valuing actuality" so that names and actualities might correspond.

Because he abhorred false names, Xu Gan also elaborated Wang Chong's ideas on verification. In "Valuing Proof" he clearly enunciates the principle of verification: "In affairs, nothing is more valuable than having proof, while in speech nothing is more futile than lacking evidence. There has never yet been a benefit in speaking when one lacks proof, nor has there ever yet been any loss in remaining silent." "Lacking evidence" refers to deceitful and exaggerated speech: empty words that neither correspond with actualities nor are supported by evidence and proof. In "A Rebuke of Social Connections," he writes: "There are cases where the name is the same but the actuality is different, and there are cases where the name is different but the actuality is the same. Thus, in regard to the principle we have been discussing, the gentleman concerns himself with the actuality rather than criticize the name." Being "concerned with the actuality" expresses a similar emphasis on verification. In "An Examination of Disputation," the argument that the aim of disputation has to be to provide actual proof again expresses Xu Gan's idea of "valuing actuality":

What the vulgar person calls "disputation" is not disputation at all. That he should call it disputation, though it is not, is probably because he has heard the name "disputation" but does not know its actuality. Accordingly, we regard this as preposterous. The person who is commonly referred to as "disputer" is, in fact, a glib person. The glib person readily embellishes the tone of his voice and multiplies his retorts, just like the onset of a gale, or the downpour of a rainstorm. Disregarding the inherent rights and wrongs of a situation, and not understanding the principles of truth and distortion, he fixes it so that he is never at a loss for words and works at securing certain victory. Hence those of shallow understanding who marvel at the unusual, seeing him like this insist that it is disputation. They do not realize that although those who have attained the way by virtue of their simplicity and reticence may be verbally defeated, yet in their hearts they do not give in.

Disputation is about persuading people in their hearts; it is not about verbal submission. Hence disputation is to articulate distinctions, and also to separate and distinguish different categories of affairs skillfully, so as to arrange them clearly. Disputation does not mean being quick-witted in one's words and speech to talk over people's heads.

Here Xu Gan uses the function of disputation as an analytical tool. This function is to "give due regard to the inherent rights and wrongs of a situation, and to understand the principles of truth and distortion" rather than simply to indulge in disputation for its own sake with no regard to being able to come to an informed conclusion.[19]

Human talent. Another important topic discussed in *Discourses* concerns standards for assessing human talent, and how people should be selected for and used in office. In the Eastern Han period, the "teaching of names" was implemented as a tool of government, and from the central administration right down to local administration, the system of selecting men for office consisted of recommendation and summons. Such categories as "filially pious and deferential to elder males" and "filially pious and incorrupt" formed the basis for assessing human talent and for selecting officials. Virtuous conduct was accorded great value, so it became customary to encourage people to build reputations based on their moral integrity and to hold moral behavior in high esteem. Abuse of these practices, however, often resulted in excessive affectation, even to the point of becoming a sham. In the period of transition between the Han and the Wei, a new trend of thought appeared which differed from traditional views on the worth of human talent. According to the new trend, talent and wisdom were to be valued over virtuous behavior. This thinking is expressed in such chapters as "Ordering Learning," "Wisdom and Deeds," and "Examining the Selection of High Officials." According to Xu Gan, in matters of assessing human talent and employing officials, a man's talent or ability should be the foremost consideration. In "Wisdom and Deeds" he clearly answers the question of which is the more important: virtuous behavior or wisdom. He objects to the traditional view in which virtuous behavior is uniformly seen to be more important than wisdom. For Xu Gan, the sage is a sage "not due to mere empty deeds; rather, it is due to his wisdom." The sage is one who is able "to realize fully his intelligence." Sages, in turn, most value men of uncommon wisdom because such men are able to render achievements and services of benefit to the world. Although such men might be faulted for failings in their virtuous conduct, it is of little consequence. Indeed, although the virtuous conduct of those men who withdraw from society to live in solitude is admirable, yet it is of no benefit to society at all. Of course, if a man's behavior is morally reprehensible and he has only a little talent and wisdom, and if he is responsible for excessive disturbances and renders insufficient achievements, then such a man is also undesirable. In "Wisdom and Action" the point that Xu Gan particularly emphasizes is that wisdom is able to make the people wealthy and alleviate their suffering, fully affirming the positive significance that wisdom has in advancing social development. In valuing talent over virtue, Xu criticized the outmoded views of the hereditary elite families in regard to human talent and the standards for selecting men to office, as he prepared the way for scholar-literati of commoner family background to ascend to the political stage. This played an important role in liberating people's thinking. The significance of his affirmation of the role of talent and the role of talented and wise scholar-literati was to arouse people's attention. It constituted a severe assault on traditional Confucianism and helped free people from the bonds of the hypocritical and ossified teachings contained in the Confucian classics and in Confucian dogma. Moreover, it subsequently provided an intellectual and theoretical basis for promoting Cao Cao's policy of "selecting only men of talent for office" and selecting men who were "neither humane nor filial but who had the skills both to put the country in order and to run military operations."[20] As such, it played a positive role in the development of history.

Learning. Xu Gan placed great emphasis on the importance of learning. In "Ordering Learning" he specifically discusses the significance and methods of learning. For Xu, learning is the sole path by which the gentleman might cultivate his body and his virtue, as well as establish meritorious achievements and a commendable legacy, so that "when he died, his name would not be forgotten." Learning has this important value because it enables one "to channel one's spirit, to be penetrating in one's reflections, to be at ease with one's emotional responses, and to order one's innate tendencies." Thus the sage considers learning to be the "highest undertaking." Xu Gan identifies with the Confucian tradition in regarding the so-called "three teachings" to be the main content of education: the six virtues, the six types of moral behavior, and the six arts.[21] In his own words, "With these three teachings in place, the human way is complete."

Xu Gan was opposed to the two main styles of commentary practiced by Han classical scholars: *zhangju* (section and sentence), and *xungu* (glossing of old terms). In "Ordering Learning" he ridicules the so-called "broad learning" of those "debased literati" because it is devoted merely "to nomenclature, meticulous in its accounts of utensils and weapons, and painstaking in glossing old terms." In his opinion, this type of learning is unable to draw "together those matters which are fully revealed in the fundamental meaning." It is, moreover, unable to "to capture the mind of the former kings" with the consequence that "scholars toil in thought yet fail to understand the way, wasting days and months with no achievement." Because of this, Xu Gan emphasized that the primary purpose of learning should be to recognize and to seek "the way." He further points out that "in all matters of learning, it is the

fundamental meaning which is of foremost importance; nomenclature is secondary. When the fundamental meaning
has been revealed, the nomenclature will follow thereafter."
Moreover, "in matters of learning, the gentleman is unremitting in his efforts, just like the movements of heaven above
and the motions of the sun and moon. He perseveres throughout his life, stopping only when he dies." For Xu Gan, a successful outcome in learning depends upon whether one has a
steadfast commitment: "Commitment is the commander of
learning, while ability is the foot soldier of learning. In learning, one does not worry about a lack of ability but rather that
one's commitment is not steadfast." Having a steadfast commitment means aspiring to be diligent and hard working, resolute, and unremitting in one's learning. In this way, the goal of
ensuring both an enlightened mind and cultivated body,
of possessing wisdom and virtuous behavior, should be
realizable.

Introduction by John Makeham

Balanced Discourses (*Zhonglun* 中論; hereafter *Discourses*) is the representative writing of the late Han philosopher-literatus, Xu Gan 徐幹 (170–217 CE).[1] Xu Gan lived at a nodal point in the history of Chinese thought, when Han (206 BCE–220 CE) scholasticism had become ossified and the creative and independent thinking that characterized Wei-Jin (220–420) thought was just emerging. *Discourses* is a collection of essays which embraces topics ranging from Confucian cultivation to calendrical calculation. Taken as a whole, the collection constitutes a wide-ranging polemical inquiry into the causes of political and social breakdown, while also proposing various remedies. Xu Gan's argumentation frequently appeals to the authority of classical Confucian ethical values; indeed, the work is classified under *Rujia* 儒家 in all bibliographical lists of the standard histories, except for that of *Songshi* 宋史, where it is listed among miscellaneous writers.

Although Xu Gan is not a major figure in Chinese intellectual history, traditionally he has been accorded a place as a representative philosopher of the Eastern Han (25–220) period. *Discourses* offers modern historians of Chinese thought a unique contemporary account of a range of social, intellectual, and cosmological factors that Xu Gan identified as having precipitated the demise of the Han order. His perspectives on these issues are also of philosophical interest as they reveal his belief in a special correlative bond that should obtain between names (*ming* 名) and actualities (*shi* 實), and his understanding of the consequences of that bond's being broken.[2]

Despite Xu Gan's having been accorded a lasting reputation in Chinese literary culture as one of the "Seven Masters of the Jianan 建安 reign period (196–220)," we know little about his life.[3] What is known derives largely from the unsigned "Preface to Xu Gan's *Balanced Discourses*."[4] That account is worth translating in full:

I consider Master Xun Qing 荀卿 [Xunzi; c. 335–c. 238] and Meng Ke 孟軻 [Mencius; fourth century BCE] to have possessed sagely talents second only to those of Confucius [551–479]. Celebrated as models of original learning and for continuing to elucidate the work of the sage Confucius, each of them recorded his own surname and personal name in his writings. While their surnames and personal names are still passed on today, their styles [*zi*], however, are not.[5] If one reflects on the reasons for this, they all stem from the situation in the Warring States period [475–221], in which too few people delighted in worthy men, leading to their contemporaries' failing to record their styles sufficiently early. Would it not be even more likely that a similar fate might await Master Xu's book, *Balanced Discourses*, should his surname and personal name not be included in the title? I fear that, after a long time, perhaps not even his personal name would be transmitted, and for this reason his talents would not be accorded their due weight. This would be a matter to make one sigh with regret. First of all, I will evaluate his virtue so as to lend prominence to his surname and personal name, and then describe his noble and enduring actions. I write these comments in a preface at the head of the chapters [*pian*] that follow. The words of the preface read:

In this age, there lived a gentleman of noble character and penetrating understanding: his surname was Xu, his personal name was Gan, his style was Weichang 偉長, and he was a native of Ju 劇 Prefecture in Beihai Kingdom.[6] His forebears had already established the family's reputation as uncorrupted and honest, and discriminating in matters of right and wrong. Each subsequent generation enhanced the family's excellence in this regard and did not detract from its store of virtue. By Xu Gan's time, this had been so for ten generations. Xu Gan embodied a pure and clear vital energy, as bestowed by the primordial vastness, and possessed the spontaneous tendency of the paragon and wise man, as bestowed by creative transformation. When, as a child, he opened his mouth to speak, he would delight in reciting the text of the nine virtues.[7] When he heard something he would remember it; once taught something he did not have to be told again. Before he was fifteen, he could probably already recite several tens of thousands of words of text.[8] When he was fourteen, he began to learn to read and recite the Five Classics.[9] He became so immersed in his studies that he would forget to eat his meals. He would pull down the curtains and devote himself entirely to reflection, working on into the night. Worrying that he might become ill, his father would often forbid him to work so hard. Due to such diligence, before he had reached the age of twenty, he had learned the Five Classics so well that he was able to recite them entirely. He read a wide range of commentaries and scholarly notes. He formed compositions of his own words and penned his own writings. This was during the last years of Emperor Ling's 靈 [r. 168–89] reign, when the canons of the state had broken down and fallen into disuse.[10] The younger male members of families who held official positions formed cliques that were aligned to powerful families. They formed social connections with these families and supported them so that they might make a name for themselves, and they vied to better one another in the attainment of ranks and titles.[11] Xu Gan was deeply upset by the confused dimness of his vulgar contemporaries and so, cutting off his contacts with others to guard his own integrity, he had nothing to do with them, deriving his pleasures solely from the Six Classics.[12]

The penetrating understanding of a gentleman is such that in his learning he has no constant teacher; rather he learns from whoever is superior to him in a particular area of undertaking.[13] Only after making sure that he has thoroughly learned every aspect of what a particular teacher knows does he leave that teacher. If he hears a saying of excellence, he does not let it pass without note but makes sure to commit it to memory. His aim is to assemble the best of all sayings and draw together the subtleties of the way and its powers. He is humiliated if there is one thing that he does not know, and ashamed if there is one art that he is unable to master. For this reason, day and night he is tireless in his diligence. When the sun is setting he does not take time off to eat, and at night he does not undress. During the day he meticulously studies the classics and their apocrypha, and at night, one by one, he observes the constellations. He investigates the primordial chaos before it takes form and repairs those areas where sagely virtue is lacking or deficient. He makes long-term considerations with regard to the boundless future, and signals that the subtle words of the ancient sages are about to fall to the ground.[14] What time does he have to make a clamor about philological matters, or to work at securing a hollow reputation by joining the ranks of vulgar men? Men of superficial and insufficient learning understand being spurred on only by honor and profit. Would they be likely to understand the root of the great way? As for the others, in what they neglect they take simplicity too far, never even experiencing sadness or happiness.[15] Those classical scholars who practice only book learning are not worth looking up to, yet there are many who are only too happy to look up to them and only a few who argue against doing so. This is why Xu Gan's reputation did not stand out in his local township and village early in his career. Nevertheless, by holding on to what was correct, through his own efforts he came to stand on his own two feet. His aim being focused, then matters of either breaking or making a reputation—matters that were of concern to his vulgar contemporaries—were like the passing clouds to him.[16] If there were individuals who awoke from their folly and sought to return to the proper way, he would advance them along the way, and forget about how they had slandered him in the past. Even though he had been wronged, he did not mind.[17] All of the above examples belong to the category of learning from human affairs to attain what is above.[18]

At the time when Dong Zhuo 董卓 [d. 192] rebelled, and removed the emperor [Emperor Xian 獻, r. 190–220] to Chang'an 長安 [190], wicked men and heroes filled the countryside, and the empire was without a ruler; the way of the sages waned, and perverse and duplicitous affairs thrived; and fellows bent on honor and profit secured reputations while worthy men who safeguarded their constancy did not stand out.[19] For these reasons Xu Gan's reputation did not reverberate throughout the land, nor were jade, silk, or easy carriages presented at his door as inducements to take up office. Yet when one examines his virtuous actions and his cultivation in literature and the arts, he was, in fact, well qualified to serve as an assistant to emperors and kings.[20] Is it not lamentable that the way was not operative in these times?

During this period of upheaval, Xu Gan withdrew to an area bordering the sea.[21] Upon his return to this area from the old capital, Luoyang, the regional governor and commandery chiefs approached him in a slow, respectful manner to present him with summonses to office, even though in their eagerness to secure his service they really wished to stride directly up to him. Xu Gan, however, maintained that in an age of plotting and intrigue, even the first sage, Confucius, had encountered straitened circumstances. Was it not still more likely that a similar fate might befall himself? He was critical of Mencius for failing to keep a proper sense of his limited capacities, emulating as he did the sage's efforts to put the way into practice by traveling about the country as the guest of one feudal lord after another. By contrast, he had a profound admiration for the way that Yan Yuan and Xun Qing had acted.²²

For these reasons, he disappeared into the mountains and valleys, living in a secluded dwelling, investigating the workings of incipience. The profound and subtle ways in which he applied his reflections led to the onset of painful bouts of malaria. Yet hidden away, he managed to prolong his years. One day, he met Cao Cao 曹操 [155–220] who was quelling a rebellion, and for the first time the way was opened up for him to enter into the service of a king.²³ Thereupon, despite his illness, he forced himself to respond to the call of Cao Cao's summons, serving in the garrison and in campaign expeditions.²⁴ After five or six years, however, his illness gradually worsened and he could no longer endure the rigors of service to the king. He thus hid himself away in a back street, nurturing his true aim and keeping his genuineness intact. Leading a plain, simple existence, and initiating no unnecessary action, he devoted himself solely to investigating the correct way. Despite living in a dwelling that had no more than four walls and a roof to shield his wife and children from the elements, and despite having to make one day's food last for two, he was not depressed. He nourished his floodlike vital energy and practiced the arts of longevity.²⁵ Hearing that he was like this, some of his contemporaries would visit him. Among them there were those who, having formed quite an appreciation of his genuineness, sought to follow him. Xu Gan agreed to see them all. He motivated them with the sound of his voice and his facial expressions, appraised the commitments they felt most keenly about, and led them in discussion and discourse.²⁶ Knowing which of them could grow in accord with the way, with subtlety he drew them on. Those who benefited from his instruction remained unaware as the great transformation worked its effects undisclosed. There were many who were saved by him.

Xu Gan did not believe in short-term friendships; in each of his friendships, he sought a long-term relationship. Moreover, he chose his friends with considerable acumen. He seldom fully expressed the depth of his friendship with someone, nor was he fond of going out of his way to display amicability and concern. He drew together the sage's undertakings of balance and harmony, and trod the path by which the worthy man and the wise man abide by the limits of what is appropriate. Profoundly silent and difficult to fathom, he was truly a great and precious vessel.

His natural inclination was such that he constantly wanted to reduce that of which the age had a surplus and increase that in which the ordinary people of the day were deficient. He saw men of letters follow one another in the contemporary fad of writing *belles lettres*, but there was never one among them who elucidated the fundamental import of the classics to disseminate the teachings of the way; or who sought the sages' point of balance to dispel the confusion of popular contemporary mores. For this reason, he abandoned the literary genres of the ode (*shi*), rhyme-prose (*fu*), eulogy (*song*), inscription (*ming*), and encomium (*zan*), and wrote the book *Balanced Discourses* in twenty chapters. Only a fraction of what Xu Gan had originally intended was selected for inclusion. Before his plans for the whole book were finished, he succumbed to an epidemic and passed away in the second month of spring in the twenty-third year of the Jianan reign period [218], at the age of forty-eight. It was bitterly painful to lose him like this. I had often sat in attendance upon him. When I contemplated his words, I was constantly fearful because, even though my intentions were in earnest and I fully exerted myself, I lacked confidence in myself. As for my talent and aims, they were vastly inferior to those of Xu Gan. Nevertheless, I still revered and esteemed him as my teacher. Since his passing, there have been those who have behaved as Zigong did on the ridge of the mountain.[27] For this reason I have reviewed the affairs of his life, crudely citing only such matters as are manifest and can be readily known. As for those matters that are obscure, subtle, profound, and vast, I leave them to gentlemen of meticulous and thorough understanding to assist in elucidating them.

According to this preface, Xu Gan's official career was frustrated early and more or less permanently. Yet other sources relate that he had served in a number of posts. Chen Shou 陳壽 (233–97) records that he held positions on the staff of the minister of works and as instructor to the leader of court gentlemen for miscellaneous purposes.[28] He must have served in the former position some time between 197 and 208, as Cao Cao was appointed minister of works in 197 and abolished the position in 208, and his instructorship must have been some time after 211, when Cao Pi was appointed as leader of court gentlemen for miscellaneous purposes. He also served as instructor of Linzi 臨菑 District some time during or after 214, when Cao Zhi 曹植 (192–232) was made marquis of Linzi.[29] Xu Gan was also offered two other positions. In his commentary to *Sanguozhi*, Pei Songzhi 裴松之 (372–451) cites *Xianxian xingzhuang* 先賢行狀, which records that Cao Cao offered him an unspecified special appointment, and that later he was also offered the post of magistrate of Shang'ai 上艾 District.[30] In both cases, he declined on the grounds of ill health.

The essays which constitute *Discourses* make no specific reference to events in his official career, although several read as if they could be memorials to a ruler (in particular, "Attend to the Fundamentals," "Be Careful of the Advice One Follows," "Destruction of the State," "Reinstitute the Three-Year Mourning Period," and "Regulate the Allotment of Corvée Laborers"). Similarly, Xu Gan's argument that a man's ability is fundamentally more important than his deeds (see "Wisdom and Deeds") resonates clearly with the substance of a series of edicts issued by Cao Cao between 203 and 217.[31]

Equally, however, there is reason to suspect that Xu's official career was punctuated by periods of dismissal or disengagement, such as the account in the preface which describes the period of his last years as one of self-imposed isolation and impoverishment. Hints that this isolation and impoverishment were self-imposed are forthcoming from Xu Gan himself and in a poem by Cao Zhi. At the end of the essay "Examining Falsity" we find the following account, which may well describe his state of mind when he decided to withdraw from official life in his later years:

> Fame seeking is certainly not something the gentleman is capable of doing. Rather, the gentleman is capable of perfecting his heart and mind. When his heart and mind are perfected, then internally he becomes settled. When internally he is settled, then things cannot upset him. When things cannot upset him, then he takes singular pleasure in his way. When he takes singular pleasure in his way, then [he is indifferent to obscurity and fame]. Hence the *Book of Rites* says:
>
>> The way of the gentleman is dim, yet grows more conspicuous by the day. The way of the small man is dazzling, yet fades by the day. The way of the gentleman is plain yet not tiresome, simple yet cultivated, understated yet patterned. He knows that what has a far-reaching effect starts with what is nearby; that tendencies derive from an origin; and that the subtle will become manifest. One may join such a person in entering into virtue.[32] . . . Is it only in regard to those aspects into which people do not see that the gentleman excels?[33]
>
> Would one such as this be likely to engage in underhand practices in a chaotic world so that he might change from being an unheard of ordinary person?"

If the final office in which Xu Gan served before retirement was as instructor of Linzi under Cao Zhi, then the following passage from a poem by Cao Zhi, dedicated to Xu, may suggest that his final retirement was motivated by his frustration at not being awarded higher office (by Cao Cao or Cao Pi?):

> I think of this fellow in his thatched cottage,
> So pitiful in his poverty and lowly condition.
> The ferns and leaves he eats do not fill his emptiness,
> Nor do the coarse garments he wears cover him.
> Behind his indignation is a melancholy heart,
> Which, given written expression, naturally comes to
>   form essays.
> Who should be blamed for casting the treasure aside?
> Master He himself is not without fault.[34]
> Awaiting a true friend so that you may brush your
>   cap,[35]
> Yet who among them does not harbor the same
>   hope?[36]

If this poem was written in the final period of Xu Gan's life, then the reference to his writings is presumably to the essays which make up *Discourses*. As noted above, it is unlikely that all of these essays were written during this period, but some of the more polemical (such as "Ordering Learning," "Examining Falsity," and "A Rebuke of Social Connections") may well have been.

The preface states that the twenty chapters included for selection in *Discourses* represent "only a fraction of what Xu Gan had originally intended." The comment of Li Shan 李善 (c. 630–89) to Cao Pi's letter to Wu Zhi also states that *Discourses* comprised only twenty chapters, but there is reason to believe that these accounts may have been mistaken.[37] For example, in his letter to Wu Zhi, Cao Pi refers to "more than twenty chapters."[38] In addition, Wei Zheng's 魏徵 (580–643) *Qunshu zhiyao* 群書治要 includes substantial parts of two chapters that have been transmitted nowhere else. There are also statements recorded in Wu Jing's 吳競 (670–749) *Zhenguan zhengyao* 貞觀正要 (see entry for 643; 6.20a) and Chao Gong-wu's 晁公武 (d. 1175) *Jun zhai dushuzhi* 郡齋讀書志 (10.17a) to the effect that the book included two chapters, entitled, re-spectively, "Fu san nian 復三年" [*sic*] and "Zhi yi 制役." The material which is included in *Qunshu zhiyao* is of a nature that matches these two titles and may be regarded as being abridged versions of the two chapters. It is not possible to determine whether the book had originally contained more than twenty-two chapters.

Regardless of how many chapters were originally included in the book, it would seem that the task of editing the essays into a book was undertaken only after Xu Gan's death. One might therefore suspect that the title *Zhonglun* may also have been chosen by the author of the unsigned preface. The inspi-ration for the title may have been the following passage in "An Examination of Disputation":

> The gentleman engages in disputation because he wants to use it to illuminate the point of balance of the great way. This being so, would he be likely to look for vic-tory in a single round of argument? The minds of people, in regard to right and wrong, are like that of taste to flavor. It is not the case that one regards only the dishes prepared by oneself as excellent and the dishes prepared by others as not so. Thus the gentleman's attitude to the way is such that when the balance resides in another, it is as though it resided in himself. So long as the balance is attained, then one's heart will be delighted by it; so why selectively distinguish it as being "his"? If the bal-ance is missed then one's heart will be saddened by it; so why insist on choosing only what is "mine"? Hence, in discoursing with others, when one encounters a point on which the other person is correct, then one should stop arguing that point.[39]

As the author of the preface explains, Xu Gan "drew together the sage's undertakings of balance (*zhong*) and harmony, and trod the path by which the worthy man and the wise man abide by the limits of what is appropriate." In titling the collection *Balanced Discourses* the preface writer may have been suggesting that the particular mode of expression common to the individual essays was itself ordered in accordance with the mode of ordering which the collection as a whole expounded. In other words, he understood these essays to display that same quality of balance which motivated Xu Gan to write them in the first place: as catalysts to restore the point of balance, or centered equilibrium, of the way.

Balanced Discourses

中論

《治學第一》

1. Ordering Learning

Like a number of Confucian writings edited in the Han dynasty, *Discourses* opens with an essay on the importance of learning. Other books in this tradition include the *Analects*, Xun Qing's (c. 335–c. 238) *Xunzi*, Yang Xiong's (53 BCE–18 CE) *Fayan*, and Wang Fu's (c. 90–165) *Qianfulun*.[1]

Xu Gan's model curriculum for the aspiring gentleman (*junzi*) is based on a range of standard Confucian virtues and the six arts (*liu yi*), the cultivation of which is essential to the development of his virtue. In *Mencius*, although it is implicit that education plays a role in the development and refinement of a person's incipient moral tendencies, the more explicit claim is that innate moral tendencies sprout of their own accord without having to be learned (*Mencius*, 2A.6). Xu Gan, on the other hand, maintains that only through learning are people able to recognize that they have these "treasures" in their heart. The place that he gives to learning draws on the views presented in Xun Qing's essay "Exhortation to Learn." Xu Gan also stresses the importance of the teacher, commitment (*zhi*), the sagely models, and a willingness to be eclectic in one's approach to learning. Of particular interest in this essay is his critique of the two main styles of commentary practiced by Han classical scholars: *zhangju* (section and sentence) and *xungu* (glossing of old terms). Xu Gan rejects the pedantic philological concerns of the Han scholiasts who had concentrated on the written word at the expense of the message behind it, seeking instead to elucidate the "fundamental mean-

昔之君子成德立行，身沒而名不朽，其故何哉？學也。學也者、所以疏神達思、怡情理性。聖人之上務也。

民之初載，其矇未知。譬如寶在於玄室，有所求而不見。白日照焉，則群物斯辯ª矣。學者、心之白日也。故先王立教官，掌教國子。教以六德。曰：智、仁、聖、義、中、和。教以六行。曰：孝、友、睦、婣、任、恤。教以六藝。曰：禮、樂、射、御、書、數。三教備而人道畢矣。

學猶飾也。器不飾則無以為美觀。人不學則無以有懿德。有懿德，故可以經人倫；為美觀，故可以供神明。故《書》曰：「若作梓材，既勤樸斲，惟其塗丹臒。」

ª Following the *Guang Han Wei congshu, Bozi,* and *Jianan qizi ji* versions of *Discourses,* as well as *Xu Hou Hanshu,* 69B.5a, in reading *bian* 辨 instead of *bian* 辯.

ing" (*dayi*). For him, it was the original and essential meaning of the classics that was of foremost importance.

The gentleman of past times would perfect his virtue and establish good deeds through his conduct; when he died, his name would not be forgotten.[2] What is the reason he was able to do this? Learning! Learning is the means to channel one's spirit, to be penetrating in one's reflections, to be at ease with one's emotional responses, and to order one's innate tendencies. Learning is the highest undertaking of the sage.

At the beginning of their lives, people are ignorant, not yet possessing understanding. Take the analogy of some valuables inside a dark room; although they are sought after, they cannot be seen. If daylight illuminates the room, however, then all the things therein can thereby be discerned. Learning is the daylight of the heart. For this reason the former kings established teaching officials, making them responsible for educating the sons of the ruler's kinsmen and high officials.[3] They instructed them in the six virtues: wisdom, humaneness, sageliness, rightness, balance, and harmony; in the six types of virtuous conduct: filial respect, friendship, maintaining harmony with one's nine degrees of relatives, being close to one's in-laws, trusting one's friends, and commiserating with the sufferings of others; and in the six arts: ritual, music, archery, charioteering, writing, and arithmetic.[4] With these three teachings in place, the human way is complete.

Learning is like an adornment. If a vessel is not adorned, then there will be no way for it to be beautiful to contemplate. Similarly, if one does not learn, then there will be no way for one to be endowed with exemplary virtue. Being endowed with exemplary virtue thus enables the bonds of human relationships to be put in good order; being beautiful to contemplate, the vessel can thus be used to make offerings to the spirits of heaven and earth.[5] Accordingly the *Book of Documents* says: "It is just like fashioning something from catalpa wood; when the bark has been chopped off and the raw wood exposed, then some red lacquer should be applied."[6] It is only

夫聽黃鐘之聲，然後知擊缶之細。視袞龍之文，然後知被褐之陋。涉庠序之敎，然後知不學之困。故學者如登山焉。動而益高。如寤寐焉。久而愈足。顧所由來，則杳然其遠。以其難而懈之，誤且非矣！《詩》云：「高山仰止，景行行止。」好學之謂也。

(倦)〔倚〕[b]立而思遠，不如速行之必至也。矯首而徇飛，不如(循雌)〔修翼〕[c]之必獲也。孤居而願智，不如務學之必達也。故君子心不苟願，必以求學。身不苟動，必以從師。言不苟出，必以博聞。是以情性合人，而德音相繼也。孔子曰：「弗學，何以行？弗思，何以得？小子勉之。」斯可謂師人矣。

馬雖有逸足，而不閑輿，則不為良駿。人雖有美質，而不習道，則不為君子。故學者求習道也。若有似乎畫

---

[b] Following the original reading of *juan* 倦. Alternatively, following *Yilin*, 5.14b, in emending *juan* 倦 to *yi* 倚: "… rather than just stand there resting against something."

[c] Following the original reading of *xun ci*. Alternatively, following *Yilin*, 5.14b, and *Taiping yulan*, 607.7b, in emending *xun ci* 循雌 to *xiu yi* 修翼: "… if one practices how to fly."

after hearing the sound of the yellow bell pitch pipe that one realizes how thin the sound of an earthenware pot is; and it is only after seeing the dragon design on the ritual vestments of the emperor and senior officials that one realizes how coarse a hemp cloak is.[7] Similarly, it is only after making one's way through proper schooling that one realizes the hindrance caused by not learning. Learning is thus like climbing a mountain: as one moves the higher one ascends. It is also like sleeping: as time passes the more one is satisfied. If one looks back to where one came from, it seems hazy in its remoteness. It is mistaken and wrong to slacken off should learning become difficult. The *Book of Odes* says: "The high mountains, I look up at them / The high road, I travel it."[8] This is what is meant by a love of learning.

One will be more certain of arriving if one travels swiftly rather than just standing there in a slump, thinking how far one still has to go. One will be more certain of catching a bird if one tracks down the female [on her nest], rather than craning one's neck to gaze up at birds flying in the air. Similarly, one will be more certain of reaching one's goal if one applies oneself to learning rather than residing in solitude, hoping for wisdom.[9] Therefore, so as not to be distracted in his heart by idle wishes, it is necessary that the gentleman pursues learning; so as not to be careless in his personal behavior, it is necessary that he follows a teacher; and so as not to utter baseless comments, it is necessary that he listens widely to others. In this way, his emotional responses and his nature will accord favorably with others, and so his reputation will be passed on. Confucius said: "If one does not learn, how can one act? If one does not reflect, how can one achieve? Mark these words, my disciples!" This is what is meant by learning from others![10]

Even if a horse is fleet of foot, if it is not trained to pull a carriage then it is not a good horse.[11] Likewise, even if a man has a good natural endowment yet does not practice the way, then he will not become a gentleman.[12] Learning is thus seeking to practice the way. In some respects it is like painting

采。玄黃之色既著，而純皓之體斯亡。皾而不渝。孰知其素歟？子夏曰：「日習則學不忘。自勉則身不墮。亟聞天下之大言則志益廣。」故君子之於學也，其不懈，猶上天之動，猶日月之行，終身亹亹，沒而後已。故雖有其才，而無其志，亦不能興其功也。志者、學之師[d]也。才者、學之徒也。學者、不患才之不贍，而患志之不立。是以為之者億兆，而成之者無幾。故君子必立其志。《易》曰：「君子以自強不息。」

大樂之成，非取乎一音。嘉膳之和，非取乎一味。聖人之德，非取乎一道。故曰：學者、所以總群道也。群道統乎己心，群言一乎己口。唯所用之故。出則元亨。處則利貞。默

---

[d] Following Yu, *Zhuzi pingyi bulu,* 74, in emending *shi* 師 to *shuai* 帥.

with colors. When the black and yellow have been applied, the pure whiteness of the original surface disappears. Although only covered and not changed, yet who can tell what the original color is? Zixia said: "If one practices daily, then one will not forget what one has learned; if one forces oneself to be diligent, then one will not fall by the way. If one often listens to the great teachings of the world, then, increasingly, one's vision shall be broadened."[13] Hence, in matters of learning, the gentleman is unremitting in his efforts, just like the movements of heaven above and the motions of the sun and moon. He perseveres throughout his life, stopping only when he dies. Thus even if he had natural ability, he would still be unable to succeed in applying it to good effect if he lacked the requisite commitment. Commitment is the commander of learning, while ability is the foot soldier of learning.[14] In learning, one does not worry about a lack of ability but rather that one's commitment is not steadfast. It is because of this that although those who are engaged in learning are legion, there are only a few who actually learn. The gentleman must, therefore, be steadfast in his commitment. The *Book of Changes* says: "The gentleman untiringly strengthens himself."[15]

The creation of a great work of music is not the product of just one note, and the harmonious blend of a fine dish is not the product of just one flavor. Similarly, a sage's virtue does not just draw on one way. Thus it is said that "learning is the means whereby the multitude of ways are brought together." When the multitude of ways are combined in one's heart, then the multitude of teachings will be united in one's speech. It is then up to the individual how he uses this learning. If he goes forth, there will be greatness and endurance; if he stays put, there will be benefit and stability.*[16] If he is si-

---

* The four mantic qualities of greatness, endurance, benefit, and stability are associated with the Qian hexagram, a symbol of strength and tenacity. Historically, there has been considerable variation in how individual commentators understood each of these terms.

則立象。語則成文。述千載之上，若共一時。論殊俗之類，若與同室。度幽明之故，若見其情。原治亂之漸，若指已效。故《詩》曰：「學有緝熙于光明。」其此之謂也。

夫獨思則滯而不通；獨為則困而不就。人心必有明焉，必有悟焉。如火得風而炎熾。如水赴下而流速。故太昊觀天地而畫八卦。燧人察時令而鑽火。帝軒聞鳳鳴而調律。倉頡視鳥跡而作書。斯大聖之學乎神明而發乎物類也。

賢者不能學於遠，乃學於近。故以聖人為師。昔顏淵之學

lent, he will establish an image; if he speaks, his words will form a pattern.[17] He will be able to describe events that occurred more than one thousand years ago as though he were living at the same time, and discourse on the differences between particular types of customs as though he were in the same room as the practitioners of those customs. He will be able to fathom the reasons underlying both the obscure and the evident, as though seeing into the true conditions of things.[18] He will be able to trace the gradual development of order and chaos as though their orientation had already come to be realized. Therefore, the *Book of Odes* says: "Through learning, one will advance energetically to enlightened understanding."[19] This is what is meant.

If one reflects in solitude, one will remain blocked and hence fail to comprehend. If one acts in isolation, one will be impeded and so fail to achieve. The human heart must have clarity and illumination in it. Then, like a fire fanned by a breeze, it blazes brightly, or like water moving downward, it flows swiftly. This is why Tai Hao was able to draw the eight trigrams only after observing heaven and earth; why Sui Ren was able to make fire with drilling sticks only after studying the seasons; why Emperor Xuan was able to harmonize pitch pipes only after hearing the song of the phoenix; and why Cang Jie was able to create writing only after observing the prints of birds.*[20] These are all examples of great sages' learning from the spirits of heaven and earth and expressing it in phenomenal categories.

Being unable to learn from the remote, the worthy learns from what is close at hand. For this reason, he takes the sages as his teachers. In the past, when Yan Yuan learned from the

---

*Tai Hao was another name of the legendary cultural hero Fu Xi 伏犧. Sui Ren was a contemporary of the legendary Yellow Emperor and is credited not only with the discovery of how to make fire but also of cooked food. Emperor Xuan, or Xuan Yuan 軒轅, was the legendary Yellow Emperor. Cang Jie was said to have been the scribe of the Yellow Emperor.

聖人也，聞一以知十。子貢聞一以知二。斯皆觸類而長之，篤思而聞之者也。非唯賢者學於聖人，聖人亦相因而學也。孔子因於文、武；文、武因於成湯；成湯因於夏后；夏后因於堯、舜。故六籍者、群聖相因之書也。其人雖亡，其道猶存。今之學者勤心以取之，亦足以到昭明而成博達矣。

凡學者、大義為先，物名為後。大義舉而物名從之。然鄙儒之博學也，務於物名，詳於器械，矜於詁訓，摘其章句，而不能統其大義之所極以獲先王之心。

sage, Confucius, although he had heard only one point, he could infer the rest about a subject. Even Zigong could infer a second point after hearing one point.*[21] These are both examples of worthies who extended meanings by extrapolating on the basis of analogy, and who learned through assiduous reflection. Yet it is not only worthies who learn from sages; sages, too, learn successively from one another. Confucius followed on from Wen [r. 1099/56–1050] and Wu [r. 1049/45–1043], Wen and Wu from Cheng Tang, Cheng Tang from the lord of Xia, and the lord of Xia from Yao and Shun.†[22] Thus the Six Classics have been passed on successively from one sage to another.[23] Although these men have gone, their way still exists.[24] If students of today adopt this way with diligence, this will be enough for them to achieve illumination and so become widely accomplished.‡

In all matters of learning, it is the fundamental meaning which is of foremost importance; nomenclature is secondary. When the fundamental meaning has been revealed, the nomenclature will follow thereafter. The "broad learning" of the debased literati, however, is devoted to nomenclature, meticulous in its accounts of utensils and weapons, and painstaking in glossing old terms.‡[25] They select "sections and sentences" for commentary yet are unable to capture the mind of the former kings by drawing together those matters which are fully revealed in the fundamental

---

*Yan Yuan and Zigong were disciples of Confucius's; see *Shiji*, 67.2195–2201, for biography.
†King Wen was the first of Zhou kings. His son, King Wu, succeeded him to the throne. Cheng Tang, or Tang 湯, was the first ruler of the Shang dynasty. The lord of Xia is another name for the traditional first ruler of the Xia dynasty, Great Yu 禹, who was most celebrated for his work in controlling floods. Yao is the most famous of the legendary emperors; Shun succeeded him upon his death.
‡*Guxun* or *xungu* means "to gloss old and ancient terms using contemporary terminology" and is characterized by its focus on semantic glossing rather than extended exegesis.

此無異乎女史誦詩、內豎傳令也。故使學者勞思慮而不知道，費日月而無成功。故君子必擇師焉。

meaning.*[26] This is no different from a female scribe's intoning odes or a junior eunuch's passing on messages.[27] Such a state of affairs thus has scholars toil in thought yet fail to understand the way, wasting days and months with no achievement. Thus, in selecting from whom he should learn, the gentleman needs to choose well.

---

*It is generally accepted that with *zhangju* commentaries, once section and sentence divisions in a piece of writing had been determined, the meaning of individual sentences would be explained by appending a commentary immediately after the sentence (or sentences) and a summary of the import of each section would similarly be appended after each section.

《法象第二》

2. *Establishing Models and Exemplars*

This essay continues the theme of the gentleman's cultivation. Xu Gan attaches greatest importance to the first of the six arts, ritual. For the gentleman, the purpose of ritual is to enable him to cultivate his virtue and so act as a model for others to emulate. He argues that ritual behavior begins with one's demeanor, in particular one's countenance: "The countenance is the external side of one's tally. The external side of one's tally being rectified, therefore one's emotional responses and spontaneous tendencies will be properly ordered. One's emotional responses and spontaneous tendencies being in proper order, therefore humaneness and rightness will be maintained. Humaneness and rightness being maintained, therefore replete virtue is manifest. When replete virtue is manifest, one can be a model and exemplar [literally, an image]. This is what is known as a gentleman." The tally is a metaphor for the gentleman. His inner self, his heart, is the inner side of his tally, and his external appearance is the outer side of his tally. Each side should match the other. The gentleman's countenance, the external side of his "tally," works both in an inward and an outward direction. Inwardly a dignified countenance puts his emotional responses and spontaneous tendencies in order, thereby giving rise to the particular virtues of humaneness and rightness. As to the outward function played by the countenance, the existence of humaneness and rightness is said to lead to replete virtue in the heart which then becomes manifest on the gentleman's countenance and demeanor, his exter-

夫法象立、所以為君子。法象者、莫先乎正容貌、慎威儀。是故先王之制禮也，為冕服采章以旌之，為珮玉鳴璜以聲之。欲其尊也。欲其莊也。焉可懈慢也？

　夫容貌者、人之符表也。符表正，故情性治。情性治，故仁義存。仁義存，故盛德著。盛德著，故可以為法象。斯謂之君子矣。

　君子者、無尺土之封，而萬民尊之。無刑罰之威，而萬民畏之。無羽籥之樂，而萬民樂之。

nal tally. The manifestation of virtue on his countenance and in his demeanor in turn functions as a model for the edification of others. Xu Gan then proceeds to stress the need to maintain an awe-inspiring demeanor, to be cautious when alone, and to be careful about what one says and does. He cautions how a thoughtless jest or a careless action can cause others to lose their respect and reverence for the gentleman, thus resulting in his ruin. The essay concludes with a series of historical illustrations of the importance of observing ritual propriety by maintaining one's respectfulness.

It is by establishing models and exemplars that the gentleman is made. Of the various models and exemplars, none is more primary than preserving an upright countenance and taking care to maintain an awe-inspiring demeanor.[1] For this reason, in the rites instituted by the former kings, gentlemen wore colored and patterned caps and garments so that they would stand out, and they dangled jade ornaments to sound their presence.[2] The kings wanted it to be seen and heard that these men should be respected and dignified. How can one be lax in regard to such matters?

The countenance is the external side of one's tally.[3] The external side of one's tally being rectified, therefore one's emotional responses and spontaneous tendencies will be properly ordered. One's emotional responses and spontaneous tendencies being in proper order, therefore humaneness and rightness will be maintained.[4] Humaneness and rightness being maintained, therefore replete virtue is manifest. When replete virtue is manifest, one can be a model and exemplar. This is what is known as a gentleman.

The gentleman may lack investiture in even a scrap of land, yet the myriad people will respect him. He may lack the authority to mete out punishments, yet the myriad people will hold him in awe. He may be unable to provide musical entertainment, yet the people will all delight in him.[5] He may

無爵祿之賞，而萬民懷之。其所以致之者一也。故孔子曰：「君子威而不猛，泰而不驕。」《詩》云：「敬爾威儀，惟民之則ᵃ。」

若夫墮其威儀，怳其瞻視，忽其辭令，而望民之則我者，未之有也。莫之則者，則慢之者至矣。小人皆ᵇ慢也，而致怨乎人！患己之卑而不知其所以然，哀哉！故《書》曰：「惟聖罔念作狂；惟狂克念作聖。」

人性之所簡也，存乎幽微；人情之所忽也，存乎孤獨。夫幽微者、顯之原也。孤獨者、見之端也。胡可簡也？胡可忽也？是故君子敬孤獨而慎幽微。雖在隱蔽，鬼神不得見其隙也。《詩》云：「肅肅兔罝，施於中林。」處獨之謂也。

又有顛沛而不可亂者，則成王、季路其人也。昔者

ᵃ Translation follows the Mao text reading of *jing shen wei yi wei min zhi ze* 敬慎威儀維民之則.
ᵇ Following *Qunshu zhiyao*, 46.10a, *Xu Hou Hanshu*, 69B.2b, and the Qian Peiming and *Jianan qizi ji* versions of *Discourses* in reading *jian* 見 for *jie* 皆.

lack the power to bestow titles and emoluments, yet the people will all regard him with affection. In all cases, the gentleman is able to achieve this by virtue of the same thing.* Thus Confucius said, "A gentleman commands authority but is never fierce," and he is "Grand but never insolent."[6] *The Book of Odes* says: "Careful of his awe-inspiring demeanor/He becomes a model for the people."[7]

It has never been the case that a man can drop his awe-inspiring demeanor, dim his gaze, or be careless in his speech, and still expect the people to model themselves on one such as he. If no one takes him as their model, then there will appear those who are disrespectful toward him. When the small man is treated with disrespect, he puts the blame on other people. He despairs at his lowness, and yet does not know why it is so. How pathetic! Thus the *Book of Documents* says: "Even the wise, when not thinking, become foolish, and the foolish, by thinking, become wise."[8]

That which human nature neglects is to pay attention to oneself when in obscurity. That which the emotional dispositions overlook is to pay attention to oneself when alone. Yet obscurity is the origin of the obvious, and solitariness is the starting-point of the manifest. So how can they be neglected and overlooked? For this reason, the gentleman is cautious when alone and vigilant when in obscurity. Even if he were in a secluded and concealed place, ghosts and spirits would not find the slightest fault with him.[9] The *Book of Odes* says: "We beat down the pegs of the rabbit nets/And place them in the middle of the forest."[10] This is what is meant by "abiding in solitariness."†

There are, moreover, men who cannot be flustered even in the most harrowing circumstances. King Cheng [r. 1042/35–1006] and Ji Lu are two such men.‡ Long ago, when King

---

*That is, his charismatic virtue, as revealed through his demeanor.

†Here, Xu Gan is simply employing the idea of a net in the middle of the forest to suggest the need for vigilance in solitariness.

‡King Cheng was the son of King Wu 武 (r. 1049/45–1043) of the Zhou dynasty. Ji Lu was one of Confucius's inner circle disciples.

成王將崩，體被冕服，然後發《顧命》之辭。季路遭亂，結纓而後死白刃之難。夫以〔崩亡〕^c之困、白刃之難，猶不忘敬，況於遊宴乎！故《詩》曰：「就其深矣，方之舟之。就其淺矣，泳之游之。」言必濟也。

　　君子口無戲謔之言，言必有防；身無戲謔之行，行必有檢。〔言必有防，行必有檢〕^d，故雖妻妾不可得而黷也；雖朋友不可得而狎也。是以不慍怒而德行行於閨門；不諫諭而風聲化乎鄉黨。傳稱大人正己而物自正者，蓋此之謂也。〔徒〕^e以匹夫之居猶然，況得意而行於天下者乎！

c Following the *Longxi jingshe* version of *Discourses* in reading *mi liu* 彌留 not *beng wang* 崩亡.

d Not following the *Concordance to Zhonglun* in introducing the characters *yan bi you fang xing bi you jian* 言必有防行必有檢 into the text.

e Not following the *Concordance to Zhonglun* in introducing this character into the text.

Cheng was dying, he made his last will and testament only after he had his cap and robe put on him.[11] When calamity befell Ji Lu, he submitted to the misfortune of being speared to death only after he had fastened his cap strings.[12] If dignity is not neglected—even when one is in such dire straits as being confronted by imminent death brought on by illness or suffering the misfortune of being speared to death—then how much more should it be preserved when sporting or banqueting? Thus the *Book of Odes* says:

> Where the water was deep,
> I crossed it by raft or a boat.
> Where it was shallow,
> I waded or swam across it.[13]

This says that the water must be crossed.*

The gentleman utters no word in thoughtless jest, for one must be guarded in one's speech. He ventures no frivolous conduct, for one must keep one's conduct in check. Thus even with one's wife or concubine one should not take liberties; even with one's friends one should not be too familiar. Thereby, without recourse to anger, virtuous conduct will be practiced in the women's quarters of the gentleman's house, and without recourse to admonishments, his reputation will have a transforming effect throughout his local community. One tradition states that things will rectify themselves of their own accord when the great man rectifies himself.† This is probably what this saying means. If this is the case with an ordinary person, how much more will it be so with one who, having realized his aims, holds sway throughout the realm? Emperor

---

* The idea being expressed is that death is inevitable and so should be faced with equanimity.
† *Mencius*, 7A.19: "There are the great men; they rectify things by rectifying themselves." In Han times, such subcanonical writings as the *Analects* and *Mencius* were often referred to as "traditions," *chuan* 傳, in contradistinction to the "classics," *jing* 經.

〔故〕^f唐堯之帝允恭克讓而光被四表。成湯不敢怠遑而奄有九域。文王祇畏而造彼區夏。《易》曰：「《觀》盥而不薦。有孚顒若。」言下觀而化也。

禍敗之由也，則有媟慢以為階。可無慎乎！昔宋敏碎首於棊局。陳靈被禍於戲言。閻邬造逆於相詬。子公生弑於嘗黿。是故君子居身也謙，在敵也讓，臨下也莊，奉上也敬。四者備而怨咎不作，福祿從之。《詩》云：「靖恭爾位，正直是與。神之聽之，式穀以汝^g。」

故君子之交人也，歡而不媟，和而不同；好而不佞詐，學而不虛行；易親而難媚，多怨而寡非。故

^f Not following the *Concordance to Zhonglun* in introducing this character into the text.
^g Following Qu, *Shijing shiyi*, 176, in reading *shen* 神 as the verb shen 慎. The modern text of *Odes* reads 共 for 恭 and 女 for 汝.

Yao was "sincere, courteous and deferential, and so his light spread over the four quarters of the realm."[14] Cheng Tang did not dare to be remiss in obeying heaven's purpose, and so he came to possess the nine regions.*[15] King Wen respected heaven, holding it in awe, and established the Zhou dynasty in the central plains region.[16] The *Book of Changes* says: "The 'Looking up' hexagram. To wash the hands but not to offer in sacrifice. There is sincerity with the head held high."[17] This is saying that those below, looking up, are transformed.[18]

Disaster and ruin are ushered in by disrespect and irreverence. Can one afford not to be vigilant? In the past, Duke Min of Song [r. 691–682] was killed at the chessboard.[19] Duke Ling of Chen [r. 613–599] met with misfortune all because of a joke.[20] Yan Zhi and Bing Chu rebelled against their master for mistreating them.[21] Zigong killed his ruler after he had tasted some of his ruler's turtle soup.[22] For this reason, in his personal behavior, the gentleman is modest; when dealing with enemies, he is deferential; when presiding over those below him, he is strong; when serving those above him, he is respectful. When all four types of conduct are fully present, enmity and misfortune will not arise; rather, fortune and emoluments will follow one. The *Book of Odes* says:

> Be careful and respectful in office;
> Associate with the correct and upright.
> Take care, heed these words,
> And good fortune will come to you.[23]

Thus in dealing with other people, the gentleman is pleasant but not familiar; he seeks harmony but not conformity; is friendly but not sycophantic and deceitful; is learned but also one whose learning is not practiced in vain; easy to draw close to but difficult to flatter; frequently puts himself in the other person's position and is seldom critical of others.[24] Because of

---

*The "nine regions" may simply refer to the empire as a whole.

無絕交，無畔朋。《書》曰：「慎始而敬終^h，以不困。」

　夫禮也者，人之急也。可終身蹈，而不可須臾離也。須臾離，則惰慢之行臻焉。須臾忘，則惰慢之心生焉。況無禮而可以終始乎！夫禮也者、敬之經也。敬也者、禮之情也。無敬無以行禮，無禮無以節敬。道不偏廢，相須而行。是故能盡敬以從禮者，謂之成人。過則生亂，亂則災及其身。

　昔晉惠公以慢端ⁱ而無嗣。文公以肅命而興國。郤犨以傲享徵亡。冀缺以敬妻受服。子(圉)〔圍〕以《大明》昭^j亂。蘧罷以《既醉》保祿。良霄以《鶉奔》喪家。子展以《草蟲》昌族。君子感凶德之如彼，見吉德之如此。故立必磬折，

^h Following Qian, "Reading Notes," 1b, and Yu, *Zhuzi pingyi bulu*, 75, in reduplicating the character *zhong* 終.

ⁱ Following Yu, *Zhuzi pingyi bulu*, 75, and *Xu Hou Hanshu*, 69B.4b, in reading *rui* 瑞 instead of *duan* 端.

^j Following the *Jianan qizi ji* version of *Discourses*, p. 138b, and *Xu Hou Hanshu*, 69B.4b in reading *wei* 圍 for *yu* 圉 and *zhao* 招 for *zhao* 昭.

this, people do not break off their relationships with him, nor does he turn his back on his friends. The *Book of Documents* says: "If one is careful of the beginning and watchful of the end, to the end there will be no difficulties."[25]

Ritual is of critical importance to man; it is a path that should be trodden to the end of one's life; one should not stray from it even momentarily.[26] If one were to stray from it momentarily, then dissolute behavior would appear in one's actions. If it were forgotten momentarily, then dissolute thoughts would be born in one's heart. Moreover, if there were no ritual, would it be possible to complete that which one had started? The practice of ritual is the main thread by which reverence is advanced. Being respectful is an essential condition of ritual. Without respectfulness there is no way to practice ritual, and without ritual there is no way to modulate respectfulness. The way proceeds by relying on both, not by abandoning one in favor of the other. For this reason, one who is capable of being thoroughly respectful in complying with the requirements of ritual is called a "fully developed person."[27] If, however, one transgresses the codes of ritual, then chaos will issue; if chaos issues, then calamities will befall one's person.

In the past, Duke Hui of Jin [r. 650–637] received the jade symbol of his rank with an air of indifference, and so he had no successor.[28] Duke Wen of Jin [r. 636–628] respectfully followed orders and so brought success to his country.[29] Because of his arrogance and self-gratification, Xi Chou [d. 574] brought death upon himself.[30] Because Xi Que of Ji treated his wife with respect, he was rewarded with office.[31] By singing "Great Brilliance," Prince Wei courted chaos.[32] Wei Pi secured his position by singing "Drunk with Wine."[33] Liang Xiao destroyed his household by singing "Bickering Quails," while Zizhan brought prosperity to his clan by singing "Insects in the Grass."[34] The gentleman discerns the difference between malevolent and benevolent potency as plainly as he distinguishes between each of these pairs of examples. Thus, when standing in court, one should incline oneself respectfully, and when

坐^k必抱鼓。周旋中規，折旋中矩。視不離乎結繪之間，言不越乎表著之位。聲氣可範，精神可愛，俯仰可宗，揖讓可貴，述作有方，動靜有常，帥禮不荒，故為萬夫之望也。

k Following Han, "Zhonglun shize," 42, in emending *zuo* 坐 to *gong* 拱.

saluting, one should hold one's arms as though embracing a drum. When turning around, one should make a full-about turn, and when turning to either side, one should do so at a right angle.[35] When speaking to one's superior, one's gaze should remain within the area of his waist sash and collar, and one's speech should not exceed the boundary of what has been marked out and determined.[36] Thereby, the force of one's words will become a model; one's spirit will become an object of fond admiration; one's physical movements will become an object of veneration; and one's deference will become praiseworthy. There will be a standard in what one transmits and creates, and a constancy in one's movement and rest. In leading others on the path of ritual there will be no dissoluteness. One will thus be a paragon for the people.

《脩本第三》

3. Cultivating the Fundamental

This essay continues the theme of individual cultivation. The opening paragraph introduces a pair of concepts that also features in several other essays in *Discourses*, the fundamental and the peripheral. Here the fundamental refers to personal cultivation of the way, for unless one unremittingly cultivates one's own virtue (*de*) in accordance with the way, it will not grow. Virtue for Xu Gan is of primary importance because it is the actuality of the gentleman and requires sustained cultivation for it to be purified and made whole. In "Ordering Learning" and "Establishing Models and Exemplars," the cultivation of virtue starts with the training of one's commitment and demeanor. In this essay its continued cultivation involves paying attention to the accord between words and deeds, steadfastness of purpose, and the refinement of one's spontaneous tendencies, or "nature" (*xing*). Xu Gan understood *xing* to be an inborn quality tainted with impurities which require unremitting cultivation to remove them and so refine virtue: "Pearls contain tiny impurities and jade harbors flaws. This is their nature. A good craftsman works on them to purify their natures, making them appear as if they had always been thus. Thus in view of the fact that these two things can be purified, one can know that the virtue of humaneness can be refined."

Xu Gan also introduces the concept of "departures from regularities" (*bianshu*): "Someone said to me, 'This way, do you really believe in it?' I replied, 'Why should I not believe in it? In times of order, practitioners of good reap good fortune, while wrongdoers meet with misfortune. In times of chaos,

人心莫不有理道。至乎用之則異矣。或用乎己，或用乎人。用乎己者謂之務本，用乎人者謂之近末。君子之理〔之〕[a]也，先務其本，故德建而怨寡。小人之理〔之〕[b]也，先近其末，故功廢而讎多。孔子之制《春秋》也，

---

[a] Not following the *Concordance to Zhonglun* in introducing this character into the text.
[b] Not following the *Concordance to Zhonglun* in introducing this character into the text.

however, practitioners of good do not reap good fortune and wrongdoers do not meet with misfortune. This is caused by departures from regularities. The wise man does not doubt the constant way because of departures from regularities.'" By the Han dynasty, people believed that all things in the cosmos were correlated with the binary numerological categories of *yin* and *yang*, and by extension, the five processes, the eight trigrams, and the sixty-four Hexagrams. It was held that if the natural succession of the *yin* and *yang* cosmological forces were disturbed, this would affect other numerical relationships by way of correlative resonance or sympathetic magic. When, however, the way changes from its regular course, because of some upset to this interrelated system of correlations, the usual pattern of retribution and reward may be temporarily suspended. "Departures from regularities" is the term Xu Gan uses for the process which leads to this changed state of affairs. Other themes in the essay include reputation, being vigilant over oneself when alone, and the relation between an individual's fortunes and "circumstances which lay beyond human control" (*ming*).

The minds of men are all endowed with a path to order their affairs; there are, however, differences in how they use it. Some use it for the sake of self-improvement, while some use it with an eye to others.[1] Using it for the sake of self-improvement means to apply one's energies to what is fundamental; using it with an eye to others means to apply one's energies to what is peripheral. The gentleman orders his affairs by first attending to what is fundamental. Hence virtue is established and little enmity is aroused. The small man, however, orders his affairs by first attending to what is peripheral. Hence effort is wasted and much antagonism is generated. In creating the *Spring and*

詳內而略外，急己而寬人。故於魯也，小惡必書；於眾國
也，大惡始筆。

　　夫見人而不自見者謂之矇。聞人而不自聞者謂之聵。慮人
而不自慮者謂之瞀。故明莫大乎自見。聰莫大乎自聞。睿莫
大乎自慮。此三者、舉之甚輕，行之甚邇，而〔人〕^c莫之知
也。故知者舉其輕之事以任天下之重，行甚邇之路以窮天下
之遠。故德彌高而基彌固，勝彌眾而愛彌廣。

^c Not following the *Concordance to Zhonglun* in introducing this character
into the text.

Autumn Annals, Confucius was detailed in his criticisms of the internal affairs of Lu yet gave few details in his criticisms of the affairs of other states; he was scrupulously demanding of his own state, yet lenient with other states. Thus with regard to Lu, he took the view that even small wrongs had to be recorded, while, in the case of the various other states, he recorded only major wrongs.*

One who sees others but not himself is called blind. To hear others but not oneself is called deafness. To give consideration to others but not to oneself is called dimness. Thus there is no greater clear-sightedness than seeing oneself; there is no greater acuteness of hearing than listening to oneself; and there is no greater wisdom than being thoughtful with regard to oneself.[2] Although it is easy to carry out these matters, and the means to implement them lies directly at hand, no one understands this. For this reason the wise are able to undertake the most difficult of tasks by employing very simple methods, and they are able to reach the most distant of places by taking the route nearest to hand. Hence the more noble the virtue, the more solid is its foundation; the larger the number of victories, the broader the application of concern to

---

*Cf. *Gongyang Commentary*, Yin 10, 3.15a–b: "The *Spring and Autumn Annals* records the internal affairs of the state of Lu but only briefly mentions the affairs of other states. It records the major wrongs of other states but does not record their minor wrongs." That is, Confucius set for his own state of Lu a particularly high standard, regarding its minor shortcomings to be equally deserving of blame and censure (and hence of being recorded) as the major shortcomings of other states. The Gongyang commentator He Xiu (129–82) explains that the purpose in treating the major wrongs committed in the state of Lu as matters of taboo, while making the major wrongs within other states a matter of public record, was to ensure that "when a clear-sighted ruler arises, he should first rectify matters in Lu. Only when there were no major wrongs internally could the major wrongs of the other feudal lords be remedied." As to why only the minor wrongs committed in Lu are recorded, while those of other states are not, He continues: "Only when there are minor wrongs internally can the major wrongs of the other feudal lords be remedied." (He's commentary to *Gongyang Commentary*, Yin 10, 3.15b). That is, other states would have been likely to take notice of Lu only if it had no identifiable wrongs except for minor ones.

《易》曰：「《復》亨。出入無疾。朋來無咎。」其斯之謂歟！

　　君子之於己也，無事而不懼焉。我之有善，懼人之未吾好也。我之有不善，懼人之未吾惡也。見人之善，懼我之不能脩也。見人之不善，懼我之必若彼也。故其嚮道，止則隅坐，行則驂乘。上懸乎冠緌，下繫乎帶珮。晝也與之遊，夜也與之息。此《盤銘》之謂「日新」，《易》曰：「日新之謂盛德。」孔子曰：「弟子勉之。汝毋自舍。人猶舍汝，況自舍乎？人違汝其遠矣。」故君子不恤年之將衰，而憂志之有倦。不寢道焉，不宿義〔焉〕。言而

others.* The *Book of Changes* says: "The 'Return' hexagram. Endurance. Exiting and entering are done without flaw. When friends come there is no trouble."[3] Is this not what is meant?†

As far as his own person is concerned, there is no matter about which the gentleman is not extremely cautious. "If one has good qualities, one fears that others will not approve of me. If one has bad qualities, one fears that others will not disapprove of me. On seeing the good qualities in others, one fears being unable to cultivate those qualities. On seeing the bad qualities in others, one fears that one is surely like them." For this reason, the gentleman aspires to the way. When staying at home, he sits on the lower edge of the mat. When traveling, he sits on the right-hand side of the carriage.‡ Above, his capstrings hang down, and below, jade ornaments are suspended from his waistband. By day, he travels with these ornaments, and at night he rests with them. This is what Tang's bathing inscription meant by daily renewal.[4] *The Book of Changes* says: "Daily renewal is what is meant by replete virtue."[5] Confucius said: "Take care, my disciples, not to be self-rejecting. No matter what, there will be some who will reject you regardless. Yet if you reject yourself then they will distance themselves from you even more."[6] Thus the gentleman is not anxious about the approaching frailty of old age but rather is worried that he might weary in his commitment to his purpose. He does not grow dormant in steering a course to that purpose, nor does he stop to rest from acting in a manner appropriate to that purpose. To say that one will do some-

---

* The first half of this sentence possibly refers to the support a ruler enjoys from his people, while the second half possibly alludes to the concern that a ruler has for newly subjugated subjects.

† Although it is not clear just how Xu Gan understood this passage, the key word is *return*. He seems to be alluding to the notion of returning to draw on the fundamental quality which constitutes the moral character of the gentleman, his virtue. For Xu Gan, the last line in the quotation possibly alludes to the charismatic quality of virtue.

‡ Both the corner of a mat and the right-hand seat on a carriage were considered to be humble positions.

不行，斯寢道矣。行而不時，斯宿義^d矣。夫行異乎言，言之
錯也，無周於智。言異乎行，行之錯也，有傷於仁。是故君
子〔之〕^e務以行前言也。

人之過、在於哀死，而不在於愛生；在於悔往，而不在於
(懷)〔慎〕^f來。喜語乎已然，好爭乎遂事。墮於今日，而懈於
後旬。如斯以及於老。故野人之事，不勝其悔。君子之悔，
不勝其事。

孔子謂子張曰：「師，吾欲聞彼將以改此也。聞彼而不
〔以〕^g改此，雖聞何益？」故《書》舉穆公之誓，善變也。《春
秋》書衛北宮括伐秦，善攝也。

夫珠之含礫，瑾之挾瑕，斯其性與。良工為之以純其性，
若夫素然。

^d Following *Qunshu zhiyao*, 46.11a, in including the sixteen characters, *yan yan er bu xing si qin dao yi xing er bu shi si su yi* 焉言而不行斯寢道矣行而不時斯宿義.
^e Not following the *Concordance to Zhonglun* in introducing this character into the text.
^f Not following the *Concordance to Zhonglun* in emending *huai* 懷 to *shen* 慎.
^g Not following the *Concordance to Zhonglun* in introducing this character into the text.

thing, yet fail to do so, is to let one's commitment to steering that course grow dormant. To act when it is untimely is to stop to rest from acting appropriately. When one's actions and words are at odds, it is the words that are at fault if they are incompatible with wisdom, and the actions that are at fault if they are harmful to being humane. For this reason, the gentleman makes it his task to act before speaking about his actions.

The mistake men make is to lament death rather than to cherish life; to regret the past rather than to embrace the future. They enjoy discussing things as they already are and are fond of arguing about events that have already come to pass. Being bogged down in the present they are neglectful of the future. They are like this unto their old age. Hence the boorish person cannot help but regret his affairs. The gentleman, however, keeps regrets about his affairs in proper perspective.

Confucius said to Zizhang: "Shi, I want to learn from others so that I can change myself. If having learned from others I still do not change myself, then of what use is it to have learned from them?"*[7] Thus the *Book of Documents* cites the vow of Duke Mu [r. 659–621] of Qin. He was good at changing.† The *Spring and Autumn Annals* records Beigong Kuo of Wei's taking part in the campaign against Qin because of his active participation in the campaign.‡

Pearls contain grains of sand and jade harbors flaws. This is their nature.[8] A good craftsman works on them to purify their natures, making them appear as if they had always been thus.

---

* Zizhang was one of Confucius's disciples.
† The vow (delivered as a speech) comprises the last chapter of *Documents*. According to *Documents*, 20.10b–11a, the vow was given by Duke Mu to his ministers when three of his defeated and humiliated officers returned from Jin as captives. He made the vow because he had failed to listen to his counselors and consequently was defeated. See *Zuo Commentary*, Xi 32, 33; Legge, *Chinese Classics*, 5:220–26.
‡ See *Zuo Commentary*, Xiang 14, 32.13a: "Beigong Kuo of Wei was not recorded [in the *Spring and Autumn Annals*] as having taken part in the meeting at Xiang, but was recorded as having participated in the campaign against Qin. This is because he actively participated in the latter."

故觀二物之既純，而知仁德之可粹[h]也。

優者取多焉。劣者取少焉。在人而已。孰禁我哉？乘扁舟而濟者，其身也安；粹大道而動者，其業也美。故《詩》曰：「追琢其章，金玉其相。勉勉我王，綱紀四方。」

先民有言：明出乎幽，著生乎微。故宋井之霜，以基昇正[i]之寒。黃蘆之萌，以兆大中[j]之暑。事亦如之。故君子修德，始乎筓丱，終乎鮐背；創乎夷原，成乎喬嶽。《易》曰：「《升》元亨。用見大人，勿恤。南征吉。」積小致大之謂也。

小人朝為而夕求其成，坐施而立望其反。行一日之善，而求終身之譽。譽不至，則曰：「善無益矣。」遂疑聖人之言，背先王之教；存其舊術，順

[h] Emending *cui* 粹 to *cui* 萃.
[i] Following Gu Guangqi in emending *sheng zheng* 昇正 to *dou zheng* 斗正. See Han, "Beitu cang Zhonglun," 279–80.
[j] Following Gu Guangqi in emending *da zhong* 大中 to *huo zhong* 火中. See Han, "Beitu cang Zhonglun," 279–80.

Thus when one sees these two things after they have been purified, one can know that the virtue of humaneness is able to be refined.

Of superior things, one should take more of them, but of inferior things one should take less of them. It is entirely up to the individual, for who will stop me choosing one or the other? Just as those who ride in a flat-bottomed boat to cross a river will be safe, those who act by taking hold of the great way will be praiseworthy in their enterprises. Thus, the *Book of Odes* says:

> Engraved and chiseled is his appearance,
> Golden and jadelike is his visage.
> Vigorous is our king,
> Effecting order and rule in the four quarters.[9]

Our forebears had a saying, "The obvious comes from the obscure, and the manifest is born of the subtle." Thus when frost forms on the wells in the state of Song, this signals the beginning of the cold time of the year, and when the shoots of the yellow reed sprout, this heralds the arrival of the hottest time of year.[10] This is indeed the case. Thus the gentleman's cultivation of virtue begins in his youth and ends in his old age; it starts on the flat plains and becomes whole in the lofty peaks. The *Book of Changes* says: "The 'Ascending' hexagram. Greatness and endurance. Use these qualities to realize the great man. Do not worry. For southern campaigns this is auspicious."[11] This is what is meant by "little by little he becomes lofty and great."[12]

The small person, however, expects to reap in the evening what he has sown in the morning. If he gives something one minute, he expects repayment the next. If he does good for one day, he expects a good reputation for a lifetime. If this good reputation is not forthcoming, then he says that there is nothing to be gained from doing good. Thereupon, he doubts the words of the sages, turns his back on the teachings of the former kings, keeps up his old ways and follows what

其常好。是以身辱名賤而不免為人役也。孔子曰：「小人何以
壽為？一日之不能善矣。久惡，惡之甚也。」

　蓋人有大惑而不能自知者，舍有而思無也，舍易而求難
也。身之與家、我之有也。治之誠易，而不肯為也。人之與
國、我所無也。治之誠難，而願之也。雖曰：「吾有術，吾
有術。」誰信之歟？故懷疾者人不使為醫。行穢者人不使畫
法。以無驗也。子思曰：「能勝其心，於勝人乎何有？不能
勝其心，如勝人何？」故一尺之錦，足以見其巧；一仞之
身，足以見其治。是以君子慎其寡[k]也。

　道之於人也，其[k]簡且易耳。其修之也，非若採金攻玉之
涉歷艱難也。非若求盈司利之競逐囂煩也。不要而遭。不徵
而盛。四時嘿而成。不言而信。德配乎

[k] Following *Xiao wanjuanlou congshu* and *Longxi jingshe* versions of *Discourses*
in reading *shen* 甚 not *qi* 其.

has always pleased him. Thus with his person demeaned and his name despised, he is certain to be ordered about by others. Confucius said: "What purpose would it serve for a small person living a long life when he cannot be good for even one day. Persistent wrongdoing is detestable in the extreme."[13]

A person who is so badly confused that he is unable to know that he is confused will abandon what belongs to him as he longs for what does not, and give up what is easy as he seeks for what is difficult. My person and my home belong to me. To keep them in order is really simple, yet the confused person is unwilling to do so. Other people and the country do not belong to me. To keep them in order is difficult, yet the confused person wishes to do so. Even if he were to say, "I have the method to bring about order," who would believe him? Thus people do not allow one who is suffering from illness to heal them, or a perpetrator of obscenities to draw up laws, because they have no proof of their efficacy. Zisi said: "If one can control one's heart, then what difficulty will there be in controlling others? If, however, one cannot control one's heart, then how will one be able to control others?"*[14] Hence a piece of embroidery one foot long is enough to see a person's handiwork skills, and a physical person six feet tall is enough to see a person's well-ordered constitution. It is for this reason that the gentleman is vigilant over himself when alone.[15]

For man, the way is very simple and easy. Its cultivation is not like the arduous undertakings involved in mining for metals or cutting and polishing jade, nor is it like the hustle and bustle involved in competing for wealth and profit. Without requesting it, one comes across it, and without one's summoning it, it presents itself in all its fullness. The four seasons come to completion in silence; they do not say anything, yet they faithfully fulfil their cycle.[16] The gentleman's virtue matches that of

---

*Zisi [Kong Ji 孔伋] was Confucius's grandson. He is recorded as having written a work in twenty-three chapters, *Zisi*, which is no longer extant. *Zhongyong* (Doctrine of the Mean) is traditionally attributed to him.

天地。功侔乎四時。名參乎日月。此虞舜、大禹之所以由匹
夫登帝位，解布衣、被文采者也。故古語曰：「至德之貴，
何往不遂？至德之榮，何往不成？」後之君子，雖不及行，
亦將至之云耳。

　　琴瑟鳴，不為無聽而失其調；仁義行，不為無人而滅其
道。故絃絕而宮商亡；身死而仁義廢。曾子曰：「士[1]任重而
道遠。仁以為己任，不亦重乎？死而後已，不亦遠乎？」夫
路不險，則無以知馬之良；任不重，則無以知人之德。君子
自強其所重以取福；小人日安其所輕以取禍。

　　或曰：「斯道豈信哉？」

　　曰：「何為其不信也？世之治也，行善者獲福，為惡

---

[1] Standard versions of *Analects*, 8.7, read *bu ke yi bu hong yi* 不可以不弘毅
after *shi* 士 and before *ren* 任.

heaven and earth; his accomplishments are equal to that of the four seasons; and his name forms a trinity with the sun and moon.[17] By cultivating virtue, Shun of Yu and the Great Yu were able to rise from the rank of commoner to that of emperor, shed their cloth jackets and don richly embroidered vestments.* An ancient saying thus says, "The nobility of supreme virtue, to which quarter does it not extend? The glory of supreme virtue, in which quarter does it not bring things to completion?" Although gentlemen of later ages may not have been able to match the behavior of Shun and the Great Yu, yet they have surely aspired to do so!

The sounds of the *qin* and *se* do not go out of key if they are not heard.† Similarly, the way of practicing humaneness and rightness is not destroyed if there are no other people.‡ Thus it is only when the strings break that the notes disappear, and only at death that humaneness and rightness are abandoned. Zengzi said: "The responsibility of a man of standing is heavy and his road is long.[18] He takes humaneness to be his responsibility. Is that not heavy? Only with death does his road come to an end. Is that not long?"[19] If the road is not rough, then there is no way to test how good a horse is. Similarly, if a man's responsibilities are not heavy, then there is no way to judge his virtue. The gentleman daily strengthens himself in matters which he finds burdensome, and so reaps good fortune. The small man, however, daily takes the easy way out, and so reaps misfortune.

Someone said to me, "This way, do you really have faith in it?"

I replied, "Why should I not have faith in it? In times of order, practitioners of good reap good fortune, while wrong-

者得禍；及其亂也，行善者不獲福，為惡者不得禍。變數也。知者不以變數疑常道。故循福之所自來，防禍之所由至也。遇不遇，非我也，其時也。」夫施吉報凶謂之命。施凶報吉謂之幸。守其所志而已矣。《易》曰：「君子以致命遂志。」然行善而不ᵐ獲福猶多，為惡而不得禍猶少。總夫二者，豈可舍多而從少也。曾子曰：「人而好善，福雖未至，禍其遠矣。人而不好善，禍雖未至，福其遠矣。」故《詩》曰：「習習谷風，惟山崔巍，何木不死？何草不萎ⁿ？」言盛陽布德之月，草木猶有枯落而與時謬者，況人事之應報乎！故以歲之有凶穰而荒其稼穡者，非良農也。

ᵐ Following Yu, *Zhuzi pingyi bulu*, 76, in omitting *bu* 不.
ⁿ The Mao recension reads *wei shan cui wei wu cao bu si wu mu bu wei* 維山崔巍無草不死無木不萎.

doers meet with misfortune. In times of chaos, however, practitioners of good do not reap good fortune and wrongdoers do not meet with misfortune. This is caused by departures from regularities. The wise man does not lose faith in the constant way because of departures from regularities. Hence, we should seek where good fortune comes from and guard against that whereby misfortune arrives. Whether I meet with good fortune is not determined by me, but rather upon the variability of circumstances."[20] To do something that should bring good fortune, yet be repaid with misfortune, is called circumstances beyond human control; to do something that should bring misfortune yet be repaid with good fortune is called luck.[21] One need only be concerned about being steadfast of purpose. The *Book of Changes* says: "The gentleman is prepared to lay down his life in the pursuit of his purpose."[22] Nevertheless, doing good will mostly result in good fortune; only seldom will wrongdoing not result in misfortune. Combining these two observations, how could one reject what is so most of the time, in favor of that which seldom happens? Zengzi said: "If a person is fond of doing good, then, although good fortune may not be forthcoming, misfortune will be far away. If a person is not fond of doing good, then, although misfortune may not be forthcoming, good fortune will be far away."[23] Hence, the *Book of Odes* says:

> Gust after gust blows the valley breeze,
> And on the scraggy heights
> There is no tree which is not dying,
> No grass which is not wilting.[24]

This says that in the months when the sun is at its fullest, bestowing its virtue, there are still plants and trees that wither and fall, quite out of keeping with the season. Such anomalies are in even greater evidence when it comes to the way in which the affairs of man are responded to and repaid. Thus, should a farmer decide to neglect his farming because he has both lean and prosperous years, then he is not a good farmer; and should

以利之有盈縮而棄其資貨者，非良賈也。以行之有禍福而改
其善道者，非良士也。《詩》云：「顒顒卬卬，如珪如璋。
令聞令望。愷悌°君子，四方為綱。」舉珪璋以喻其德貴不變
也。

°The Mao recension reads *gui* 圭 for *gui* 珪, and *qi di* 豈弟 for *kai ti* 愷悌.

a merchant decide to throw away his capital and goods because his profits fluctuate, then he is not a good merchant. Similarly, if a man of social standing decides to change his commitment to follow the path of goodness because his conduct brings both misfortune and good fortune, then he is not a good man of social standing. The *Book of Odes says*:

> Grand and imposing
> Like a jade scepter or a token of jade.
> Renowned and distinguished,
> Pleasant lord,
> The four realms are guided by you.[25]

The scepters are employed as metaphors for his noble and unchanging virtue.

《虛道第四》

人之為德，其猶虛器歟？器虛則物注，滿則止焉。故君子常虛其心志，恭其容貌，不以逸群之才加乎眾人之上。視彼猶賢，自視猶不足也。故人願告之而〔不厭，誨之而〕[a] 不倦。《易》曰：「君子以虛受人。」《詩》曰：「彼姝者子，何以告之？」

　　君子之於善道也，大則大識之，小則小識之。善無大小，咸載

[a] Following *Qunshu zhiyao*, 46.12a, in adding the characters *bu yan hui zhi er* 不厭誨之而.

4. *The Way of Humility*

Xu Gan describes the way of humility as a means of enhancing the gentleman's virtue. By adopting a humble attitude, the gentleman encourages other people to share what they know with him, and so, even though "his vision does not extend beyond the walls of his house, nevertheless he sees beyond the frontiers of his country. [And] although his hearing does not extend past the threshold of his house, nevertheless he hears about matters that occur more than a thousand miles away." In the second half of the essay Xu Gan uses historical examples to illustrate that humility is just as important for the rulers of men. By heeding the remonstrances of their ministers and the complaints of their people, wise rulers in ancient times were able to avert calamity and enable their countries to prosper.

In acting virtuously, can one be likened to an empty vessel?[1] If a vessel is empty then things will flow into it, ceasing when it is full.[2] Thus the gentleman constantly empties his heart's ambitions and makes his demeanor respectful. He does not use his superior talents to be condescending toward the multitude of people; rather, he looks upon others as though they were gifted, and upon himself as though he were insufficiently endowed. Thus others are only too happy to tell him what they know, never tiring of instructing him. The *Book of Changes* says: "The superior person receives others with emptiness."[3] The *Book of Odes* says: "That admirable gentleman / What do I have that I could tell him?"[4]

In regard to the way of goodness, the gentleman remembers both major and minor good ways for what they are. Yet irrespective of a good way's relative importance, he will bear it

於心，然後舉而行之。我之所有既不可奪，而我之所無又取於人。是以功常前人而人後之也。故夫才敏過人，未足貴也。博辯過人，未足貴也。勇決過人，未足貴也。君子之所貴者，遷善懼其不及，改惡恐其有餘。故孔子曰：「顏氏之子，其殆庶幾乎！有不善，未嘗不知；知之，未嘗復行。」

夫惡猶疾也。攻之則〔日〕[b]益悛，不攻則日甚。故君子〔之〕[c]相求也，非特興善也。將以攻惡也。惡不廢則善不興。自然之道也。《易》曰：「《(比)〔否〕[d]》之匪人。不利君子貞。大往小來。」陰長陽消之謂也。

先民有言：人之所難者二。樂攻其惡者難。以惡告人者難。夫惟君子然後能為己之所難，能到人之所難致[e]。既能其所難也，猶恐舉人惡之輕，而舍己惡之重。君子患其如此也。故反之、復之、鑽之、核之，然後彼之所懷者竭。始盡知己惡之重矣。既知己惡之重者，而不能取彼，

[b] Not following the *Concordance to Zhonglun* in introducing this character into the text.
[c] Not following the *Concordance to Zhonglun* in introducing this character into the text.
[d] Emending *bi* 比 to *pi* 否 after the *Changes* text. The characters *bi* and *pi* being closely homophonous would appear to have led to a copyist mistakenly writing *pi* as *bi*.
[e] Following *Qunshu zhiyao*, 46.12b, in emending *neng dao ren zhi suo nan zhi* 能到人之所難致 to *neng zhi ren zhi suo nan ye* 能致人之所難也.

in mind, one day drawing upon it and making use of it. What one knows cannot be taken away, while what one does not know will be learned from others. Thereby he regularly excels others in his achievements while they lag behind him. Hence not superior talent and keen-wittedness, or superior learning and polemical skills, or superior courage and decisiveness are worth attaching importance to. The gentleman attaches importance to movement toward the good, fearing that he might never attain it, and correcting his faults, fearing that there are still some remaining.[5] Thus Confucius said: "That son of the Yan family, he's almost there. He never fails to know when he has done something that is not good. Knowing this, he never does it again."[6]

Wrongdoing is like a disease. If treated, one's condition will increasingly improve, and if not, it will worsen daily. Hence gentlemen seek one another's company in order to combat wrongdoing rather than specifically to promote the good. For if what is bad is not eliminated then what is good cannot thrive. This is simply the way that things are. *The Book of Changes* says: "The bad men associated with 'Obstruction' make it an unfit time for the gentleman to practice rectitude. Thus the great depart and the petty arrive."[7] This is what is meant by the waxing of the *yin* and the waning of the *yang*.[8]

People in former times had a saying: "There are two things which people find difficult: it is difficult to take pleasure in combating one's own wrongdoing, and it is difficult to tell others of theirs." Only the gentleman is able to do that which he himself finds difficult and to bring himself to do what others find difficult. Yet even though able to do that which it is difficult to do, he still fears that he may have dismissed the gravity of his own bad qualities while singling out the pettiness of other people's bad qualities. Perturbed that this may indeed be the case, the gentleman reflects, reviews, investigates, and checks. Only then will the suspicions of others cease and he begin to become fully aware of the gravity of his own bad qualities. If, however, despite having become aware of the gravity of his own bad qualities, he is unable to accept that

又將舍己，況拒之者乎！

夫酒食、人之所愛者也，而人相見莫不進焉。不吝於所愛者，以彼之嗜之也。使嗜〔忠言〕者[f]甚於酒食，人豈〔其〕愛之〔乎〕[g]？故忠言之不出，以未有嗜之者也。《詩》云：「匪言不能，胡斯畏忌？」

目也者，能遠察〔天際〕，而不能近見。其〔皆〕心[h]亦如之。君子誠知心之似目也，是以務鑒於人以觀得失。故視不過垣墻之裏，而見邦國之表；聽不過閫閾之內，而聞千里之外。因人〔之耳目〕[i]也。人之耳目盡為我用，則我之聰明無敵於天下矣。是謂人一之，我萬之；人塞之，我通之。故知其高不可為員，其廣不可為方。

先王之禮，左史記事，右史記言，師瞽誦詩，庶僚箴誨，器用載銘，筵席書戒，月考其為，歲會其行，所以自供正也。

[f] Following *Qunshu zhiyao*, 46.12b, in emending *zhe* 者 to *zhong yan* 忠言. Not following the *Concordance to Zhonglun* in retaining *zhe*.
[g] Following *Qunshu zhiyao*, 46.12b in emending *ren qi ai zhi* 人豈愛之 to *ren qi qi ai zhi hu* 人豈其愛之乎.
[h] Reading *neng yuan cha er bu neng jin jian. Qi xin* 能遠查而不能近見。其心 and not following the emendations in *Concordance to Zhonglun*.
[i] Following *Qunshu zhiyao*, 46.13a, in adding the characters *zhi er mu* 之耳目.

fact, then not only will others stop associating with him, but, moreover, they will positively refuse to have anything to do with him.⁹

People like wine and food and are always served them on social occasions. There is no parsimony when it comes to offering guests things that they like, precisely because they prefer these things. Now suppose that people preferred sincere words to wine and food, would people like to have the food and wine instead? Sincere words are not forthcoming because no one has developed a preference for them. The *Book of Odes* says: "It is not that his words are inadequate / Then why should I fear them?"¹⁰

The eye can discern things at a distance yet fail to see what is close by. The heart is also like this. Because the gentleman truly understands that the heart is like the eye, he endeavors to use people as a mirror to see into matters of loss and gain. Hence although his vision does not extend beyond the walls of his house, nevertheless he sees beyond the frontiers of his country. Although his hearing does not extend past the threshold of his house, nevertheless he hears about matters that occur more than a thousand miles away.¹¹ He does this by relying on the eyes and ears of other people. If the ears and eyes of others are completely at one's disposal, then no one in the world will be able to match the acuteness of one's hearing and the keenness of one's vision. This is what is meant by "where others are one, I am a multitude; where others become stuck, I break through." Hence such a person understands that the height and breadth of his knowledge cannot be circumscribed within the limits demarcated by the compass and square.

In the ritual practices of the former kings, the left scribe recorded events and the right scribe recorded speech; the music master and blind musicians recited odes and the various officials admonished; vessels and implements carried inscriptions and mats had admonishments written on them.¹² Each month the king would examine his actions and each year he would review his behavior so as to help correct himself. In the

昔衛武公年過九十，猶夙夜不怠，思聞訓道。命其群臣曰：「無謂我老耄而舍我。必朝夕交戒〔我〕[j]。」又作《抑》詩以自儆也。衛人誦其德，為賦《淇澳》。且曰「睿聖。」凡興國之君，未有不然者也。故《易》曰：「君子以恐懼修省。」

下愚反此道也。以為己既仁矣、智矣、神矣、明矣。兼此四者，何求乎眾人？是以辜罪昭著，腥德發聞，百姓傷心，鬼神怨痛，曾不自聞，愈休如也。若有告之者，則曰：「斯事也，徒生乎子心，出乎子口。」於是刑焉、戮焉、辱焉、禍焉。不能免。則曰：「與我異德故也，未達我道故也。又安足責？」是己之非，遂初之繆，至於身危國亡，可痛矣〔已〕[k]夫。《詩》曰：「誨爾諄諄，聽之[l]藐藐，匪用為教，覆用為虐。」

[j] Not following the *Concordance to Zhonglun* in introducing this character into the text.

[k] Not following the *Concordance to Zhonglun* in introducing this character into the text.

[l] The Mao text reads *wo* 我 for *zhi* 之.

past, although Duke Wu of Wei (r. 812–758) was more than ninety years old, nevertheless from early morning to night he remained resolute in his desire to listen to admonishments. Commanding his assembled ministers, he said, "Don't stop admonishing me on the pretext that I am well on in years; from first light to dusk you must set forth your reprimands." He also composed the ode "Restraint" to admonish himself.[13] The people of Wei eulogized his virtue and presented him directly in the ode "Qi River Cove" and, moreover, said that he was perspicacious and sagely.[14] Indeed, all rulers who made their countries flourish acted likewise. Therefore the *Book of Changes* says: "Beset with fear, the gentleman cultivates and examines himself."[15]

The most benighted rulers act contrary to this way, believing that they are already humane, wise, godlike and clear-sighted. "Being in possession of these four qualities, what then is left to be sought from other people?" Hence their crimes are glaringly obvious and their foul-smelling "virtue" reeks; the people are grieved and the ghosts and spirits are resentful. Completely oblivious to this, these rulers grow ever more pleased. If anyone should inform them about this, they would say, "Such a thing is a mere product of your imagination and your mouth." Thereupon it would be impossible for them to escape corporal punishment, execution, humiliation or misfortune. Such a ruler will accordingly say, "Things have come to this because this person's virtues are different from mine and because they have not attained my way. So what is there to reproach me about?" He portrays his wrongdoings as justified and persists in his original error until ultimately his person is placed in danger and the country is destroyed. It pains me to think about it! The *Book of Odes* says:

> I instructed you with earnestness,
> But you listened with contempt.
> You do not use my words to edify;
> Instead you treat them as a joke.[16]

蓋聞舜之在鄉黨也，非家饋而戶贈之也。人莫不稱善焉。象之在鄉黨也，非家奪而戶掠之也。人莫不稱惡焉。由此觀之，人無賢愚，見善則譽之，見惡則謗之。此人情也，未必有私愛也，未必有私憎也。今夫立身不為人之所譽，而為人之所謗者，未盡為善之理也。盡為善之理，將若舜焉。人雖與舜不同，其敢謗之乎？故語稱救寒莫如重裘。止謗莫如修身。療暑莫如親冰 [m]。信矣哉。

[m] Following *Yilin*, 5.15a, in moving *liao shu mo ru qin bing* 療暑莫如親冰 to before *jiu han mo ru chong qiu* 救寒莫如重裘.

Indeed, I have heard that when Shun was still living in his local community, it was not because he gave presents to every family and household that everyone approved of him.* Similarly when Xiang was still living in his local community, it was not because he stole from every family and household that everyone disapproved of him.† In view of this, both wise and foolish people praise good when they see it and criticize bad when they see it. It is only natural that people act in this way and is not necessarily because personal likes and dislikes are involved. Nowadays, establishing oneself successfully in society attracts people's criticism rather than their praise. The reason for this is that people have not yet fully realized the principle for doing good. If they were to do so, then should there be someone like Shun amongst them, even though other people were not the same as he, would they dare to criticize him? According to an old saying, "There is nothing better than applying ice to stop feeling hot and there is nothing better than putting on a fur jacket to stop feeling cold. Similarly, there is nothing better than cultivating oneself to stop criticism." How true!

* Shun was a predynastic sage ruler; see Allan, *Heir and Sage*, chapter 2.
† Xiang was Shun's younger brother.

《貴驗第五》

事莫貴乎有驗，言莫棄乎無徵。言之未有益也，不言未有損也。水之寒也，火之熱也，金石之堅剛也。此數物未嘗有言，而人莫不知其然者，信著乎其體也。使吾所行之信，若彼數物，而誰其疑我哉？今不信吾所行，而怨人之不信〔己〕ᵃ也，猶教人執鬼縛魅，而怨人之不得也。惑亦甚矣！孔子曰：「欲人之信己也，則微言而篤行之。篤行之，則用日久。用日久，則事著明。事著明，則有目者莫不見也。有耳者莫不聞也。其可誣哉！」故根深而枝葉茂，

ᵃ Not following the *Concordance to Zhonglun* in introducing this character into the text.

5. *Valuing Proof*

This essay takes its title from the proof provided by the gentleman that his deeds consistently match his words. In establishing good faith (*xin*), the gentleman gradually establishes a reputation (*ming*) commensurate with the growth of his virtue. Other themes in the essay include the ability to appreciate good advice, learning from one's mistakes, and being careful about one's choice of friends.

In affairs, nothing is more valuable than having proof, while in speech nothing is more futile than lacking evidence. There has never yet been a benefit in speaking when one lacks proof, nor has there ever yet been any loss in remaining silent. Water is cold, fire is hot, metal and stone are hard. These several things have never spoken and yet there is no one who does not know that they are so. Our faith comes from the embodiment of these qualities in these objects. If the faith I inspire in others through my actions could be like the case of these several things, then who would doubt one? Now if I do not have faith in what I do and yet resent the fact that others have no faith in me, this would be like resenting the fact that someone had been unable to take hold of a ghost or tie up a demon after I had instructed them to do so. How utterly confused! Confucius said, "If you want people to have faith in you, then be sparing in what you say and diligently carry out what you have undertaken to do. If one diligently carries out what one has undertaken to do, then the usefulness of having done so will endure. Lasting a long time, then the matter will become evident and clear. It being evident and clear, then anyone with eyes cannot fail to see it and anyone with ears cannot fail to hear about it. How could it be lied about?"[1] Hence, if the roots are deep, the branches and leaves will flourish, and if one

行久而名譽遠。《易》曰：「《恆》亨。無咎。利貞。」言久於其道也。

伊尹放太甲，展季覆寒女，商、魯之民不稱淫篡焉，何則？積之於素也。故染不積則人不觀其色；行不積則人不信其事。子思曰：「同言而信，信在言前也。同令而化，化在令外也。」

謗言也，皆緣類而作、倚事而興、加其似者也。誰謂華岱之不高、江漢之不長與？

works at something for a long time, one's name and reputation will travel far. The *Book of Changes* says: "The 'Constancy' hexagram. Perseverance. Good omens and benefit."[2] This is saying that "there is duration in the path he takes."[3]

Yi Yin banished Taijia, and Zhan Ji covered a freezing woman with his clothes.*[4] The people of Shang and Lu did not say that Taijia was a usurper or that Zhan Ji was licentious. Why? Because over the course of their lives they had already established reputations as good men. Hence if the dye does not build up, then people will not be able to see what color the cloth is, and if one does not build up a reputation for one's behavior, people will have no faith in what one does. Zisi said: "Two different people may utter the same words but only one will be believed, because faith had been established before he had spoken. Two different people may issue the same order but only one will have a transforming effect, because the efficacy of the transformation lies outside the order itself."†[5]

Words of criticism all arise when one makes generalizations on the basis of analogy or relies on the similarity of particular matters. Who would deny that Mount Hua and Mount Dai are tall, or that the Chang and Han rivers are long?‡ If the

---

*Yi Yin is said to have been instrumental in banishing Jie, the wicked last ruler of the Xia, thus paving the way for Tang to found the Shang dynasty (*Mencius*, 5A.7). He is also said to have banished Tang's grandson, Taijia, for having upset the laws instituted by Tang, but later allowed him to return after he had reformed. See *Mencius*, 5A.6; *Shiji*, 3.99; Allan, *Heir and Sage*, 96–101. Zhan Qin 展禽 was posthumously know as Hui 惠. He was also known as Hui of Liuxia 柳下惠. He is best known as an upright minister of Lu. See *Analects*, 18.2, 18.8. His behavior with the woman was considered by many to have contravened the rules of propriety.

†That is, it depends on the virtue—the power to transform—of whomsoever issued the order.

‡Mount Hua or Mount Taihua 太華 was formerly also known as the Western Peak. It is situated in today's Huaying County, Shaanxi Province, and stands 1,997 meters above sea level. Mount Dai is the alternative name for Mount Tai 泰, and was formerly also known as the Eastern Peak. It is situated in the middle of Shandong Province, with its main peak in Tai'an County. It is 1,524 meters above sea level. The 5,800 km. Changjiang or Yangtze is the longest river in China. The 1,532 km. Hanjiang is the longest tributary of the Yangtze.

君子修德，亦高而長之，將何患矣？故求己而不求諸人，非自強也，見其所存之富耳！

　　子思曰：「事、自名也。聲、自呼也。貌、自眩也。物、自處也。人、自官也。無非自己者。」故怨人之謂壅，怨己之謂通。通也、知所悔。壅也、遂所誤。遂所誤也、親戚離之。知所悔也、疏遠附之。疏遠附也、常安樂。親戚離也、常危懼。自生民以來，未有不然者也。殷紂為天子而稱獨夫，仲尼為匹夫而稱素王，盡此類也。故善釣者不易淵而殉魚，君子不降席而追道。治乎八尺之中而德化光矣。古之人謌曰：「相彼玄鳥，止于陵阪。仁道在近，求之無遠。」

gentleman similarly cultivates his virtue, elevating and extending it, then what will he have to worry about? For this reason, one should seek from within oneself, not from others.[6] This is not a matter of regarding oneself as possessing superior strengths but simply a question of discovering the wealth of one's inherent qualities.

Zisi said: "Affairs name themselves, sounds call themselves, appearances show themselves off, things place themselves and men determine their own office. All do this by themselves."[7] Thus to blame others is what is meant by "being obstructed," while to blame yourself is what is meant by "getting through." To get through is to realize the cause for repentance; to be obstructed is to pursue what is mistaken. If you pursue what is mistaken, then even family and relatives will leave you. If you realize what you should repent, then even those without close ties will come and join you. If those without close ties come and join you, you will constantly be content and happy. If, however, your family and relatives leave you, you will constantly be in danger and fear. Ever since our people were given birth it has always been like this. Although Zhou of the Yin dynasty was the son of heaven, he is called only a "solitary fellow," while Confucius, although an ordinary person, is referred to as an "uncrowned king."*[8] Zhou and Confucius are each ultimate examples of these two types. Thus just as a good angler does not need to change his pool to go after fish, so too the gentleman does not need to leave his mat to pursue the way, for by cultivating his person, his virtue will be transformed and illuminated. In antiquity, people used to sing:

> Look at that swallow
> Resting on the slope of that mound.
> The way of humaneness is nearby,
> It is not far to seek.[9]

*The epithet applied to Zhou, the infamous last ruler of the Shang-Yin dynasty, suggests that he was without the legitimating support of heaven or other people.

人情也、莫不惡謗，而卒不免乎謗。其故何也？非愛致力而不已之也。已之之術反也。謗之為名也，逃之而愈至，距之而愈來，訟之而愈多。明乎此，則君子不足為也。闇乎此，則小人不足得也。帝舜屢省，禹拜昌言，明乎此者也。厲王蒙戮，吳起刺之，闇乎此者也。〔夫人也〕[b]、皆書名前策，著形列圖，或為世法，或為世戒，可不慎之！

曾子曰：「或言予之善，予惟恐其聞。或言予之不善，惟恐過而見予之鄙色焉。」故君子服過也，非徒飾其辭而已。誠發乎中心，形乎容貌。其愛之也深，其更之也速。如追兔，惟恐不逮。故有進業、無退功。《詩》曰：

---

[b] Following *Qunshu zhiyao*, 46.146, in introducing the three characters *fu ren ye* 夫人也 into the text.

The common emotional response of people is to detest criticism, yet in the end, no one can avoid being criticized. What is the reason? Is it not precisely because people love to make every effort to put a stop to it that they fail to do so? The real method for stopping it is just the opposite. Take the example of being criticized for making a reputation for oneself. The more one tries to avoid a criticism, the more it pursues one; the more one tries to distance oneself from it, the more it comes along; and the more one tries to dispute it, the more one is criticized. Understanding this, the gentleman realizes that it is not worthwhile making a reputation; failing to understand this, no amount of reputation is enough for the small man. Emperor Shun frequently examined himself, and when Yu heard good advice he bowed to the speaker.[10] They understood this. King Li (r. 857/53–842/28) met with his death and Wu Qi was killed by arrows.[11] They failed to understand this. All these people have their names recorded in the documents of previous times, and their images are illustrated in paintings. Some have become models for posterity, while others have become object lessons for later generations. Can one afford not to be careful?

Zengzi said: "When someone says something about my good points, my single fear is that it will become known.* When someone says something about my bad points, my single fear is that in passing this person the wretched expression on my face will be revealed to him."[12] Thus the gentleman bears his faults. He does not just make a show of words and then have done with it. His sincerity issues from within his heart and manifests itself on his countenance. The deeper he cherishes the good, the quicker he will be able to change. It is just like chasing a rabbit; his single fear is that he will not catch it. Hence he advances his affairs and suffers no setback to his achievements. The *Book of Odes* says:

---

* Presumably out of embarrassment caused by modesty or a fear of succumbing to vanity.

「相ᶜ彼脊令,載飛載鳴。我日斯邁,而月斯征。」遷善不懈之謂也。夫聞過而不改,謂之喪心。思過而不改,謂之失體。失體、喪心之人,禍亂之所及也。君子舍游。

　　《周書》有言:「人毋鑒於水,鑒於人也。」鑒也者、可以察形,言也者、可以知德。小人恥其面之不及子都也;君子恥其行之不如堯、舜也。故小人尚明鑒,君子尚至言。至言也、非賢友則無取之。故君子必求賢友也。《詩》曰:「伐木丁丁,鳥鳴嚶嚶。出自幽谷,遷于喬木。」言朋友之義務,在切直以升於善道者也。故君子不友不如己者,非羞彼

ᶜ *Odes* reads *ti* 題 where *Discourses* reads *xiang* 相.

Look at that *jiling* bird,
Singing as it flies.
We stride forth every day,
And advance every month.[13]

This is what is meant by untiringly moving toward the good. To learn that one has been wrong and yet not change is called "losing one's heart." To reflect upon one's mistakes and yet not change is called "losing one's bearings." People who have lost their hearts and bearings are susceptible to the spread of misfortune and chaos. The gentleman rejects being like them.

There is a saying in the Zhou texts section of the *Book of Documents* that "people should not seek a reflection of themselves in water but in other people."[14] With a mirror one can examine one's external form, but from words one can come to know a person's virtue. The small man is ashamed that his face is not as handsome as Zidu's face, while the gentleman is ashamed that his virtuous behavior does not measure up to that of Yao or Shun.* Hence the small man values a bright mirror while the gentleman values frank words. Unless they are the frank words of a worthy friend, however, he will not heed them. Thus the gentleman must choose worthy friends. The *Book of Odes* says:

They hew the trees, *ding ding*,
And the birds cry out, *ying ying*.
From out of the dark valley,
They move to the lofty trees.[15]

This says that the duty of friends is to be forthright so as to enable one to rise to the ranks of those who are skilled at the way. Hence the reason "the gentleman does not befriend those inferior to himself" is not that he is embarrassed by such people

---

*Zidu was a classical paragon of male beauty. *Mencius* 6A.7: "As for Zidu, everyone recognizes him to be a beauty. Those who fail to do so have no eyes."

而大我也,不如己者須己而植者也。然則扶人不暇,將誰相我哉?吾之儥也亦無日矣!故(儥)〔墳庫〕則〔水〕縱(多)[d]。友邪則己僻也。是以君子慎取友也。孔子曰:「居而得賢友,福之次也。」

　　夫賢者、言足聽,貌足象,行足法,加乎善獎人之美,而好攝人之過。其不隱也如影。其不諱也如響。故我之憚之,若嚴君在堂,而神明處室矣。雖欲為不善,其敢乎?故求益者之居遊也,必近所畏而遠所易。《詩》云:「無棄爾輔,員于爾輻。屢顧爾僕,不輸爾載。」親賢求助之謂也。

[d] Following the *Qunshu zhiyao*, 46.15a, reading of *fen bei ze shui zong* 墳庫則水縱.

and has a high opinion of himself but that those inferior to himself need to rely on themselves if they are to stand upright.[16] As such, if I spend all my time supporting others, then who will assist me? My own downfall will be imminent! If the banks are too low the waters will overflow; if one's friends are crooked, then one will bend. For this reason the gentleman takes care when choosing his friends. Confucius said, "If when disengaged from office one has worthy friends, this is the second-best kind of good fortune."[17]

A worthy man is one whose words are worth heeding, whose countenance is worth imitating and whose behavior is worth emulating. On top of this, he is good at praising the good qualities of other people yet is also fond of seizing upon their faults. Like a shadow, he does not conceal their faults from them, and like an echo, he does not avoid talking about their shortcomings. Hence I fear him as I would a stern father in the hall or a spirit presence in the room of the ancestral temple. Even if I wanted to do wrong, would I dare do so? Hence if one seeks the company of a truly beneficial friend, one must be prepared to draw near to that which one fears and stay far away from that which is easy. The *Book of Odes* says:

> Do not throw away your sideboards
> Lest you fall on the spokes.
> Often pay attention to the driver,
> And do not let your cargo fall off.[18]

This is what is meant by being close to the worthy and soliciting their help.

《貴言第六》

君子必貴其言。貴其言則尊其身。尊其身則重其道。重其道所以立其教。言費則身賤。身賤

6. Valuing Words

For Xu Gan, it is important not only that a gentleman's deeds match his words but also that he be circumspect about what he says and with whom he speaks. Accordingly, the gentleman speaks only about matters that are pertinent to a given topic or suitable for a particular audience and adjusts his words to a level appropriate to the understanding of the person with whom he is speaking. The effect of using his words sparingly is that he will be respected by others, who will then prize his way. In a passage reminiscent of the account in the preface describing how Xu Gan tested and instructed those who came to learn from him during the period of his final years of retirement, this essay similarly outlines the process by which the gentleman instructs those who come to learn from him. A stylistic feature of this essay is the use of the question-and-answer dialogue form with a nameless interlocutor. This technique is employed in several essays. Here it is used to address the question of the gentleman's engaging in disputation with "vulgar persons," with Xu Gan arguing that the vulgar person often sets out to engage the gentleman in pointless discussion. The final part of the essay employs a series of historical precedents to argue that sometimes it is better for a gentleman to act expediently rather than apply principles too rigidly.

The gentleman must value his words. If he values his words, then he brings honor to his person; if he brings honor to his person, then his way will be prized. His way being prized is the means by which he establishes his teachings. If he is verbose, then his person will be demeaned; if his person is de-

則道輕。道輕則教廢。故君子非其人則弗與之言。若與之言，必以其方。農夫則以稼穡，百工則以技巧，商賈則以貴賤，府史則以官守，大夫及士則以法制，儒生則以學業。故《易》曰：「艮其輔，言有序。」不失事中之謂也。

　　若夫父慈子孝，姑愛婦順，兄友弟恭，夫敬妻聽，朋友必信，師長必教，有司日月慮知乎州閭矣。雖庸人則亦循循然與之言此，可也。過此而往，則不可也。故君子之與人言也，使辭足以達其知慮之所至，事足以合其性情之所安，弗過其任而強牽制也。苟過其任而強牽制，則將昏瞀委滯，而遂疑君子以為欺我也。不則曰「無聞知」矣。非故也。明偏而示之以幽，弗能照也。聽寡

meaned, then his way will not be valued; if his way is not valued, then his teachings will be cast aside. Hence if someone is not an appropriate person, then the gentleman will not speak with him.[1] Should he speak to someone, it will certainly be on matters in which the other person is expert. With farmers, it will be about planting and harvesting; with artisans of every sort, it will be about techniques and skills; with traders or merchants, it will be about prices; with office scribes, it will be about the responsibilities of office; with counselors and men of social standing, it will be about laws and regulations; and with students of classical learning, it will be about the task of learning. Hence the *Book of Changes* says: "Restraining the cheeks so that one's words are ordered."[2] This is what is meant by not straying from what is to the point in a given situation.[3]

A constant concern of minor officials is how best to exercise well-informed control over the local districts of their region in regard to such matters as: fathers being kind, sons being filial, mothers-in-law being loving, daughters-in-law being submissive, elder brothers being friendly, younger brothers being deferential, husbands being respectful, wives being obedient, friends necessarily being trustworthy, and teachers and elders necessarily giving instruction. Even when such officials are men of modest talent, it is permissible to explain these matters to them step by step.[4] It would not be appropriate, however, for the gentleman to go beyond this. Hence, when speaking to others, the gentleman phrases his words in a manner sufficient to reach the extent of that person's understanding and concern, and discusses matters congenial to their nature and emotional responsiveness. He does not make them exceed their assigned duties and compel them to follow whatever he stipulates. Should he do so, they will become confused and frustrated and begin to grow suspicious of the gentleman, thinking that he has tried to deceive them; otherwise they will say that he is ignorant. It is not that he has purposely set out to deceive them. It is, rather, as if someone has only partial vision and you show him something that is indistinct, he will not be able to see what it is; or if someone is hard of hearing

而告之以微，弗能察也。斯所資於造化者也。雖曰無訟，其如之何？故孔子曰：「可與言而不與之言，失人。不可與言而與之言，失言。知者不失人，亦不失言。」

夫君子之於言也，所致貴也。雖有夏后之璜，商湯之駟，弗與易也。今以施諸俗士，以為志誣而弗貴聽也。不亦辱己而傷道乎！是以君子將與人語大本之源，而談性、義之極者，必先度其心志，本其器量，視其銳氣，察其墮衰，然後唱焉以觀其和，導焉以觀其隨。隨、和之徵發乎音聲，形乎視聽，著乎顏色，動乎身體，然後可以發〔幽〕[a]而步遠，功察而治微。於是乎闓張以致之，因來以進之，審諭以明之，雜稱以廣之，立準以正

[a] Following the *Liangjing yibian* and *Longxi jingshe* versions of *Discourses* in reading *er* 邇, and not the *Concordance to Zhonglun* reading of *you* 幽.

and you tell him something in a hushed voice, he will not be able to hear. These capacities are natural endowments.[5] Although one might say that there is nothing reproachful in being born like this, yet still there is nothing that can be done about it. Hence Confucius said: "To fail to speak to someone to whom one should speak is to fail that person. To speak to someone to whom one should not speak is to fail one's words. A wise man fails neither other people nor his own words."[6]

The gentleman places the utmost value on speech. He would not exchange it even for the lord of Xia's semicircular piece of jade (*huang*) or the four-horse chariot of Tang of the Shang.[7] Nowadays, however, if a gentleman were to address a vulgar person, the latter would think that the gentleman was intending to deceive him and so consider it not worth listening to what the gentleman has to say. Does this not shame the gentleman himself and injure his way? For this reason, before the gentleman tells someone about the source of the "great fundament" and talks about the ultimate scope of human nature and rightness, he needs first to appraise what sort of aspiration that person harbors in his heart; gauge his caliber; examine his mettle; and check whether he is degenerate or weak.* Only then will the gentleman sing to him so as to observe how well he sings in response, and lead him so as to observe how well he follows. Proof that a person is able to follow and to sing in response starts in his voice, is given form in the way he looks and listens, becomes manifest on his face, and is animated in the movements of his body. Only then may he develop from what is closest and move on to more distant tasks. As his accomplishments grow more evident, his mastery grows more refined. Thereupon the gentleman sets out matters openly so as to bring him forward; shows him how to build on what he has learned so as to draw him on; draws detailed analogies so as to make him understand; cites a wide variety of accounts so as to broaden his knowledge; sets him standards so as to cor-

---

*"Great fundament": another name for the way.

之，疏煩以理之。疾而勿迫，徐而勿失，雜而勿結，放而勿逸，欲其自得之也。故大禹善治水，而君子善導人。導人必因其性，治水必因其勢。是以功無敗而言無棄也。荀卿曰：「禮恭、然後可與言道之方。辭順、然後可與言道之理。色從、然後可與言道之致。」「有爭氣者、勿與辨也[b]。」孔子曰：「惟君子、然後能貴其言，貴其色。小人能乎哉？」仲尼、荀卿先後知之。

問者曰：「或有周乎上哲之至論，通乎大聖之洪業，而好與俗士辨者何也？」

曰：「以俗士為必能識之故也。」

「何以驗之？」

「使彼有金石絲竹之樂，則不奏乎聾者之側。有

[b] The order of the two quoted sentences is reversed in *Xunzi*. My translation follows (i) *Discourses* in reading *ran* 然 where the *Xunzi* text reads *er* 而 (three occurrences); and (ii) the *Xunzi* text in reading *bian* 辯 where *Discourses* reads *bian* 辨.

rect him; and disentangles the complex so that he may keep things in order. The gentleman shows him how to act swiftly yet not in haste; how to act slowly yet not miss anything; how to deal with complicated matters without becoming enmeshed; and how to be unfettered without becoming dissolute. The gentleman wants him to be able to find the capacity within himself to do these things. Thus just as the Great Yu was skilled at controlling flood waters, the gentleman is skilled at leading men. In leading men, the gentleman acts in accordance with their natures, just as in controlling floods one acts in accordance with the configurations of the situation.* Thereby his efforts will not be defeated and his words will not be rejected. Xun Qing said: "Only when a person's ritual actions are imbued with reverence may one speak to him about the norms of the way; only when his speech shows deference may one speak to him about the principles of the way; and only when his demeanor is complaisant may one speak to him about the fine points of the way. . . . Do not engage in disputation with one who is argumentative."[8] Confucius said: "Only the gentleman is able to value his words and countenance. Is the small man?"[9] Zhongni and Xun Qing, in their respective times, knew this.

Someone asked me, "There is a person who is fully conversant with the supreme discourses of the wisest men and comprehends the grand enterprise of the great sages, yet he still enjoys engaging in disputation with vulgar persons. What do you make of this?"

I replied, "This is because he thinks the vulgar person must be able to understand the subject of the disputation."

"How can you prove this?"

"Suppose that he had bells, musical stones, string and woodwind instruments—he would not play them beside the deaf. Or suppose he had clothing embroidered with designs of

---

*Depending, that is, upon whether the nature of the flooding calls for dredging, constructing dikes, or other measures.

山龍華蟲之文，則不陳乎瞽者之前。知聾者之不聞也。知瞽者之不見也。於己之心分數明白。至與俗士而獨不然者。知分數者不明也。」

「不明之故何也？」

「夫俗士之牽達人也，猶鷃鳥之欺孺子也。鷃鳥之性善近人，飛不峻也，不速也。蹲蹲然似若將可獲也。卒至乎不可獲。是孺子之所以跔膝跛足而不以為弊也。俗士之與達人言也，受之雖不肯，拒之則無說。然而有贊焉，有和焉，若將可寤[c]。卒至乎不可寤。是達人之所以乾唇竭聲而不舍也。斯人也、固達之蔽者也，非達之達者也。雖能言之，猶夫俗士而已矣。

「非惟言也，行亦如之。得其所則尊榮，失其所則賤辱。昔倉梧丙娶妻美，而以與其兄。欲以為讓也。則不如無讓焉。尾生與婦人期於水邊。水暴至不去而死。

mountains, dragons, and the decorative fowl—he would not display them in front of a blind person. He knows that the deaf do not hear and the blind do not see. In his own mind he is perfectly clear about how to divide up regular matters. The matter of engaging the vulgar person in disputation, however, is uniquely different. The one who knows how to divide up regular matters did not understand this."

"What is the reason for this lack of clarity?"

"The way in which the vulgar person leads on the man of understanding is analogous to the way in which quail deceive young children. By their nature, quail are quite capable of drawing close to people, flying neither high nor swiftly. As they move slowly about on the ground, they seem an easy catch, and yet, right to the end, they cannot be caught. This is how young children end up scraping their knees and twisting their ankles yet still not feel hurt. When a vulgar person is talking with the man of understanding, even though he is not prepared to accept what the man of understanding has to say, he nevertheless has nothing to say to refute him. In this manner, sometimes he agrees with the gentleman and sometimes he echoes the gentleman's words. It is just as if the vulgar person were about to be enlightened, yet in the final analysis he remains incapable of enlightenment. This is how the man of understanding ends up with his lips drying out and losing his voice, yet still not giving up. He would have to be a benighted man of understanding and not a true man of understanding! Although he is able to speak about matters of understanding, he remains, after all, just a vulgar person.

A gentleman is not just talk; his deeds must also be like those of a gentleman. Whoever attains his proper place will be respected and honored; whoever fails to attain his proper place will be despised and humiliated. "In the past, Cangwu Bing took a wife, but because she was beautiful, he gave her to his elder brother.[10] It would have been better not to have deferred to him at all than to have been deferential in this manner. Wei Sheng arranged to meet his wife at the edge of a river. When the water suddenly rose, he would not leave and so drowned.[11]

欲以為信也。則不如無信焉。葉公之黨，其父攘羊而子證之。欲以為直也。則不如無直焉。陳仲子不食母兄之食。出居於陵。欲以為潔也。則不如無潔焉。宗魯受齊豹之謀。死孟摯之難。欲以為義也。則不如無義焉。故凡道、蹈之既難，錯之益不易。是以君子慎諸己以為往鑒焉。」

It would have been better not to have kept his word to her at all than to have done so in this manner. In the community of the governor of She, a father stole a sheep and his son bore witness against him.* It would have been better not to have been honest with his community at all than to have been honest in this manner. Chen Zhongzi would not eat the food provided by his mother and elder brother, and left to go and live elsewhere in Wuling.† It would have been better not to have remained pure to his principles at all than to have been pure in such a manner. Zong Lu knew of Qi Bao's plot, yet died in the troubles with Gongmeng Zhi.‡ It would have been better not to have done his duty to him at all than to have done so in such a manner. Thus while all of these ways are already difficult to walk, it is even more difficult to know when to put them to one side. Because of this the gentleman is watchful over himself and uses the past to reflect on himself."

---

* The *locus classicus* of this story is *Analects*, 13.18: "Speaking to Confucius, the governor of She said, 'In my community we have an upright man who turned in his father for stealing a sheep.' Confucius said, 'Those who are upright in my community are quite different: fathers cover up for their sons and sons cover up for their fathers.'"

† Regarding his elder brother's wealth as ill gotten, Chen Zhongzi refused to share in any part of it, and so he left the house of his mother and brother to live elsewhere in Wuling. On another occasion he unwittingly ate some of the goose that his brother had brought home. Upon learning that his brother had provided it, he regurgitated the food. Mencius concludes the anecdote by ridiculing him for being so uncompromising in his principles. See *Mencius*, 3B.10; Lau, *Mencius*, 116.

‡ Zong Lu was a charioteer for Gongmeng Zhi (the heir to the state of Wei), but had been able to enter into his service only because of Qi Bao's introduction. Because he was plotting an attempt on Gongmeng Zhi's life, Qi Bao first warned Zong Lu of his plan. Zong Lu, however, still insisted on driving the doomed chariot (the target of the ambush) because he felt that such an action would undermine the recommendation given by Qi Bao to Gongmeng Zhi which had enabled him to obtain his position as charioteer for Gongmeng Zhi. In the end, he was killed with Gongmeng Zhi. See *Zuo Commentary*, Zhao 20; Legge, *Chinese Classics*, 5:681–82.

《藝紀第七》

7. The Fundamental Principles of the Arts

The roots-branches (*benmo*) metaphor is here employed to present the relationship between virtue and the six arts as a structure-application (*tiyong*) relationship. Analogous to the roots of a tree, virtue sustains and supports man. The arts, like the branches and leaves of a tree, in turn give formal expression to man's virtue and also enable him to cultivate virtue and foster its growth. Only in this way can unadorned simplicity (*zhi*) and refinement (*wen*) be integrated as a balanced whole. Xu Gan equates the innate virtues (including humaneness, wisdom, and artistic ability) with unadorned simplicity, and external appearance, conduct, and the six arts with refinement.

Although it was the sages who had first used wisdom to create the arts, Xu Gan also refers to "artistic ability," a capacity closely aligned to the wisdom each person possesses in his or her nature. This innate ability enables people to become adept in the arts and so "highlight wisdom, … embellish ability." The "replete virtue" of the gentleman is a product both of realizing humaneness and rightness in his virtuous conduct and of realizing his artistic abilities in his virtuous conduct.[1] Conversely, artistic abilities are measured by virtuous conduct, just as the gentleman's aims correspond with his speech and countenance.

Xu Gan laments that few of those qualities which inform the genuine impulses and actuality of the six arts remain; in-

藝之興也，其由民心之有智乎？造藝者將以有理乎？民生而心知物。知物而欲作。欲作而事繁。事繁而莫之能理也。故聖人因智以造藝。因藝以立事。二者近在乎身，而遠在乎物。藝者所以旌智飾能統事御群也。聖人之所不能已也。藝者、所以事成德者也。德者、以道率身者也。藝者、德ᵃ之枝葉也。德者、人之根幹也。斯二物者不偏行，不獨立。木無枝葉則不能豐其根幹，故謂之瘣；人無藝則不能成其德，故謂之野。若欲為夫君子，必兼之乎！

先王之欲人之為君子也，故立保(民)〔氏〕，掌教六藝：一曰五禮，二曰六樂，三曰五射，

ᵃ Emending *de* 德 to *ren* 人 on the basis of the tree analogy.

stead, only their trappings are attended to. The essay concludes with a description of the merits of each of the six arts.

Did the arts arise because there is wisdom in people's hearts, or because the creators of the arts intended thereby to have ordered patterns?* When our people first came into existence, their hearts came to know things; as they came to know things, they desired to do things; as their desires to do things were realized, their affairs proliferated; and as their affairs proliferated no one was able to put them in order. Hence the sages relied on their wisdom to create the arts, and they relied on the arts to establish affairs properly. These two things [wisdom and the arts] are examples of "nearby, residing in what lies close at hand [themselves], and far away, residing in other things."[2] The arts are the means by which to highlight wisdom, to embellish ability, to bring affairs under control and to steer the masses. The sage cannot do without them. The arts are the means by which to bring one's virtue to realization in concrete affairs, while virtue is that by which one guides oneself in accordance with the way. The arts are the leaves and branches of man, while virtue is the roots and trunk of man.[3] These two things [the arts and virtue] do not function separately and are not independent of one another. If a tree is without leaves and branches, then it would be unable to make the roots and trunk thrive, and so it would be called malnourished; if men did not have the arts, then they would be unable to perfect their virtue, and so would be called uncultivated. If one wants to become a gentleman, then does one not need both to perfect one's virtue and to cultivate the arts?

Because the former kings wanted people to become gentlemen, they established the office of the palace protector and put him in charge of the teaching of the six arts: the five types of ritual, the six types of music, the five skills of archery, the

*That is, the six arts.

四曰五御，五曰六書，六曰九數。教六儀：一曰祭祀之容，二曰賓客之容，三曰朝廷之容，四曰喪紀之容，五曰軍旅之容，六曰車馬之容。大胥掌學士之版。春入學，舍采，合萬[b]舞。秋班學，合聲，諷誦講習，不解於時。故《詩》曰：「菁菁者莪，在彼中阿。既見君子，樂且有儀。」美育〔群〕[c]材，其猶人之於藝乎！

　既脩其質，且加其文。文質著然後體全。體全然後可登乎清廟，而可羞乎王公。故君子非仁不立。非義不行。非

---

[b] Following the *Zhouli* text in treating the character *wan* 萬 as excrescent.
[c] Following the *Longxi jingshe* version of *Discourses* in reading *ren* 人, not the *Concordance to Zhonglun* reading of *qun* 群.

five skills of driving a carriage, the six classes of script, and the nine ways of reckoning.[4] He was also responsible for teaching the six types of demeanor: the demeanor for offerings, the demeanor for guests, the demeanor for court, the demeanor for funerals and commemorative occasions, the demeanor for armies, and the demeanor for driving chariots and riding horses.[5] The senior dancing master was in charge of the student register. In spring, when the students began school, he would pick and choose from among them and have them dance as a group.*[6] In autumn, he would put them into different classes according to their talents and also get them to sing as a group.† The chanting, recitation, lectures and study went on without let-up throughout the year. Hence, the *Book of Odes* says:

> Luxuriantly grow the tarragon,
> In the middle of that sloping hill.
> Since I have seen my lord,
> I am delighted and of attentive demeanor.[7]

This is praising the cultivation of talent.[8] Is man's relation to the arts not like this?

If a man has already cultivated his unadorned simplicity and, moreover, enhanced it with refinement, then both refinement and unadorned simplicity will be manifest. Only then will his physical body be fully developed. Only when his physical body is made whole will he be able to ascend the ancestral temple and be recommended to the king and the highest officials. Thus if it were not for the gentleman's humaneness, he would not be able to stand upright. If it were not for his rightness, he would not be able to act. If it were not for his ability

---

*To see how well each one danced so that they could be graded and then instructed on how to dance rhythmically to music.
†To hear how well each one sang so that they could be graded and then instructed on how to sing.

藝不治。非容不莊。四者無怠，而聖賢之器就矣。《易》曰：「富有之謂大業。」其斯之謂歟！

　　君子者、表裏稱而本末度者也。故言貌稱乎心，志藝能度乎德行。美在其中，而暢於四支。純粹內實，光輝外著。孔子曰：「君子恥有其服而無其容。恥有其容而無其辭。恥有其辭而無其行。」故寶玉之山，土木必潤；盛德之士，文藝必眾。

　　昔在周公，嘗猶豫於斯矣。孔子稱「安上治民，莫善於禮」；「移風易俗，莫善於樂。」存乎六藝者，著其末節也。謂夫陳籩豆，置尊俎，執羽籥，擊鐘磬，升降趨翔，屈伸

in the arts, he would not be cultivated. If it were not for his demeanor, he would not be able to command authority.* If he cannot be faulted in these four respects, then he has the makings of a sage or worthy. The *Book of Changes* says: "To have it in rich abundance is what is meant by 'great enterprise.'"⁹ Is this not what is meant?

The gentleman is one whose fundamental and peripheral are both in due measure, having attained a balance between his interior and exterior. Hence his speech and countenance correspond with his heart's aims, and his artistic ability is measured by his virtuous conduct.¹⁰ When "beauty resides within it will flow freely to the four limbs"; when purity fills the inside, its aura will shine outward.¹¹ Confucius said: "The gentleman finds it shameful to have the attire but not the proper countenance; to have the proper countenance but not the right speech; and to have the right speech but not the behavior."¹² Hence mountains of precious gems are sure to have well-watered soil and forests.¹³ Similarly, men of standing whose virtue is replete are sure to be widely skilled in the arts.

Long ago, the duke of Zhou was undecided about how best to foster the arts.† Confucius declared that "there is nothing better than ritual to make the ruler secure and to order the people, and there is nothing better than music to change *mores* and alter customs."¹⁴ Yet all too often, those responsible for the six arts emphasize only their trivial details.¹⁵ What is meant by this is that the techniques involved in setting out *bian* and *dou*, arranging *zun* and *zu*, holding *yu* and *yue*, striking bells and stone chimes, raising and lowering the leg, taking small steps and striding, bending the arm in and out,

---

* Here simplicity is equated with humaneness and rightness (innate virtues of the nature), while refinement is equated with the arts and demeanor.
† Son of King Wu and the most famous of the Confucian paragons. For a summary account of his depiction in the early literature, see Knoblock, *Xunzi*, 2:38–42.

俯仰之數也。非禮樂之本也。禮樂之本也者，其德音乎！
《詩》云：「我有嘉賓，德音孔昭。視民不恌，君子是則是
效。我有旨酒，嘉賓式宴以敖ᵈ。」此禮樂之所貴也。

　　故恭恪廉讓、藝之情也。中和平直、藝之實也。齊敏不
匱、藝之華也。威儀孔時、藝之飾也。通乎群藝之情實者，
可與論道。識乎群藝之華飾者，可與講事。事者、有司之職
也。道者、君子之業也。先王之賤藝者，蓋賤有司也。君子
兼之則貴也。故孔子曰：「志於

ᵈ The modern text of *Odes* reads *tiao* 桃 for *tiao* 恌, *xiao* 傚 for *xiao* 效,
and *yan* 燕 for *yan* 晏.

and moving the head up and down do not constitute the roots of ritual and music.*[16] Are not the roots of ritual and music the virtuous teachings of the former kings?[17] The *Book of Odes* says:

> I have a fine guest,
> His virtuous teachings are grand and brilliant.
> He does not regard the people with disdain;
> Gentlemen take him as a their model, imitating him.
> I have fine wine,
> Which my admirable guest drinks, enjoying himself.[18]

It is this that is prized in ritual and music.

Hence being reverent, respectful, honest, and deferential are the genuine impulses of the arts, and being balanced, harmonious, even, and upright are its actuality. Maintaining an orderly appearance and an alert disposition on a consistent basis is an embellishment to the arts, and bearing an awesome demeanor on a regular basis is an adornment to the arts. One may discuss the way with someone who fully understands the impulse and actuality of the various arts. One may explain details to someone who has knowledge only of the ornaments and embellishments to the various arts. Details are the occupation of minor officials, while the way is the vocation of the gentleman.[19] Those former kings who looked down upon the arts probably did so because they looked down upon minor officials. If the gentleman fully understands both that which is fundamental to the arts and that which is ornamentation, he will be esteemed. Hence Confucius said: "Set one's aims on

---

*The *bian* was a splint bamboo basket with cover, used to contain fruit offered in worship; the *dou* was a platterlike vessel used for holding food in offerings. The *zun* was a wine vessel used in offerings; the *zu* was a stand for meat used at feasts or offerings. The *yu*, a feathered fanlike object, and the *yue*, a three-holed flute, were two props or dancing accessories, and distinguished the type of dances they were used in as being cultural (*wen*) or martial (*wu*). See *Zhouli*, 24.6a–b, and subcommentary.

道，據於德，依於仁，游於藝。」

藝者、心之使也，仁之聲也，義之象也。故禮以考敬，樂以敦愛，射以平志，御以和心，書以綴事，數以理煩。敬考則民不慢。愛敦則群生悦。志平則怨尤亡。心和則離德睦。事綴則法戒明。煩理則物不悖。六者雖殊，其致一也。其道則君子專之，其事則有司共之。此藝之大體也。

the way, support oneself on virtue, rely on humaneness, and move freely in the arts."[20]

The arts are the servants of the heart, the voice of humaneness, and the model of rightness. Thus ritual is used to perfect respect, music to encourage concern, archery to balance one's aim, driving to bring harmony to one's heart, writing to link affairs, and calculation to bring order to confusion. If respect is perfected, then the people will not be remiss; if concern is encouraged, then the myriad living forms will be content; if one's aim is balanced, then resentment and blame will disappear; if one's heart is harmonious, then there will be concord between the separate virtues; if affairs are linked, then laws and prohibitions will be clear. If confusion is brought to order, then things will not be at odds. Although each of these six is unique, they ultimately reach a state of unity. As for the way, it is the gentleman's preoccupation, while responsibility for details is shared jointly by minor officials. Such, then, are the fundamental principles of the arts.

《覈辯第八》

8. An Examination of Disputation

In Warring States times, disputation was a method of argumentation particularly associated with thinkers who in pre-Qin times were sometimes known as *bianzhe*, "those who argue out alternatives." Some of these were later retrospectively identified as *mingjia*, "school of names." The Mohists explain proper disputation in the following terms: "One calling it 'ox' and one calling it 'not ox'; this is to contend over 'that.' This being the case, they are not both appropriate. Since they are not both appropriate, one is necessarily appropriate" [Summa, A 74, Explanation]. Unlike Greek dialectic, in which further dialectical development remains an open possibility, in disputation there must be a winner. In stipulating that something must be either "X" or "not X" there is no room for shades of gray because once a standard has been established, an object either conforms to it or does not. Like the Mohists, Xunzi, and Han Fei, *mingjia* thinkers excelled in the art of disputation, but, unlike them, for the *mingjia* disputation was an opportunity to shock and disconcert by arguing for propositions that defied common sense. It was their perceived frivolous attitude to disputation—in which victory was valued above real understanding—rather than any lack of argumentative skill, which led many contemporary and later critics to hold them in disdain.

In this essay, Xu Gan argues that what in his day passed under the name of disputation was not the actuality of disputation. For him genuine disputation is about understanding.

俗士之所謂辯者，非辯也。非辯而謂之辯者，蓋聞辯之名，而不知辯之實。故目之妄也。俗之所謂辯者，利口者也。彼利口者，苟美其聲氣，繁其辭令。如激風之至，如暴雨之集。不論是非之性，不識曲直之理。期於不窮，務於必勝。以故淺識而好奇者，見其如此也，固以為辯。不知木訥而達道者，雖口屈而心不服也。

　　夫辯者、求服人心也，非屈人口也。故辯之為言別也。為其善分別事類而明處之也。非謂言辭切給而以陵[a]蓋人也。故《傳》稱《春秋》「微而顯、婉而辯」者。然則

[a] Following *Yilin*, 5.15a, in emending *qie* 切 to *jie* 捷 and *ling* 陵 to *ling* 淩.

In particular, "the gentleman engages in disputation because he wants to use it to illuminate the point of balance of the great way." Having described how proper disputation should proceed, he contrasts the gentleman's attitude to disputation with that of some of his own contemporaries who used false disputation to retain their positions by gaining a hold over the minds of stupid and ignorant people. The sole concern of false disputation is to secure "verbal submission" and not "persuading people in their hearts."

What the vulgar person calls "disputation" is not disputation at all. That he should call it disputation, though it is not, is probably because he has heard the name "disputation" but does not know its actuality. Accordingly, we regard this as preposterous. The person who is commonly referred to as "disputer" is, in fact, a glib person. The glib person readily embellishes the tone of his voice and multiplies his retorts, just like the onset of a gale, or the downpour of a rainstorm. Disregarding the inherent rights and wrongs of a situation, and not understanding the principles of truth and distortion, he fixes it so that he is never at a loss for words and works at securing certain victory. Hence those of shallow understanding who marvel at the unusual, seeing him like this insist that it is disputation. They do not realize that although those who have attained the way by virtue of their simplicity and reticence may be verbally defeated, yet in their hearts they do not give in.[1]

Disputation is about persuading people in their hearts; it is not about verbal submission. Hence disputation is to articulate distinctions, and also to separate and distinguish different categories of affairs skillfully, so as to arrange them clearly. Disputation does not mean being quick-witted in one's words and speech to talk over people's heads.[2] Hence the [*Zuo*] *Commentary* states that the *Spring and Autumn Annals* are "subtle yet evident, tactful yet discriminating."[3] Given this, then the

辯之言必約以至，不煩而諭。疾徐應節，不犯禮教，足以相稱。樂盡人之辭，善致人之志，使論者各盡得其願，而與之得解其稱也。無其名其理也不獨顯。若此則可謂辯。故言有拙而辯者焉，有巧而不辯者焉。

君子之辯也，欲以明大道之中也。是豈取一坐之勝哉！人心之於是非也，如口於味也。口者、非以己之調膳則獨美，而與人調之則不美也。故君子之於道也，在彼猶在己也。苟得其中，則我心悅焉。何擇於彼？苟失其中，則我心不悅焉。何取於此？故其論也，遇人之是則止矣。

words used in disputation should be terse and to the point, unconvoluted and informative. The tempo of their delivery needs to be modulated so that they do not transgress the ritual teachings and their import should serve as a worthy complement to those teachings. Deriving pleasure from letting the other person complete what he has to say, and being skilled at bringing forth the intention behind the other person's words enables each discussant to achieve fully his wishes, and each interlocutor to understand what the other speaker is saying. After all, without names principles would not be made individually evident.* If things are such, then it can be deemed to be disputation. Hence some speech is artless yet there is genuine disputation in it; some speech is cunning yet there is not genuine disputation in it.†

The gentleman engages in disputation because he wants to use it to illuminate the point of balance of the great way. This being so, would he be likely to look for victory in a single round of argument? The minds of people, in regard to right and wrong, are like that of taste to flavor. It is not the case that one regards only the dishes prepared by oneself as excellent and the dishes prepared by others as not so. Thus the gentleman's attitude to the way is such that when the balance resides in another, it is as though it resided in himself. So long as the balance is attained, then one's heart will be delighted by it; so why selectively distinguish it as being "his"?‡ If the balance is missed then one's heart will be saddened by it; so why insist on choosing only what is "mine"? Hence in discoursing with others, when one encounters a point on which the other person is correct, then one should stop arguing that point. If,

---

* "Names" here refers to the words used in disputation; "principles" (or more literally, "patterns") refers to the thread of reasoned argument.
† The point is that it is the use of appropriate words which constitutes proper disputation. When words are articulated into speech, the speech may be guileless, but if what is said is to the point then it will provide the vehicle for genuine disputation.
‡ And, on this ground alone, reject it out of hand.

遇人之是而猶不止，苟言苟辯，則小人也。雖美說，何異乎鵙之好鳴、鐸之喧譁哉？故孔子曰：「小人毀訾以為辯，絞急以為智，不遜以為勇。」斯乃聖人所惡，而小人以為美。豈不哀哉？

夫利口之所以得行乎世也，蓋有由也。且[b]利口者、心足以見小數，言足以盡巧辭，給足以應切問，難足以斷俗疑。然而好說而不倦，諜諜如也。夫類族辯物之士者寡，而愚闇不達之人者多，孰知其非乎？此其所〔以〕[c]無用而不見廢也，至賤而不見遺也。

先王之法：(折)〔析〕[d]言破律、亂名改作者，殺之。行僻而堅、言偽而辯、記醜而博、順非而澤者，亦殺之。為其疑眾惑民，而潰亂至道也。孔子曰：「巧言亂德。惡似而非者也。」

[b] Following *Qunshu zhiyao*, 46.15a, in reading *fu* 夫 in place of *qie* 且.
[c] Following *Qunshu zhiyao*, 46.15a, in reading *yi* 以 after *suo* 所.
[d] Not following the *Concordance to Zhonglun* in emending *zhe* 折 to *xi* 析.

however, one persists in arguing the point, speaking and disputing just to keep the argument going, then one is a small man. Although one might adduce some fine arguments, yet how different is it from the shrike's fondness for squawking, or the din of a signal bell used in battle? Hence Confucius said: "The small man thinks that slander is disputation, that a flurry of words is wisdom, and that insolence is courage."[4] These are things which the sage detests but which the small man regards as excellent. Is it not pathetic?

There is surely a reason why glibness has been able to gain currency in our age. The minds of those who are glib are perceptive enough to see minor regular principles; their verbal skills are good enough to look to make maximum use of cunning phrases; they are eloquent enough to respond to incisive questioning; and the objections they raise are convincing enough to dispel common doubts. In this manner, they enjoy discoursing tirelessly and loquaciously. Those men who "distinguish things by placing them in their proper categories and kinds" are few, while foolish, ignorant, and unenlightened men are many.[5] Yet who realizes that the many are wrong? This is why they are able to retain their positions, despite being useless and utterly contemptible.

According to the laws of the former kings, those who split words to break the written statutes or who confounded names to alter what had been recorded were put to death.[6] Those who persisted obstinately in aberrant behavior, who defended duplicitous speech with disputation, who recorded what is scandalous and spread it widely, or who followed what is wrong as if it were beneficial were also put to death.[7] This is because, by sowing doubt among the masses and confusing the people, they caused disorder to spill over the supreme way. Confucius said: "Cunning words confound virtue.... I detest something pretending to be what it is not."[8]

《智行第九》

或問曰：「士或明哲窮理，或志行純篤。二者不可兼，聖人將何取？」

對曰：「其明哲乎。夫明哲之為用也，乃能殷民阜利，使萬物無不盡其極者也。聖人之可及，非徒空行也，

9. *Wisdom and Deeds*

This essay may well be an example of what Xu Gan considered to be proper disputation. It takes the form of a dialogue with a nameless interlocutor, in which Xu Gan argues that wisdom, intellectual brilliance, and ability are fundamentally more important than being committed to aspirations (*zhi*) and deeds (*xing*), for while the former can be translated into action independently of the latter, without wisdom, brilliance, or ability, aspirations and deeds (including virtuous conduct) cannot be realized.[1] Furthermore, "the sage values the special abilities of the talented and wise: they are able to render achievements and services of benefit to the world," a theme that is also central to the series of edicts issued by Cao Cao between 203 and 217. Xu Gan discusses the examples of the sages of antiquity, Confucius's disciple Yan Yuan and Guan Zhong, to argue his case. The final part of his argument addresses the importance of intelligence and wisdom in situations in which it is necessary to adjust priorities (*quan*).

Someone asked me, "Suppose that a man of social standing could either be intelligent and wise enough to fathom all principles, or purely committed to the realization of his aspirations and deeds, but not both. Which would a sage choose?"

I replied, "Would it not be intelligence and wisdom? Only through the practical application of intelligence and wisdom can the people prosper and profit be increased, such that the myriad creatures will all realize their full capacities. That which the sage is able to attain is not due to mere empty deeds; rather,

智也。伏羲作八卦，文王增其辭。斯皆窮神知化。豈徒特行善而已乎！《易離象》稱『大人以繼明照於四方。』且大人、聖人也。其餘象皆稱君子。蓋君子通於賢者也。聰明惟聖人能盡之。大才通人有而不能盡也。《書》美唐堯，『欽』『明』為先。驩兜之舉共工。四嶽之薦鯀。堯知其行。眾尚未知。信也。若非堯則裔土多凶族。兆民長愁苦矣。明哲之功也如是，子將何從？」

或曰：「俱謂賢者耳，何乃以聖人論之？」

對曰：「賢者亦然。人之行莫大於孝，莫顯於清。曾參之孝，有虞不能易；原憲之清，伯夷不能閒。

it is due to his wisdom. Fu Xi's creation of the eight trigrams and King Wen's appending of the statements are both examples of fathoming the divine and understanding transformation.[2] Surely these are not merely and only examples of practicing goodness? The 'Commentary on the Image' of the 'Cohesion' hexagram in the *Book of Changes* says: 'The great man continues to cast his brilliance in all four directions.'[3] The great man is the sage. All of the other 'Commentaries on the Images' talk only about the 'gentleman.' This is because the 'gentleman' is equivalent to the 'worthy.' Only the sage is able to realize fully his intelligence. A greatly talented man of penetrating comprehension may possess intelligence but will be unable to realize it fully. Above all else, the *Book of Documents* praises Yao for his 'reverence and clarity of vision.'[4] When the Huan Dou recommended Gong Gong and the Four Leaders recommended Gun, Yao was well aware of the conduct of these two men, but his many officers were still not aware of their conduct and so had faith in them.*[5] Had it not been for Yao, then the presence of the many ferocious tribes on the border regions would have meant lasting sadness and suffering for the multitude of people within the borders.† Such is the merit of intelligence and wisdom. So which would you follow, intelligence and wisdom or commitment to aims and deeds?"

He said, "But what about all those who are known as worthies? Why do you discuss this matter only with regard to sages?"

I replied, "The same applies to worthies. Of man's deeds, none is greater than being filial, or more outstanding than being incorrupt. Even the sage Youyu would have been unable to look down upon Zeng Shen's filial behavior, nor would the sage Boyi have been able to find fault with Yuan Xian's incor-

---

*In contrast to Yao, that is, who did not wish to appoint them to office.
†Presumably because if Gong Gong and Gun had been successfully recommended, then they would have promoted the interests of the border tribes.

然不得與游、夏列在四行之科,以其才不如也。仲尼問子貢曰:『汝與回也孰愈?』對曰:『賜也何敢望回?回也聞一以知十,賜也聞一以知二。』子貢之行不若顏淵遠矣。然而不服其行,服其聞一知十。由此觀之,盛才所以服人也。仲尼亦奇顏淵之有盛才也。故曰:『回也非助我者也,於吾言無所不說。』顏淵達於聖人之情,故無窮難之辭。是以能獨獲亹亹之譽,為七十子之冠。曾參雖質孝,原憲雖體清,仲尼未甚嘆也。」

或曰:「苟有才智,而行不善,則可取乎?」

對曰:「何子之難喻也。水

ruptibility.* Despite this, Zeng Shen and Yuan Xian were not listed with You and Xia as exemplifying one of the four types of conduct, because their abilities were inferior.†6 When Confucius asked Zigong, 'Whom do you consider superior, yourself or Hui?' he replied, 'How would I dare to look up to Hui? Hearing one point, Hui is able to infer another nine; I, however, hear one point and can infer only a second point.'‡7 Although Zigong's deeds were far inferior to those of Yan Yuan, it was not Yan Yuan's deeds that Zigong admired; rather, it was his ability to hear one point and then infer another nine. From these examples one can see that it is abundant talent that makes others admire one. Confucius also marveled at Yan Yuan's endowment of abundant talent; thus he said: 'Hui is of no help to me. There is nothing I say in which he does not delight.'§8 Yan Yuan, having attained the emotional responses of the sage, had no disputatious words in consequence. Of all seventy disciples, he alone earned a reputation for being indefatigable. Although Zeng Shen personified filial piety and Yuan Xian embodied incorruptibility, Confucius never exclaimed in admiration as he had in the case of Yan Yuan."

The interlocutor then asked, "If a man has talent and wisdom, yet practices wrongdoing, can he be selected for office?"

I replied, "How difficult it is to instruct you! Although water can overcome fire, could one *sheng* of water be enough to pour

---

* Youyu: Style of the predynastic sage ruler, Shun. Zeng Shen (style Ziyu 子輿) was one of Confucius's best known disciples. Traditionally, he is reputed to have written *Xiaojing* (Classic of Filial Piety) and was famous for his filial piety. Yuan Xian (style Zisi 子思) was another of Confucius's best-known disciples. He was known for his integrity and modesty, and for his commitment to live by the teachings of his master, even though he lived in great poverty.
† You and Xia: Ziyou 子游 and Zixia 子夏, the styles of Yan Yan 言偃 and Bu Shang 卜商, respectively. Again, both were prominent disciples of Confucius.
‡ Zigong (style Duanmu Si 端木[沐]賜) was another of Confucius's best-known disciples. Hui: Yan Yuan.
§ Confucius's rebuke is not, of course, intended literally; he is actually proud of Yan's perspicacity.

能勝火，豈一升之水、灌一林之火哉？柴也愚，何嘗自投於井？夫君子仁以博愛，義以除惡，信以立情，禮以自節，聰以自察，明以觀色，謀以行權，智以辨物，豈可無一哉？謂夫多少之間耳。

「且管仲背君事讎，奢而失禮。使桓公有九合諸侯，一匡天下之功。仲尼稱之曰：『微管仲，吾其被髮左衽矣。』召忽伏節死難，

on a whole forest on fire?* And although Chai was stupid, he never jumped into a well.† The gentleman spreads his loving concern to others by being humane, uproots wrongdoing by acting rightly, establishes cordiality by showing good faith, modulates himself by acting in accord with ritual, examines himself closely by being keen-witted, observes people's facial expressions by being perceptive, acts expediently by planning adroitly, and distinguishes things by being wise.[9] Could he do without any one of these?‡ I say, it is only to a degree.

"Furthermore, Guan Zhong [d. 645 BCE] turned his back on his ruler to serve his ruler's enemy, and through extravagance failed to observe the rites, yet he enabled Duke Huan of Qi to achieve the merit of uniting and rectifying the whole realm by assembling the feudal lords on repeated occasions.§[10] Confucius praised him, saying: 'If it were not for Guan Zhong, today we would be wearing our hair loose and folding our upper garments on the left.'[11] Upholding his moral principles, Shao Hu died a loathsome death.‖[12] Such an act was

---

*A sheng, in Han times, was approximately two hundred cubic centimeters.
†Gao Chai 高柴 (style Zigao 子羔) was one of Confucius's disciples. *Analects*, 11.18: "Chai is stupid." There is no necessary connection between, on the one hand, a man's ability and intelligence and, on the other hand, his capacity for virtuous action. Thus in the case of Chai, even though he lacked ability and intelligence, this did not mean that he would be likely to engage in an act such as suicide (morally reprehensible because it is unfilial).
‡The point is that although the gentleman values wisdom over virtuous action when forced to choose between the two, he himself is endowed with both of these qualities, so that even if those in his employ were talented yet lacked virtue, the gentleman would not be corrupted by them.
§Guan Zhong had originally served Gongzi Jiu 公子糾, but after the latter was killed, Guan Zhong switched his allegiance to Gongzi Jiu's younger brother and enemy, Xiaobo, 小白, the future Duke Huan of Qi (r. 685–643). See *Zuo Commentary*, Zhuang 9, 8.20a; *Shiji*, 32.1485–86.
‖Originally, both Guan Zhong and Shao Hu had supported Jiu's claim to the dukedom and fled with him to Lu. Jiu's brother, Xiaobo, the future Duke Huan, however, claimed that he was the rightful ruler, and had Jiu killed. Consequently, Shao Hu took his own life.

人臣之美義也。仲尼比為匹夫匹婦之為諒矣。

　「是故聖人貴才智之特能立功立事益於世矣。如愬過多、才智少，作亂有餘，而立功不足，仲尼所以避陽貨而誅少正卯也。何謂可取乎？漢高祖數賴張子房權謀以建帝業。四皓雖美行而何益夫倒懸？此固不可同日而論矣！」

　或曰：「然則

considered to be an excellent example of a minister doing what was right in the circumstances. Confucius, however, compared it to the act of common men and women preserving a bond of faith.*

"That is why the sage values the special abilities of the talented and wise: they are able to render achievements and services of benefit to the world. If, however, a man has many faults and errors but only a little talent and wisdom, and causes excessive disturbances and renders insufficient achievements, then such are the conditions which led Confucius to avoid Yang Huo and to execute Deputy Mao.†[13] How could such a man be deemed as fit for office? On several occasions Han Gaozu relied upon Zhang Liang's tactics to establish his dynastic enterprise.‡[14] Although the four Gray Beards were of virtuous conduct, of what use were they for bringing succor?§[15] These two types of people cannot be mentioned in the same breath."

The interlocutor said, "That being the case, then what did

---

* *Analects*, 14.17: "Would Guan Zhong have behaved like common men and women, who, in order to preserve a bond of faith, take their own lives in a ditch without anybody knowing?"
† Yang Huo has traditionally been identified as Yang Hu 陽虎, a powerful official employed by the Ji 季 family who controlled the government in Lu. On the question of this identification, see Lau, *Analects*, 167–68. *Analects*, 17.1, records the incident referred to by Xu Gan as follows: "Yang Huo wanted to see Confucius, but because Confucius was not prepared to go and see him, Yang Huo sent him a present of a suckling pig. Choosing a time when Confucius knew that Yang Huo was not at home, he went to pay his respects, but by chance met him on the way."
‡ Han Gaozu: the first emperor of the Han dynasty, Liu Bang 劉邦 (r. 202–195).
§ The four Gray Beards were Master Dongyuan 東園公, Scholar Luli 甪里 (or as a variant, Jiaoli 角里), Qi Liji 綺里季 and Master Xia Huang 夏黃公. They were said to have been four old men who, in the disorder that followed in the wake of the fall of the Qin, had become recluses at Mount Shang. Liu Bang is said to have solicited their counsels, but they refused to meet him. After he became emperor, the four old men, by then in their eighties, went to serve the crown prince. See *Shiji*, 55.2044–47; *Hanshu*, 40.2033–36, 72.3056.

仲尼曰『未知焉得仁』，乃高仁耶？何謂也？」

　　對曰：「仁固大也，然則仲尼此亦有所激然，非專小智之謂也。若有人相語曰：『彼尚無有一智也，安得乃知為仁乎？』

　　「昔武王崩。成王幼。周公居攝。管、蔡啟殷畔亂。周公誅之。成王不達。周公恐之。天乃雷電風雨以彰周公之德。然後成王寤。成王非不仁厚於骨肉也。徒以不聰叡之故。助畔亂之人，幾喪周公之功，而墜文、武之業。召公見周公之既反政，而猶不知。疑其貪位。周公

Confucius mean when he said, 'Without wisdom, how can one achieve humaneness?'[16] Didn't he thus elevate humaneness over wisdom?* What do you say?"

I replied, "Humaneness is certainly important. Nevertheless, Confucius was a little agitated when he said this; he was not specifically intending to downplay wisdom. Rather, it was as if he were saying to someone, 'If you still do not have even a modicum of wisdom, how could you know how to practice humaneness?'†

"In the past, after King Wu had died, because King Cheng was still young, the duke of Zhou acted as regent.‡[17] Because Guan and Cai had incited the remaining descendants of the Yin to rebellion, the duke of Zhou had them executed.§[18] King Cheng, however, failed to understand why the duke of Zhou had so acted, and the duke of Zhou became afraid. Heaven then issued thunder, lightning, wind, and rain so as to proclaim the duke of Zhou's virtue. Only after this did King Cheng awaken to the truth.[19] It was not that King Cheng did not treat his own kinsman with humaneness and sincerity; the reason was simply that he was not intelligent and wise. Aiding those treacherous and rebellious men almost destroyed the duke of Zhou's good work and ruined the enterprise of Kings Wen and Wu. The duke of Shao had seen that the duke of Zhou had returned the running of the government to King Cheng, yet even so he still did not understand the duke of Zhou's actions and so he suspected him of coveting the king's position.||[20] Only after the duke of Zhou

*That is, although wisdom might be a necessary means, the end was humaneness.
†That is, humaneness needs at least a minimal amount of wisdom to direct its application.
‡Cheng was son of Wu and successor to the throne.
§Guan and Cai were the third and fifth sons, respectively, of King Wen, and brothers of King Wu, the second-eldest brother.
|| The duke of Shao is variously said to have been a son of King Wen or a more distant member of the Zhou clan. According to *Documents*, 18.17b, he served as guardian and was responsible for aiding King Cheng.

為之作《君奭》。然後悅。夫以召公懷聖之資，而猶若此乎！末業之士，苟失一行，而智略褊短，亦可懼矣！仲尼曰：『可與立，未可與權。』孟軻曰：『子莫執中。執中無權。猶執一也。』仲尼、孟軻可謂達於權智之實者也。

「殷有三仁：微子介於石。不終日。箕子內難而能正其志。比干諫而剖心。君子以微子為上，箕子次之，比干為下。故《春秋》大夫見殺，皆譏其不能以智自免也。

「且徐偃王知脩仁義，而不知用武。

wrote 'Lord Shi' was the duke of Shao satisfied.²¹ If even the duke of Shao can thus suspect the qualities of a sage, is it not also frightening to think what might happen should a fellow in some lowly occupation make even one mistake in his deeds when his wisdom and ability are somewhat partial and short-sighted? Confucius said: 'A man with whom one may take one's stand is not necessarily one with whom one can adjust priorities.'*²² Mencius said: 'Zimo holds on to the middle. To hold on to the middle is to lack leverage. It is like holding on to one end.'²³ Confucius and Mencius are ones who may be said to have comprehended the actuality of the wisdom of adjusting priorities.

"In the Yin dynasty there were three men who realized humaneness.²⁴ The virtue of the viscount of Wei was firmer than a rock and he acted promptly.†²⁵ Although confronted by adversity at court, the viscount of Ji was able to keep his aims squarely on course.‡²⁶ Bigan remonstrated and had his heart cut out.§²⁷ The gentleman [Confucius] considered the viscount of Wei's behavior as most exemplary, followed by that of the viscount of Ji, and Bigan's as being the least so. Hence in the *Spring and Autumn Annals*, all those counselors who were killed for remonstrating with their rulers are criticized for not using their wisdom to escape death.²⁸

"Furthermore, although King Yan of Xu [r. early seventh century BCE] knew how to cultivate humaneness and rightness, he was ignorant about the employment of arms, and so,

---

*Xu Gan is employing this *Analects* passage to make the point that while the actions of King Cheng and the duke of Shao were motivated by proper principles, they were incapable of discerning the appropriateness of the duke of Zhou's actions. This is because they lacked the wisdom needed to respond to the exigency of particular circumstances and readjust normal priorities (such as taking the exceptional step of executing Guan and Cai).
†The viscount Qi of Wei is variously identified as the elder brother of the last Shang ruler, Zhou, and as the uncle of Zhou.
‡The viscount of Ji was a relative, possibly an uncle or older brother, of the tyrant Zhou.
§Bigan was an uncle (younger brother of the father) of the tyrant Zhou.

終以亡國。魯隱公懷讓心，而不知佞偽。終以致殺。宋襄公守節，而不知權。終以見執。晉伯宗好直，而不知時變。終以隕身。叔孫豹好善，而不知擇人。終以凶餓。此皆蹈善而少智之謂也。故《大雅》貴『既明且哲、以保其身。』

　「夫明哲之士者，威而不懾，困而能通，決嫌定疑，辨物居方。穰禍於忽杪。求福於未萌。見變事則達其機。得經事則循其常。巧言不能推，令色不能移。動作可觀，則出辭為師表。比諸志行之士，不亦謬乎！」

in the end, his country was destroyed. Although Duke Yin of Lu [r. 722–712] yearned to yield the throne, he did not understand the nature of cunning words and falsehood, and so, in the end, he was killed.[29] Although Duke Xiang of Song [r. 650-637] stood by his principles, he did not understand adjusting priorities, and so, in the end, was taken as hostage.[30] Although Bozong of Jin [d. 576] was fond of being frank, he did not understand how to change with the times, and so, in the end, lost his life.[31] Although Shusun Bao [d. 538] was fond of the good, he did not understand how to select people, and so, in the end, was starved to death.[32] These are all examples of what is known as walking the path of goodness yet having little understanding. Hence, the 'Greater Court Odes' valued 'being both intelligent and wise in order to protect oneself.'[33]

"The man of social standing who is intelligent and wise is one who, when threatened, does not panic, and who, when blocked, is able to find a way through. He settles suspicions and fixes uncertainty, and 'distinguishes between things, situating them in their proper categories.'[34] He makes offerings and prays to avert misfortune when it suddenly sprouts, and looks for good fortune in places before it has yet to germinate. When he notices events in transformation, he penetrates their incipient trends; when he obtains events in their regular state, he acts in accord with their norms. Clever words are unable to sway him, just as contrived appearances are unable to move him.[35] The actions he initiates are worthy of contemplation, thus the words he utters are the mark of a teacher. Is it not absurd to compare him with a man of social standing committed to his aspirations and deeds?"

《爵祿第十》

10. Titles and Emoluments

In this essay, which again takes the form of a dialogue, Xu Gan argues that a man's reputation (*ming*) should be the visible mark and representation of his actual worth: his virtue and the meritorious services he has performed. Only when the proper conditions prevail, however, does reputation perform this role. Xu Gan appeals to the authority of the sagely paragons to justify his claim that in early times there had been a correlation between title and virtue, and also emoluments and meritorious service. Titles and emoluments serve to represent to the world that their holder is a man of virtue and merit. The bestowing of titles and ranks is regulated by the social customs and ritual behavior that are practiced in a society which is well ordered and ruled by sage rulers. When these conditions are in place, as a natural consequence, accord between name and actuality is realized. Where this accord exists, then "by observing a person's title, one [can] distinguish his virtue, and by seeing how much his emolument was, one [can] know how great a person's meritorious service had been." Leverage (*shi*), the "power vested in authority," and position (*wei*) are described as of great importance to the gentleman, for they enable rightness to be put into practice and the transforming effects of his virtue to exert a far-reaching influence on others. The essay concludes with Xu Gan's lamenting that, like Confucius and several of his disciples, he too had not been born in the right times.

或問：「古之君子貴爵祿歟？」

曰：「然。」

「諸子之書稱『爵祿，非貴也。資財、非富也。』何謂乎？」

曰：「彼遭世之亂，見小人富貴而有是言。非古也。古之制爵祿也，爵以居有德，祿以養有功。功大者〔其〕祿厚，德遠者〔其〕[a] 爵尊。功小者其祿薄，德近者其爵卑。是故觀其爵，則別其人之德也。見其祿，則知其人之功也。不待問之〔也〕。古之君子貴爵祿者，蓋以此也。非以黼黻華乎其身，芻豢之適於其口也。非以美色悅乎其目，鐘鼓之樂乎其耳也。孔子曰：『邦有道，貧且賤焉，恥也。』明王在上，序爵班祿而不以逮也。君子以為至羞，何賤之有乎？

「先王將建諸侯而錫爵祿也，

---

[a] Following *Qunshu zhiyao*, 46.15b, in reading *qi* 其 after *gong da zhe* 功大者 and again after *de yuan zhe* 德遠者.

Someone asked, "Did gentlemen in early times value titles and emoluments?"

I replied, "Yes."

"Then how do you explain that the books of the philosophers state that titles and emoluments are not of value and that goods and valuables are not wealth?"

I replied, "Living in an age of chaos, they saw petty men who were wealthy and noble and so they said these things. It was not like this in early times. In early times, titles and emoluments were regulated so that titles were bestowed to give a position to those who possessed virtue, and emoluments were granted to support those who had rendered meritorious services. On the one hand, those whose meritorious services had been great were granted a generous emolument, while those whose virtue was far-reaching were bestowed a high-ranking title; on the other hand, those whose meritorious services had been minor were granted a modest emolument, while those whose virtue was not far-reaching were bestowed a low-ranking title. For this reason, by observing a person's title, one could distinguish his virtue, and by seeing how much his emolument was, one could know how great a person's meritorious service had been—one did not have to wait to ask about them. Those gentlemen in early times who valued titles and emoluments did so because of this. They did not value them so that they might bedeck themselves with emblem-adorned robes and embroidered skirts; or to gratify their palates with fine meats; or to please their eyes with beautiful colors; or to delight their ears with the music of bells and drums. Confucius said: 'When a country has the way, being poor and holding a humble position are things to be ashamed of.'[1] When an enlightened ruler is on the throne, he awards the different ranks of title and bestows the different grades of emolument. If, however, he is not included among the recipients, the gentleman considers it to be a matter of the utmost shame. What reason is there to hold a humble position?

"When the former kings were about to invest the feudal lords and present them with their titles and emoluments, they

必於清廟之中。陳金石之樂，宴賜之禮。宗人擯相，內史作策也。其《頌》曰：『文王既勤止，我應受之。敷時繹思。我徂維求定，時周之命。於繹思！』由此觀之，爵祿者、先王之所重也，非所輕也。故《書》曰：『無曠庶官，天工，人其代之。』

「爵祿之賤也，由處之者不宜也。賤其人，斯賤其位矣。其貴也，由處之者宜之也。貴其人，斯貴其位矣。《詩》云：『君子至止，黻衣繡裳。佩玉鏘鏘，壽考不忘[b]。』黻衣繡裳、君子之所服也。愛其德，故美其服也。暴亂之君(子)[c]非

[b] The Mao version reads *xiu* 繡 for *xiu* 綉, *jiang* 將 for *qiang* 鏘, and *wang* 亡 for *wang* 忘.

[c] Translation follows *Qunshu zhiyao*, 46.16a, in reading *jun* 君, not *junzi* 君子.

invariably conducted the investiture in the ancestral temple and presented bell and stone music and a ceremonial banquet for the occasion.[2] Ancestral intendants assisted with the reception of the guests and with ceremonial procedures, and the royal secretary prepared a record of the names of the titles and emoluments to be bestowed.[3] The 'Zhou Song of Praise' says:

> King Wen labored,
> And we receive the fruit of his labors.
> He spread everywhere that abundance;
> We go and seek to establish firmly the charge of
>   Zhou.
> Oh, the abundance![4]

In view of this, it is apparent that the former kings attached great importance to titles and emoluments—they were not matters to be treated lightly. Hence the *Book of Documents* says: 'Do not wastefully appoint official positions. Man acts on behalf of heaven in carrying out its work.'[5]

"The debasement of titles and emoluments arises when those who possess them are unsuitable. If a man is looked down upon, so too will the position that he holds. Titles and emoluments become esteemed when those who possess them are suitable. If a man is esteemed, so too will be the position he holds. The *Book of Odes* says:

> The gentleman arrives.
> He wears an emblem-adorned robe and an embroi-
>   dered skirt,
> With jade pendants suspended from his waist,
>   tinkling,
> May he live on forever in our esteem.[6]

The emblem-adorned robe and embroidered skirt are what the gentleman wears. Because the people cherish his virtue, therefore they also praise his dress. It is not that the tyrannical

無此服也，而民弗美也。位亦如之。

「昔周公相王室以君天下。聖德昭聞。王勛弘大。成王封以少昊之墟。地方七百里。錫之山川土田，附庸備物，典策官司，彝器龍旗九旒，祀帝於郊。太公亮武王克商寧亂。王封之爽鳩氏之墟。東至於海，西至於河，南至於穆陵，北至於無棣。『五侯九伯，汝實征之。世祚太師撫寧東夏。』當此之時，孰謂富貴不為榮寵者乎？自時厥後，文武之教衰，黜陟之道廢。諸侯僭恣，大夫世位。爵人不以德，祿人不以功。竊國而貴者有之，竊地而富者有之。姦邪得願，仁賢失

and reckless ruler does not have such clothes, but rather that the people do not regard them as beautiful. It is the same with a man's position.

"In the past, the duke of Zhou assisted the royal house to rule the world. His sagely virtue was illustrious and renowned, and his meritorious services to the king were magnificent and great. King Cheng enfeoffed him with a domain of seven hundred square *li* centered on the old capital of Shaohao.*[7] He gave him mountains, rivers, lands and fields, dependencies, ritual objects used by rulers, books and records, ritual vessels used by officials and nine-tailed dragon banners, and the authority to sacrifice to the imperial progenitor in the suburbs.[8] Jiang Taigong aided King Wu in defeating the Shang and restoring order, so the king enfeoffed him with a domain centered on the old site where the Shuangjiu clan had been enfeoffed, which spread eastward to the sea, westward to the River, southward to Muling, and northward to Wudi.†[9] 'You indeed led punitive expeditions against the five marquises and nine earls. I grant the office of grand preceptor in perpetuity, for supporting us in pacifying the eastern regions of our county.'[10] In those times, who would have said that wealth and noble rank were not honored and cherished? Yet after this period, the teachings of kings Wen and Wu declined in influence, the paths of promotion and demotion fell into disuse, and feudal lords overstepped their position and did as they pleased, while counselors made theirs hereditary. Titles were bestowed without regard to a person's virtue, and emoluments were granted without regard to a person's meritorious services. There were cases in which men attained noble rank by stealing whole states and other cases in which men became wealthy by stealing territory. The treacherous and aberrant got what they wanted, while the humane and worthy failed to

---

*Shaohao was the dynastic appellation of Jintian 金天, a legendary ancestor/sage ruler of high antiquity.
†Jiang Taigong was also known as Grand Duke Wang 太公望 and Lü Shang 呂尚.

志。於是則以富貴相詬病矣。故孔子曰：邦無道，富且貴焉，恥也。』然則富貴美惡存乎其世也。

「《易》曰：『聖人之大寶曰位。』何以為聖人之大寶曰『位』？位也者、立德之機也。勢也者、行義之杍也。聖人蹈機握杍，纖成天地之化。使萬物順焉，人倫正焉。六合之內，各(竟)〔充〕[d]其願。其為大寶不亦宜乎！故人以無勢位為窮，百工以無器用為困。困則其資亡，窮則其道廢。故孔子栖栖而不居者，蓋憂道廢故也。《易》曰：『井渫不食，為我心惻，可用汲。王明，並受其福。』

「夫登高而建旌，則其所視者廣矣。順風而振鐸，則其所聞者遠矣。非旌色之益明，〔非〕[e]鐸聲之益遠也。

[d] Not following the *Concordance to Zhonglun* in emending *jing* 竟 to *chong* 充.

[e] Not following the *Concordance to Zhonglun* in introducing the character *fei* 非.

achieve their aims. Thereupon people were criticized for having wealth and noble rank. Thus Confucius said: 'When a country does not have the way, it is shameful to be rich and of noble rank in that country.'[11] This being the case, wealth and noble rank, excellent reputation or bad, depended on the age.

"The *Book of Changes* says, 'The sage's great treasure is position.' How is it that the sage's great treasure is said to be position? Adopting the metaphor of the loom, position is the pivot mechanism by which virtue is established, while leverage is the shuttle by which rightness is put into practice.[12] Treading on the crossboard of the loom and grasping the shuttle, the sage weaves the transformations of heaven and earth, causing the myriad things to follow in order after him, human relations to be put right by him, and all within the six enclosures to have their desires fully realized.* Is it not appropriate that it should be called a 'great treasure'? Thus the sage regards the lack of leverage and position as being in dire straits, just as the various artisans regard the lack of tools and implements as being impeded. If the various artisans are impeded, then their talents will go to waste, and if the gentleman is in dire straits, then his way will fall into disuse. Hence Confucius's itinerant existence, not ever settling in one place, was probably because he was worried that his way would fall into disuse.[13] The *Book of Changes* says: 'The well has been cleaned, yet no one drinks from it, which makes my heart sad, for people might use this opportunity to draw from it. If there were a king who was clear-sighted, then he, too, would receive its blessings.'[14]

"If one climbs to a high place and erects a flag, then the area over which it can be seen will be vast. If one strikes a bell when upwind, then the distance over which it can be heard will be far. This is not because the flag's color becomes brighter nor because the bell's sound becomes drawn out; rather, it is

---

*The six enclosures were heaven, earth, and the four directions.

所託者然也。況居富貴之地，而行其政令者也。故舜為匹夫、猶民也。及其受終於文祖，稱曰『予一人』，則西王母來獻白環。周公之為諸侯、猶臣也。及其踐明堂之祚，負斧扆而立，則越裳氏來獻白雉。故身不尊則施不光，居不高則化不博。《易》曰：『《豐》亨，無咎[f]。王假之。勿憂，宜日中。』身尊居高之謂也。

「斯事也，聖人之所務也。雖然求之有道，得之有命。舜、禹、孔子可謂求之有道矣。舜、禹得之，孔子不得之，可謂有命矣。非惟聖人，賢者亦然。稷、契、伯益、伊尹、傅說得之者也。顏淵、

[f]The two characters, *wu jiu* 無咎, do not appear in the received text of *Changes*.

because of what color and sound rely on.[15] How much more so will this principle apply if those who occupy positions of wealth and noble rank are responsible for carrying out government edicts? Thus when Shun was a commoner, he was just like other people. When he was granted rule in perpetuity in the temple of the Great Ancestors, calling himself 'I, the solitary man,' then the Queen Mother of the West came and presented a white jade bracelet as tribute.*[16] When the duke of Zhou was one of the feudal lords, he was still a subject. Yet when he acceded to the throne in the Bright Hall [as regent], standing facing south with his back to the silken screen embroidered with axe-head designs, the gentlemen of Yueshang came and presented white pheasants as tribute.†[17] Thus if one does not have an honored status, then one will not be able to bestow one's gifts extensively, and if one does not occupy a high position, then the transformations one effects will not be far-reaching. The *Book of Changes* says: 'Abundance. Endurance. There is no trouble. The king extends abundance far away. Do not worry; it is appropriate that you should be like the midday sun.'[18] This is what is meant by having an honored status and occupying a high position.

"These matters—honored status and high position—are what the sage strives for. Nevertheless, 'there is a way to be followed in seeking them and their attainment is also determined by conditions which lie beyond human control.'[19] Shun, Yu, and Confucius may all be said to have had a way to seek them. Shun and Yu's successful attainment of them and Confucius's failure to have done so may be said to have been determined by conditions lying beyond human control. This is so not only for sages; the same applies to worthies. Ji, Xie, Bo Yi, Yi Yin, and Fu Yue were successful, while Yan Yuan,

* "I, the solitary man" was the ruler's first-person term of address.
† The Bright Hall is said to have been a hall or palace used in Zhou times for many religious and civil state ceremonies, as well as the place where the Zhou king held audiences with the feudal lords.

閔子騫、冉耕、仲弓不得者也。故良農不患壖場之不修，而患風雨之不節。君子不患道德之不建，而患時世之不遇。《詩》曰：『駕彼四牡，四牡項領。我瞻四方，蹙蹙靡所騁。』傷道之不遇也。豈一世哉！豈一世哉！」

Min Ziqian, Ran Geng, and Zhong Gong were not.*[20] Hence a good farmer does not worry that the borders of his fields are not maintained, but rather that the winds and rains will not be in proper measure. Similarly, the gentleman does not worry that the way and its power are not established, but rather that he will not encounter the right times.[21] The *Book of Odes* says:

> I yoke my four stallions,
> My four stallions stretch their necks.
> I gaze to the four quarters,
> Frustrated that I have nowhere to drive.[22]

He is hurt because he has not encountered the way. Yet, is this so of one age only? Is it so of one age only?"

---

*Ji, or Hou Ji 后稷, was the title of Shun's minister of agriculture. He was also an ancestor worshiped by the Zhou. Xie was Shun's minister of education. He is also said to have been a founding father of the Shang dynasty. Bo Yi is listed, along with Ji and Xie, as someone upon whom Shun wished to bestow office. He is also said to have been the progenitor of the Qin, and to have assisted Yu in controlling the floods. Fu Yue was the sagely minister said to have rendered meritorious services to king Wu Ding 武丁 of the Yin. Yan Yuan, Min Ziqian, Ran Geng, and Zhong Gong are the four disciples listed in *Analects*, 11.3, as exemplifying virtuous conduct (*de xing*). Yan Yuan died in poverty, Min Ziqian did not take up office, and Ran Geng died of a disease. Zhong Gong, however, achieved substantive office; therefore it is not evident why Xu Gan included him in the example.

《考偽第十一》

11. *Examining Falsity*

The subject of this essay is the "falsity" of the fame seeker. "Fame seeker" is the term Xu Gan uses to describe the type of person who pursues personal fame as a means to secure position and wealth. To a significant extent, the fame seeker was the product of the excessive importance that the scholar-gentry class had come to attach to individual reputation as a result of the Eastern Han recommendation system.[1] The first part of the essay is a detailed satirical portrayal of how the fame seeker perniciously garners his reputation through duplicity. Particularly abhorrent to Xu Gan is the fame seeker's failure to match his reputation (*ming*) with his actual qualities: "If he finds himself in a situation where he can earn himself a reputation without necessarily securing an actual achievement, then he will stay put. If, however, he finds himself in a situation where he can secure an actual achievement without necessarily earning himself a reputation, then he will leave." The consequence of this breakdown between name and actuality is that "virtue is thrown into disorder" and this, in turn, through the operation of "departures from regularities (*bian shu*)," has repercussions on a cosmic scale. In the final part of the essay, Xu Gan expounds his philosophical views on the correlative relationship between name and actuality, maintaining that names have a special "natural" bond with actualities such that only the name that is inherently appropriate to a given actuality should be used with that actuality.[2]

仲尼之沒，于今數百年矣。其間聖人不作，唐虞之法微，三代之教息，大道陵遲，人倫之中不定。於是惑世盜名之徒，因夫民之離聖教日久也，生邪端，造異術，假先王之遺訓以緣飾之。文同而實違，貌合而情遠。自謂得聖人之真也。各兼說特論誣謠一世之人。誘以偽成之名，懼以虛至之謗。使人憧憧乎得亡，惙惙而不定。喪其故性，而不自知其迷也。咸相與祖述其業而寵狎之。斯術之於斯民也，猶內關之疾也。非有痛癢煩苛於身，情志慧然。不覺疾之已深也。然而期日既至，則血氣暴竭。故內關之疾，疾之中夭，而扁鵲之所甚惡也。

Confucius died several hundred years ago.* In the interim, no more sages have appeared, the laws of Tang and Yu have become effete, the teachings of the Three Dynasties have waned, the great way has fallen into decline and the point of balance for ordered human relationships has remained in an unsettled state.† Thus charlatans and seekers of undeserved fame, taking advantage of the people's long departure from the sagely teachings, have spawned aberrant heresies and created heterodox practices. They use the instructions bequeathed by the former kings to avail themselves of a facade. While externally they conform, in actuality they contravene those instructions; while in appearance they accord, yet their real condition places them at a great remove. They boast that they have attained the truth of the sages, and each of them deploys two-sided explanations and uneven arguments. They have deceived a whole era of people, enticing them with their falsely earned reputations and frightening them with fabricated slander. They cause people to become agitated about success and failure, and in their despondency, they become restless. Having lost their original natures, people are not even aware that they have fallen under their spell, and so, joining with one another, they all adopt the teachings of these men as their model, revering them and drawing close to them. The effect of these heterodox practices upon these people is similar to a the effect of a blocked artery. Because the body does not suffer any painful itching or vexations, and because the faculties will function as if still keen, one does not sense that the illness is already acute. When, the time comes, however, the blood and vital energy will suddenly cease flowing.‡³ Hence a blocked artery, once developed, will lead to premature death, and this is what those people afflicted by it detest most of all. Being such a malady, not even the

---

*The traditional date for Confucius's death is 479 BCE.
†Tang and Yu: legendary sage rulers, Yao and Shun. The Three Dynasties were the Xia, Shang, and Zhou.
‡Here vital energy refers to the activating fluids in the body.

以盧醫不能別，而遘之者不能攻也。[a]

　　昔楊朱、墨翟、申不害、韓非、田駢、公孫龍汩亂乎先王之道，譸張乎戰國之世。然非人倫之大患也。何者？術異乎聖人者易辨，而從之者不多也。

　　今為名者之異乎聖人也微，視之難見，世莫之非也。聽之難聞，世莫之舉也。何則？勤遠以自旌，託之乎疾固。廣求以合眾，託之乎仁愛。枉直以取舉，託之乎隨時。屈道以弭謗，託之乎畏愛。多識流俗之故，矙誦《詩》、《書》之文，託之乎博文。飾非

[a] Following Yu, *Zhuzi pingyi bulu*, 78, in emending the text to read *er gou zhi zhe suo shen e ye, yi Lu yi bu neng bie, er Bian Que bu neng gong ye* 而遘之者所甚惡也以盧醫不能別也而扁鵲不能攻也.

physician of Lu could diagnose it nor could Bian Que treat it.*4

In the past, Yang Zhu [fifth-fourth centuries BCE], Mo Di [fifth century BCE], Shen Buhai [fourth century (?) BCE], Han Fei [c. 280–c. 233 BCE], Tian Pian [fl. 300 BCE], and Gongsun Long [fl. first half third century BCE] brought obstruction and disorder to the way of the former kings.†5 They lied and deceived during the Warring States period. Yet despite this, no great calamity was brought to bear on ordered human relationships. Why was this? Because those methods of theirs which differed from those of the sages were easy to distinguish, hence those who followed them were few.

Nowadays, however, that which distinguishes the fame seeker from the sage is slight. Because the difference is not easy to see or to hear, nobody in the world criticizes or exposes the fame seeker. How is this so? The fame seeker exerts his energies traveling to faraway places to display himself, on the pretext of being pressingly committed to the task of remaining steadfast in his aims.6 He seeks widely for ways in which to join with the mass of people, on the pretext of being humane and caring for others. He places what is crooked over what is upright so as to secure recommendation, on the pretext of following current trends.7 He twists the way so as to quell detractions being leveled at him, on the pretext of showing fear and concern.8 Being well acquainted with the common ways of the world, he crudely recites the texts of the *Book of Odes* and the *Book of Documents*, on the pretext that he is broadening himself with culture.9 He covers up what is not right,

---

*The "physician of Lu" was a legendary physician who lived during the Spring and Autumn period. His name was Qin Yueren 秦越人 (fl. c. 500 BCE?) and he was also known as Bian Que. Given that Xu Gan distinguishes between the "physician of Lu" and Bian Que, then the Bian Que of the text would seem to refer to another physician, also known as Bian Que, who treated either the first (r. 685–643) or the second (r. 363–357?) Duke Huan of Qi.
†All famous Warring States philosophers.

而言好，無倫而辭察，託之乎通理。居必人才，遊必帝都，託之乎觀風。然而好變易姓名，求之難獲，託之乎能靜。卑屈其體，輯柔其顏，託之乎（熅）〔溫〕恭。然而時有距絕，擊斷嚴厲，託之乎獨立。獎育童蒙，訓之以己術，託之乎勤誨。金玉自待，以神其言，託之乎説道。其大抵也。

　　苟可以收名，而不必獲實，則不去也。可以獲實，而不必收名，則不居也。汲汲乎常懼當時之不我尊也。皇皇爾又懼來世之不我尚也。心疾乎內，形勞於外。然其智調足以將之。便巧足以莊之。稱託比類足以充之。文辭聲氣足以飾之。是以欲而如讓。躁而如靜。幽而如明。跛而如正。

　　考其所由來，則非堯、舜之律[b]也。核

[b] Emending *lü* 律 to *tu* 徒 on the grounds of intelligibility and parallelism with *men* 門.

saying that it is good, and in unrivaled fashion makes fine points of verbal distinction, on the pretext that he is fully conversant with the principles of things. At home he insists that he must have men of talent to associate with, and when he travels, he insists on visiting the imperial capital, on the pretext that he is observing current mores. Yet he is also fond of changing his name, making it difficult for others to track him down, on the pretext that this enables him to remain undisturbed. He scrapes and bends his frame and puts on an ingratiating countenance, on the pretext that he is being genial and courteous.[10] Yet occasionally, he also refuses to associate with certain people and he sternly cuts off his dealings with them, on the pretext of being independent. He encourages the education of the young and the novice, instructing them in his own methods, on the pretext that he is diligently guiding them. He treats himself with an exalted regard so as to give his words a godlike quality, on the pretext that he is expounding the way. Such is an overall characterization of the fame seeker.

If he finds himself in a situation where he can earn himself a reputation without necessarily securing an actual achievement, then he will stay put. If, however, he finds himself in a situation where he can secure an actual achievement without necessarily earning himself a reputation, then he will leave. In anxiety, he often fears that his contemporaries do not respect him, and, with agitation, he also dreads the prospect that in times to come he will not be honored. His heart pains within, and his physical appearance labors under the strain externally. Nevertheless, his wisdom and ability suffice to carry him through; his glib tongue and smooth talking suffice to give him a dignified air; his use of pretext and analogy suffices to pass something off as something else; and his cultivated phrases and tone of voice suffice to put a good face on matters. Thereby although avaricious, he seems deferential, although restless, he seems composed, although secretive, he seems open, and although crooked, he seems upright.

If one examines into the origins of the fame seeker, it is not found among the followers of Yao and Shun. If one checks

其所自出，又非仲尼之門也。其回遹而不度。窮涸而無源。不可經方致遠，甄物成化。斯乃巧人之雄也，而偽夫之傑也。然中才之徒，咸拜手而贊之，揚聲以和之。被死而後[c]論其遺烈，被害而猶恨己不建。悲夫！人之陷溺蓋如此乎！

孔子曰：「不患人之不己知。」[d]者，雖語我曰「吾為善」，吾不信之矣。何者？以其泉不自中涌，而注之者從外來也。苟如此，則處道之心不明，而執義之意不著。雖依先王稱《詩》《書》，將何益哉？以此毒天下之民，莫不離本趣末。事以偽成，紛紛擾擾，馳騖不已。其流于世也，至於父盜子名，兄竊弟譽，骨肉相訕，朋友相詐。此大亂之道也。故求名者，聖人至禁也。

昔衛公孟多行無禮，取憎於國人。齊豹殺之以為名。《春秋》書之曰「盜」。其《傳》曰：

[c] Following Qian's "Reading Notes" in emending *hou* 後 to *fu* 復.
[d] Following Wang Shumin 王叔岷 in reduplicating the six characters *huan ren zhi bu ji zhi* 患人之不己知. See Liang, *Zhonglun jiaozheng*, 88.

where the fame seeker has come from, it is not from among the disciples of Confucius. He is aberrant and does not constrain himself. He is dried up with no source to draw on. He cannot proceed along the correct path and journey far, nor create things and complete their transformation. He is the champion of the cunning, the hero of the impostor. Despite this, those of mediocre talent all salute and praise him, and raise their voices in song as they follow his lead. After the fame seeker meets his death, they continue to discuss his "brilliant legacy"; and even when he is killed they regret that they have not been able to save him in time. How pitiful! How is that people should be so deluded?

Confucius said: "Don't worry about the failure of others to recognize your worth."[11] As for someone who does worry about the failure of others to recognize his worth, even if he were to say to me, "I am a practitioner of good deeds," I would not believe him. Why? Because his fountain does not issue from inside and that which directs its flow is external. If one is like this, then the resolve to situate oneself in the way will not shine forth, and the intention to uphold rightness will not be manifest. Although he may comply with the former kings' precedents of extolling the *Book of Odes* and the *Book of Documents*, to what avail is it? All such ways of poisoning the people of the realm reject what is fundamental and hasten towards the peripheral. Affairs are thus brought about by contrivance and so confusion and disorder are spurred on unceasingly. When such practices become prevalent in the world, it leads to fathers stealing their sons' names, elder brothers stealing their younger brothers' reputations, family members deceiving one another, and friends cheating one another.[12] This is the way of great chaos. It is for these reasons that sages absolutely prohibited fame seekers.

In the past, Gongmeng Zhi of Wei frequently acted without propriety and so incurred the hatred of the people of Wei. Qi Bao killed him so as to become famous, and the *Spring and Autumn Annals* recorded him as being a bandit.[13] The *Zuo Commentary* says:

「是故君子動則思禮，行則思義。不為利回，不為〔不〕ᵉ義疚。或求名而不得，或欲蓋而名章。懲不義也。齊豹為衛司寇，守嗣大夫。作而不義，其書為『盜』。邾庶其、莒牟夷、邾黑肱以土地出。求食而已。不求其名。賤而必書。此二物者、所以懲肆而去貪也。若艱難其身，以險危大人，而有名章徹，攻難之士將奔走之。若竊邑叛君以徼大利而無名，貪冒之民將寘力焉。是以《春秋》書齊豹曰『盜』，三叛人名，以懲不義。數惡無禮，其善志也。」

　　問者曰：「齊豹之殺人以為己名，故仲尼惡

ᵉ Not following the *Concordance to Zhonglun* in introducing this *bu* 不 character.

It is for this reason that a gentleman is mindful of propriety in his actions and mindful of rightness in his behavior. He does not take a crooked course for gain nor bring shame in order to do what is right. Some people seek to become famous yet fail, while others want to remain obscure yet their names stand out. This is to punish improper behavior. Qi Bao was minister of justice in Wei, a rank he had inherited from his forebears who had been counselors. Because he acted improperly, he was recorded as a bandit. Shu Qi of Zhu, Mou Yi of Ju, and Hei Gong of Zhu, ceding territory, left their states.*[14] Their purpose was to seek only emoluments, not fame. Although their ranks were humble, their names nevertheless had to be recorded. These two object lessons serve to punish wanton behavior and to get rid of greed. If men can have their names made famous by placing their own lives at risk in their attempts to threaten the lives of their superiors, then those fellows who undertake hazardous affairs would rush to follow them.†[15] If men can make great profit by stealing towns and betraying their rulers, and yet not be named for their crimes, then covetous and audacious people would apply their energies to such undertakings. Therefore the *Spring and Autumn Annals* records Qi Bao, calling him a bandit, and the names of the three traitors, so as to punish improper behavior. To set out examples in which negative judgment is passed on impropriety is the excellent intention of this method of recording.[16]

A questioner asked me, "Qi Bao killed for the sake of earning a reputation for himself, thus Zhongni passed negative

*Shu Qi was a counselor from the state of Zhu, Mou Yi was a counselor from the state of Ju, and Hei Gong was a counselor from the state of Zhu; all fled, ceding territory to Lu.
†Qi Bao was responsible for the assassination of Gongmeng Zhi, the rightful heir to the Wei state.

而『盜』之。今為名者豈有殺[f]之罪耶？」

　　曰：「《春秋》之中，其殺人者不為少，然而不盜不已[g]。聖人之善惡也，必權輕重，數眾寡以定之。夫為名者、使真偽相冒，是非易位，而民有所化。此邦家之大災也。殺人者、一人之害也。安可相比也。然則何取於殺人者以書盜乎？荀卿亦曰：『盜名不如盜貨。』鄉愿亦無殺人之罪也，而仲尼惡之。何也？以其亂德也。今偽名者之亂德也，豈徒鄉愿之謂乎？萬事雜錯，變數滋生。亂德之道，固非一端而已。《書》曰：

---

[f] Following Yu, *Zhuzi pingyi bulu*, 79, in adding the character *ren* 人 between *sha* 殺 and *zhi* 之.

[g] The text is corrupt at this point. Following Yu, *Zhuzi pingyi bulu*, 79, in emending *bu dao bu yi* 不盜不已 to *bu shu dao* 不書盜.

judgment on him and classed him as a bandit.* You surely don't mean that those who presently strive for fame might also have committed the crime of murder?"

I replied, "During the Spring and Autumn period there were many murderers, but they were not recorded as bandits. In determining that which he finds commendable and that which he find detestable, the sage must weigh up what is important and what is not, and calculate what is so in the majority of cases and what is the exception. Because fame seekers cause truth and falsehood to appear as their opposites, and right and wrong to change places, this influences the people. This is a great calamity for the state. A murderer harms but one person—how can he be put on a par with the fame seeker? This being so, then what would be the point of recording as bandits only those selected from among murderers? Why should those who are recorded as bandits be given preference over murderers? Xun Qing also said, 'Those who steal names are worse than those who steal goods.'[17] The 'honest villagers' did not murder anyone, either, yet Zhongni despised them. Why? Because they threw virtue into disorder.† Nowadays, are the only fame seekers who throw virtue into disorder the honest villagers? The myriad affairs are complex and interwoven; when departures from regularities proliferate, it is certain that the path leading to the disordering of virtue has more than only one starting point. The *Book of Documents* says: 'He

---

*Confucius (Zhongni) was traditionally believed to have edited the *Spring and Autumn Annals*.
†*Analects*, 17.13: "The Master said, 'The honest villagers are the thieves of virtue.'"Also *Mencius*, 7B.37, Lau, *Mencius*, 203, mod: "'What sort of man can be described as an honest villager?' '[The man who says,] "What is the point of having such great ambition? Their words and deeds take no notice of each other, and yet they keep on saying, 'The ancients! The ancients!' Why must they walk along in such solitary fashion? Being in this world one must behave in a manner pleasing to this world. So long as one is good, it is all right." Cringingly, he tries in this way to please the world. Such is the honest villager.... Confucius said, "I dislike the honest villager, fearing that he will confound virtue."''"

『靜言庸違，象恭滔天。』皆亂德之類也。《春秋外傳》曰：『姦仁為佻，姦禮為羞，姦勇為賊。』夫仁、禮、勇，道之美者也。然行之不以其正，則不免乎大惡。故君子之於道也，審其所以守之，慎其所以行之。」

問者曰：「仲尼惡殁世而名不稱。又疾偽名。然則將何執？」

曰：「是安足怪哉？名者、所以名實也。實立而名從之，非名立而實從之也。故長形立而名之曰長，短形立而名之曰短。非長短之名先立，而長短之形從之也。仲尼之所貴者，名實之名也。貴名乃所以貴實也。夫名之繫於實也，猶物之繫於時也。物者、春也吐華。夏也布葉。秋也凋零。冬也成實。斯無

never agrees with good words, and although he appears respectful, he slights even heaven!'[18] All those who are like this fall into the category of 'confounders of virtue.' *The Outer Commentary to the Spring and Autumn Annals* says: 'To offend against humaneness is stealing, to offend against propriety is shameful, and to offend against courage is theft.'[19] Humaneness, propriety, and courage are all excellent virtues of the way. If, however, they are not practiced correctly, then great wrongdoing will be unavoidable. Thus the gentleman must be cautious how he safeguards the way and be careful how he applies it."

My interlocutor asked me, "On the one hand, Zhongni hated the prospect of dying without having established a reputation.*[20] Yet on the other hand, he despised false reputations.[21] This being so, then how could such a position be upheld?"

I replied, "What is so strange about that? A name is that which is used to name an actuality. When an actuality has been established, its name follows after it; it is not the case that a name is established and then its actuality follows after it. Thus if a long shape is established, then it will be named 'long,' and if a short shape is established, then it will be named 'short.' It is not the case that the names 'long' and 'short' are first established and then the long and short shapes follow after them. That which Confucius valued were those 'names' which truly name actualities. In so valuing names, he thereby valued actualities. Names are tied to actualities just as plants are tied to the seasons. In spring, plants blossom into flower; in summer, they are covered in leaves; in autumn, their foliage withers and falls; and in winter, they bear fruit. This is not

---

* For example, *Analects*, 15.19. This passage can be interpreted in at least two ways. Waley's translation (*Analects*, 197) captures the sense inferred by Xu Gan's interlocutor: "A gentleman has reason to be distressed if he ends his days without making a reputation for himself." Alternatively, the following translation captures the sense in which Xu Gan evidently understood the passage: "The gentleman is distressed at the prospect of dying and leaving behind him a name that does not match his actual qualities."

為而自成者也。若強為之，則傷其性矣。名亦如之。故偽名者皆欲傷之者也。人徒知名之為善，不知偽善[h]者為不善也。惑甚矣。

　　求名有三：少而求多，遲而求速，無而求有。此三者不僻為幽昧。離乎正道，則不獲也。固非君子之所能也。君子者、能成其心。心成則內定。內定則物不能亂。物不能亂則獨樂其道。獨樂其道則不聞為聞，不顯為顯。故《禮》稱『君子之道，闇然而日彰[i]；小人之道，的然而日亡。君子之道，淡而不厭，簡而文，溫而理，知遠之近，知風之自，知微之顯，可與入德矣。』『君子之[j]不可及者，其惟人之所不見乎。』夫如是者，豈將反側於亂世，而化庸人之未稱哉！」

[h] Emending *shan* 善 to *ming* 名 on the grounds of parallelism and coherence.
[i] Standard modern versions of *Liji* read *zhang* 章 not *zhang* 彰.
[j] Standard modern versions of *Liji* read *suo* 所 not *zhi* 之.

acting for a specific purpose and yet the plant completes itself. If a plant is forced, its natural tendencies will be harmed. It is the same with names. Hence false name makers are all those who would harm names. People are aware only of the good that is done by names and are ignorant of the bad that is done by false name makers. They are greatly confused!

"There are three types of fame seekers: those with little who seek more, those for whom it is slow in coming and seek faster results, and those lacking fame who seek to have it. These three do not avoid being benighted. Because they have departed from the proper way, they will not be successful. Fame seeking is certainly not something the gentleman is capable of doing. Rather, the gentleman is capable of perfecting his heart and mind. When his heart and mind are perfected, then internally he becomes settled. When internally he is settled, then things cannot upset him. When things cannot upset him, then he takes singular pleasure in his way. When he takes singular pleasure in his way, then what is not renowned becomes renowned, and what is not eminent becomes eminent. He regards a lack of renown as renown and obscurity as eminence. Hence the *Book of Rites* says: 'The way of the gentleman is dim, yet grows more conspicuous by the day. The way of the small man is dazzling, yet fades by the day. The way of the gentleman is plain yet not tiresome, simple yet cultivated, understated yet patterned. He knows that what has a far-reaching effect starts with what is nearby; that tendencies derive from an origin; and that the subtle will become manifest. One may join such a person in entering into virtue.[22] … Is it only in regard to those aspects into which people do not see that the gentleman excels?'[23] Would one such as this be likely to engage in underhand practices in a chaotic world so that he might change from being an unheard of ordinary person?"

《譴交第十二》

民之好交游也，不及聖王之世乎？古之不交游也，將以自求乎？

昔聖王之治其民也，任之以九職，（斜）〔糾〕之

12. A Rebuke of Social Connections

In this essay—which largely takes the form of a response to a nameless interlocutor—Xu Gan criticizes the late Eastern Han practice of traveling far and wide to make social contacts and to form cliques. It may have been this essay that the author of the preface to *Discourses* had in mind when he wrote of the situation during the last years of Emperor Ling's (r. 168–89) reign: "The younger male members of families who held official positions formed cliques that were aligned to powerful families. They formed social connections with these families and supported them so that they might make a name for themselves, and they vied to better one another in the attainment of ranks and titles."[1]

Having criticized his interlocutor for failing to distinguish between the name and actuality of social intercourse, Xu Gan proceeds to distinguish between an idealized type of social intercourse purportedly practiced in antiquity and the one practiced during the reigns of emperors Huan (r. 146–68) and Ling. He describes in considerable detail those distorted features of its contemporary practice that he found most objectionable.

Is people's fondness for traveling around to form social connections less now than in the time of the sage kings? Or is it the case that in antiquity people did not travel around to form social connections, instead seeking support from within themselves?

In the past, so as to give order to their people, the sage kings assigned them the nine occupations, disciplined them

以八刑，導之以五禮，訓之以六樂，教之以三物，習之以六容。使民勞而不至於困，逸而不至於荒。當此之時，四海之內，進德脩業，勤事而不暇，詎敢淫心舍力、作為非務以害休功者乎？

自王公至於列士，莫不成正畏，相厥職有恭，不敢自暇自逸。故《春秋外傳》曰：「天子大采朝日，與三公、九卿祖識地德。日中考政，與百官之政事。師尹惟[a]旅、牧相宣序民事。少采夕月，

[a] Standard modern versions of the *Guoyu* text read *wei* 維 not *wei* 惟.

with the eight punishments, guided them with the five types of ritual, instructed them in the six types of music, taught them the three matters, and trained them in the six types of countenance.*[2] This led to the people's working hard, but not to the point of exhausting themselves, and also to their relaxing, yet not to the point of dissipation. In those times people everywhere "advanced virtue, applying themselves to their professions"; being diligent in their work, they had no time to waste.[3] Would they have dared to become dissolute in their hearts, abandon their efforts, engage in matters that were not their concern and so damage work of great merit?

From the son of heaven and the feudal lords to the various lower-ranking members of the bureaucracy, all "brought things to their proper completion and respected those who assisted them. They performed the duties of their offices with respect. They did not dare to be indolent or seek their own ease."[4] Thus the *Outer Commentary to the Spring and Autumn Annals* says:

> [On the morning of the spring equinox], the Son of Heaven would wear his five-colored ceremonial costume and pay homage to the sun. With the three dukes and the nine chamberlains, he learned to understand the earth's potency.†[5] During the day, he would inspect the affairs of state. Together with the various officials he would work on matters of government. Senior officials would assemble the various lower-ranking members of the bureaucracy, regional representatives, and the administrators who served the feudal lords, comprehensively putting the affairs of the people in order.[6] At sunset on the autumnal equinox he would make offerings

---

* Five types of ritual and six types of music: see "The Fundamental Principles of the Arts," note 4. The "three matters" is another name for the "three teachings," and the "six types of countenance" is another name for the "six types of demeanor." See "The Fundamental Principles of the Arts."
† "The earth's potency" refers to the various forms of life on earth.

與太史、司載（紏）〔糾〕虔天刑。日入監九御，〔使〕[b]潔奉禘、郊之粢盛。而後即安。諸侯朝修天子之業命。畫考其國職。夕省其典刑。夜警其[c]百工，使無慆淫。而後即安。卿大夫朝考其職。畫講其庶政。夕序其業。夜庀其家事。而後即安。士朝而[d]受業。畫而講貫。夕而習復。夜而計過無憾。而後即安。」正歲使有司令於官府曰：「各修乃職，考乃法，備乃事，以聽王命。其有不恭，則邦有大刑。」

由此觀之，不務交游者，

[b] Following the *Guoyu* text in inserting the character *shi* 使.
[c] The *Guoyu* text reads *jing* 儆 for *jing* 警, and does not have the character *qi* 其 following *jing*.
[d] The character *er* 而 does not appear in the *Guoyu* text.

to the moon, wearing his three-colored ceremonial costume, and, together with the grand scribe and the astronomer official, he would reverently and respectfully study the celestial models.[7] After nightfall, he would inspect the nine groups of nine secondary concubines and have them cleanse and fill the vessels used to make offerings of grain in the imperial progenitor ceremony and in the suburban sacrifice. Only then would he rest. In the early morning, the feudal lords would deal with the business and orders entrusted to them by the Son of Heaven. During the day, they would examine the various offices in their states. In the evening, they would review the statutes and standards. At night, they would caution the various officers, and so have them free of insolence and debauchery. Only then would they rest. In the early morning, the highest officials would examine the affairs of their office. In the day they would thoroughly investigate various administrative matters. In the evening they would put their tasks in order. At night they would take care of family matters. Only then would they rest. In the early morning the lower-ranking members of the bureaucracy would receive their tasks. In the daytime they would thoroughly attend to their regular duties. In the evening they would repeatedly practice their skills. At night they would reckon their mistakes. Only if there were no matters for regret would they then rest.[8]

In the first month of the year, minor officers would be dispatched to the various offices to announce a decree. The decree said: "In order to obey the king's commands, each of you is to conduct the affairs of your office, thoroughly examine your models, and fully execute the tasks you undertake. If you fail to respect the king's commands, the state has severe punishments."[9]

In the light of this, the reason that people did not devote their energies to traveling around to form social connections

非政之惡也。心存於職業而不遑也。且先王之教，官既不以交游導民。而鄉之考德，又不以交游舉賢。是以不禁其民，而民自舍之。及周之衰，而交游興矣。

問者曰：「吾子著書，稱君子之有交，求賢交也。今稱交非古也。然則古之君子無賢交歟？」

曰：「異哉！子之不通於大倫也。若夫不出戶庭，坐於空室之中，雖魑魅魍魎，將不吾覯，而況乎賢人乎？今子不察吾所謂交游之實，而難其名。名有同而實異者矣。名有異而實同者矣。故君子於是倫也，

was not that poor government of the day made it a perilous undertaking, but rather that they were preoccupied with the duties of office— they simply had no leisure for such pursuits. Furthermore, in the teachings of the former kings, the officials did not lead the people by setting an example of traveling around to form social connections. Similarly, when local communities examined the moral qualities of candidates for recommendation, they did not select men of worth for recommendation on the strength of their having engaged in this activity. This is an example of where the people were not forbidden to engage in the practice, yet they chose to reject it anyway. It was not until the decline of the Zhou dynasty that traveling around to form social connections became popular.

A questioner asked, "My good sir, you have written a book in which you state that in his social intercourse a gentleman seeks to mix with men of worth and that what today is called social intercourse is not the same as the social intercourse of antiquity. Doesn't this then mean that in antiquity gentlemen did not mix with men of worth?"*

I replied, "How strange, sir, that you have not grasped the main principle of what I'm saying! If a man were not to leave his house but just sat inside an empty room, then even spirits of the mountains and streams would not get to see him, much less worthies.[10] Now, sir, you have raised an objection relating to the name without examining into the actuality of what I have called 'traveling around to form social connections.'†

There are cases where the name is the same but the actuality is different, and there are cases where the name is different but the actuality is the same. Thus, in regard to the principle we

---

*The implication is that contemporary gentlemen engaged in social intercourse with men of worth.

†Actually, the interlocutor's objection relates to the term *jiao* 交 (social intercourse) rather than *jiao you* 交游, "traveling around to form social connections." It will be noted that Xu Gan consistently uses the term *jiao you* in a pejorative sense, while the term *jiao*, when referring to its practice in antiquity, has no pejorative sense.

務於其實，而無譏其名。吾稱古之不交游者，不謂嚮屋漏而居也。今之好交游者，非謂長沐雨乎中路者也。

「古之君子因王事之間，則奉贄以見其同僚及國中之賢者。其於宴樂也，言仁義而不及名利。君子未命者，亦因農事之隙，奉贄以見其鄉黨同志。及夫古之賢者亦然。則何為其不獲賢交哉？非有釋王事、廢交[c]業、遊遠邦、曠年歲者也。故古之交也近，今之交也遠。古之交也寡，今之交也眾。古之交也為求賢，今之交也為名利而已矣。

「古之立國也有四民焉。執契脩版圖，奉聖王之法，治禮義之中，謂之士。竭力以盡地利，謂之農夫。審曲直形勢，飭五材以別民器，謂之百工。

[c]Following the *Jianan qizi ji* version of *Discourses* in emending *jiao* 交 to *ben* 本.

have been discussing, the gentleman concerns himself with the actuality rather than criticize the name. In saying that in antiquity people did not travel around to form social connections, I did not mean that they lived hidden away, facing an obscure corner of the house; and in saying that today people are fond of traveling around to form social connections, I did not mean that they are frequently soaked from being out on the road all the time!

"Taking advantage of breaks in the course of carrying out the king's affairs, gentlemen of antiquity would take gifts and visit fellow officials and worthies within their state. When feasting and merrymaking, they would discuss humaneness and rightness and not touch upon matters of fame and profit. As for those gentlemen who were not yet engaged in office, they would take advantage of breaks in farming to take gifts and visit those men in the local district who shared common aspirations. Similar opportunities were available to the worthies of antiquity. Thus how could it be that in antiquity gentlemen did not get to mix with men of worth? There were none, however, who set aside the king's business, neglected farming, traveled to distant states or wasted their years in idle pursuit. In antiquity social intercourse was practiced locally, but today it is practiced faraway; in antiquity those with whom one engaged socially were few, but today they are many; in antiquity the purpose of social intercourse was to seek the company of men of worth, but today it is merely for the sake of fame and advantage.

"When a state was established in antiquity, it had four types of people in it. Those who held official credentials dealt with issues concerning household registers and territorial maps, respectfully upheld the king's laws, and maintained the balance of ritual decorum and appropriateness were called members of the bureaucracy. Those who exerted their energies exploiting the benefits to be yielded from the earth were called farmers. Those who had full knowledge of curved and straight forms and fashioned the five materials so as to differentiate the utensils used by the people were called the various artisans.[11] Those

通四方之珍異以資之，謂之商旅。各世其事，毋遷其業。少
而習之，其心安之，則若性然而功不休也。故其處之也。各
從其族，不使相奪。所以一其耳目也。不勤乎四職者，謂之
窮民[f]，役諸圍土。凡民出入行止，會聚飲食，皆有其節。不
得怠荒，以妨生務，以麗罪罰。

「然則安有群行方外，而專治交游者乎？是故五家為比，
使之相保。比有長。五比為閭，使之相(憂)〔受〕[g]。閭有胥。
四閭為族，使之相葬。族有師。五族為黨，使之相救。黨有
正。五黨為州，使之相賙。州有長。五州為鄉，使之相賓。
鄉有大夫。必有聰明慈惠之人，使各掌其鄉之政教、禁令。

「正月之吉，受法

---

[f] Following *Zhouli*, 34.14a, in reading *pi min* 罷民 rather than *qiong min* 窮
民. See Yu, *Zhuzi pingyi bulu*, 79.
[g] Following *Zhouli*, 10.22b, in emending *you* 憂 to *shou* 受.

who exchanged rare and precious goods from the four corners of the earth in order to supply them were called itinerant traders. Each passed his occupation on to the next generation; they did not change their vocations. Having practiced their particular occupations from when they were young, their hearts became so familiar with those occupations that it was as if it was their nature to be so, and so they were unremitting in their efforts. Accordingly, they remained in those occupations. Because each group followed their clans in a particular occupation, the situation did not arise where one group tried to take over another's occupation. In this way, their ears and eyes became as one. Those who did not work hard in one of the four occupations were called good-for-nothings and were put in jail. All matters relating to the people's movements, assemblies, and eating and drinking practices were subject to standards. They were not allowed to become idle or wasteful, which would interfere with the task of production and so lead to sentencing and punishment being brought to bear upon them.

"This being so, how could there possibly have arisen a situation in which masses of people traveled outside their own locale specifically for the purpose of forming social connections? Hence, five families formed a neighborhood so as to make them look after one another. Each neighborhood had a head. Five neighborhoods formed a village so as to make them watch over one another. Each village had an assistant. Four villages formed a precinct so as to make them bury one another's dead. Each precinct had a mentor. Five precincts formed a ward so as to make them help one another. Each ward had a head. Five wards formed a township so as to make them aid one another. Each township had a township head. Five townships formed a district so as to make them honor each other's men of worth. Each district had a counselor.[12] The office of counselor had to be filled by intelligent and compassionate men, as each of them was made responsible for political decrees, ritual teachings, prohibitions and edicts in his district.

"On New Year's Day, having received copies of the laws

于司徒。退而頒之于其州黨族閭比之群吏。使各以教其所治之民。以考其德行，察其道藝。以歲時登其大夫[h]，察其眾寡。

「凡民之有德行、道藝者，比以告閭，閭以告族，族以告黨，黨以告州，州以告鄉，鄉以告[i]。民有罪奇衺者，比以告，亦如之。有善而不以告，謂之蔽賢。蔽賢有罰。有惡而不以告，謂之黨逆。黨逆亦有罰。故民不得有遺善，亦不得有隱惡。

「鄉大夫三年則大比而興賢能者。鄉老及鄉大夫群吏獻賢能之書於王。王拜受之。登於天府。

「其爵之命也，各隨其才之所宜。不以大司小，不以輕任重。故《書》曰：『百僚師師，百工

[h] Following *Zhouli*, 12. 1b, in emending the *da fu* 大夫 to *fu jia* 夫家. This description of the duties of the counselor is based on *Zhouli*, 12.1a–1b.
[i] Following Ikeda, "Chūron kōchū (3)," 120, n.18, in appending the two characters *si tu* 司徒.

from the minister of education, the counselors would retire and proclaim these laws to the assembled local agents from the townships, wards, precincts, villages, and neighborhoods. They would then have each of these local agents instruct the people they governed in the laws. Through the examination of the people's virtuous conduct, their accomplishments in the teachings of the former kings and the six arts could be ascertained, and by a quarterly census of families, the size of the population could be ascertained.[13]

"In all cases where, among the people, there were individuals of virtuous conduct and those accomplished in the teachings of the former kings and the six arts, the neighborhoods would report them to the village; the villages would report them to the precinct; the precincts would report them to the ward; the wards would report them to the township; the townships would report them to the district; and the districts would report them to the minister of education. If there were people who had committed crimes or who had engaged in strange or perverse practices, neighborhoods would also report them in a similar manner. If some good deed was known of but not reported, this was known as 'concealing the worthy,' which was a punishable offence. If some wrongful deed was known but not reported, this was known as 'conspiracy,' which was also a punishable offence. Thus people could afford neither to omit reporting good deeds nor to conceal wrongful deeds.

"Every three years the counselors of the various districts would conduct a major selection so as to foster the worthy and talented. The district elders as well as the counselors and assembled minions would present a record of the worthy and talented to the king. The king would respectfully receive it and register it with the keeper of the temple treasures.[14]

"Each of the titles conferred by the king's mandate followed in accord with the appropriate level of the conferee's ability. Those of formidable ability were not put in charge of petty matters, nor were those of negligible ability made responsible for weighty matters. Hence the *Book of Documents* says: 'If the various officials learn from one another, then they can all per-

惟時。』此先王取士官人之法也。

「故其民莫不反本而自求。慎德而積小。知福祚之來不由於人也。故無交游之事，無請託之端。心澄體靜，恬然自得。咸相率以正道，相屬以誠愨。姦說不興，邪陂自息矣。

「世之衰矣，上無明天子，下無賢諸侯。君不識是非，臣不辨黑白。取士不由於鄉黨，考行不本於閥閱。多助者為賢才，寡助者為不肖。序爵聽無證之論，班祿采方國之謠。民見其如此者，知富貴可以從眾為也，知名譽可以虛譁獲也。乃離其父兄。去其邑里。不脩道藝。不治德行。講偶時之說。結比周之黨。汲汲皇皇，無

form the duties of their office with excellence.'[15] This was the way the former kings selected their officials.

"Thus all the people in those times returned to that which was fundamental, seeking for it within themselves. By being careful in matters of virtue, they accumulated small things. They knew that the source of blessings did not come from other people. Hence there was no such business as traveling around to form social connections and no beginning of people asking others for favors and requests. People's hearts were pure and their bodies were composed. They were content to find it within themselves.* They led one another by way of the proper path, and encouraged one another by being sincere and honest. Slander did not arise and aberrance and perversity died out by themselves.

"When an age is in decline, above there is no enlightened emperor and below no worthy feudal lords. Rulers do not distinguish between what is right and what is wrong, and subordinates do not differentiate between black and white. The selection of men for recommendation is not conducted through the village communities, and the examination of a man's conduct is not based on his achievements and experience.[16] Those with many supporters become the 'worthy and talented,' while those with few supporters become the 'good-for-nothings.'[17] The allotment of titles is based on what is heard in unsubstantiated talk and the issuing of emoluments is based on rumors from the four corners of the country. Those people who see that this is how things are become aware that wealth and position can be brought about by doing as the majority do, and that reputation is obtainable by deceit. Thereupon such people leave their fathers and elders, depart their villages, cease cultivating the way and the six arts, and no longer practice virtuous behavior. They discuss fashionable issues and band together in cliques for selfish purposes. Frantic and frenzied, not

---

* "It" here refers to that which is "fundamental": the accumulation of virtue through the cultivation of one's moral nature.

日以處。更相歎揚,迭為表裏。

　「橋枒生華,憔悴布衣,以欺人主,惑宰相,竊選舉,盜榮寵者,不可勝數也。既獲者賢己而遂往。羨慕者並驅而追之。悠悠皆是。孰能不然者乎?

　「桓靈之世其甚者也。自公卿大夫、州牧郡守,王事不恤,賓客為務。冠蓋填門,儒服塞道。飢不暇餐,倦不獲已。殷殷沄沄,俾夜作晝。下及小司,列城墨綬,莫不相商[j]以得人,自矜以下士。

　「星言夙駕,送往迎來。亭傳常滿,吏卒傳問。炬火夜行,闇寺不閉。把臂捩[k]腕,扣天矢誓。推託恩好,不較輕重。文書委於官曹,繫囚積於囹圄,而不遑省也。詳察其為也,非欲

[j] Following Yu, *Zhuzi pingyi bulu*, 80, in emending *shang* 商 to *gao* 高.
[k] Following the *Jianan qizi ji* version of *Discourses* in reading emending *lie* 捩 to *e* 扼.

stopping even for a day, they take turns in singing each other's praises.

"There are countless numbers of evil men living in the lap of luxury, while ordinary people are suffering, as they deceive the ruler, confuse the ministers of state, usurp places rightfully belonging to others in the recommendation system, and steal honor and favor. Those who succeed regard themselves as men of worth and so onward they go. Those envious of them hasten one another on in pursuit. There were multitudes doing it—who could avoid following suit?

"During the reigns of Huan and Ling this was particularly so. At the level of the Three Excellencies, counselors, provincial governors, and commandery administrators, none was concerned about state affairs, devoting themselves instead to their retainers.[18] Officials thronged at the gates of other officials, and scholars blocked the roads traveling to see other scholars.[19] They had no time to eat when hungry and no chance to rest when tired. Their swirling multitudes turned night into day. At the level of minor officials, town after town of district magistrates all praised one another for having obtained the right men and boasted of themselves that they had enlisted the service of talented members of the bureaucracy of lower rank.

"In the morning, by starlight they would have their carriages yoked so as to hurry to see off those departing or welcome those arriving. The overnight guest inns were always full, and the functionaries who accompanied the officials would pass on gossip and seek information. Because some officials traveled at night by torchlight, the gatekeepers would not lock the gates. When making farewells, the officials would grasp each other's arms, clutch each other's wrists and, bowing to heaven, swear oaths of lasting friendship. They would plead for favors and special treatment without consideration of the commitment involved. Official documents would pile up in their offices and prisoners fill up the jails, yet they would spare no time to deal with these matters. When their behavior is carefully investigated, it is clear that they had no desire to be

憂國恤民、謀道講德也。徒營己治私、求勢逐利而已。

　「有策名於朝，而稱門生於富貴之家者，比屋有之。為之師而無以教。弟子亦不受業。然其於事也，至乎懷丈夫之容，而襲婢妾之態。或奉貨而行賂，以自固結。求志屬託，規圖仕進。然擲目指掌，高談大語，若此之類，言之猶可羞，而行之者不知恥，嗟乎！王教之敗，乃至於斯乎。[1]

　「且夫交游者、

---

[1] Following *Yiwen leiju*, 397, in inserting the following passage into the text: 林宗之時所謂交遊者也輕位不仕者則有巢許之高廢職待客者則有仲尼之稱委親遠學者則有優遊之美是以各眩其名而忘天下之亂也.

concerned about the country or to feel sympathy for the people; nor did they aspire to the way or thoroughly examine virtue. They were merely concerned with taking care of their own private affairs and interests and seeking power and profit.

"There were those who, despite holding court appointments, nonetheless called themselves 'household disciples' of rich and high-ranking families.[20] They were to be found in house after house. Those who were teachers had nothing to teach, and so their students received no learning either. When it came to conducting affairs of office, they went so far as to be manly in appearance, yet emulate the mannerisms of a servant girl. Some of them offered goods as bribes so as to consolidate their positions. Pursuing their ambitions, they would seek favors and scheme on ways to be promoted in office. Yet even to talk about such things as glaring eyes, animated gesticulation, and bombastic boasting is an embarrassment—as for those who actually behaved in such a manner, they had no sense of shame at all. Alas! The demise of the kingly teachings had come to this! In the time of [Guo] Linzong [128–69], 'traveling around to form social connections' referred to the following: describing those who despised position, refusing to serve in office, as having the noble-mindedness of Chao[fu] and Xu [You]; describing those who did not attend to their posts, devoting themselves instead to receiving guests, as having the praiseworthy qualities of Confucius; and describing those who abandoned their parents, traveling to distant places to pursue learning, as having the excellent virtue of being carefree.*[21] This is how such individuals became dazzled by personal reputation and so forgot about the chaotic state of the realm.

"When those who travel around to form social connec-

---

*Guo Linzong (=Guo Tai 郭泰) was the most celebrated of the Eastern Han "character assessors" and a leading figure in the "pure faction." Xu You 許由 was a legendary hermit who refused the throne when Yao offered it to him. Chaofu 巢父 was another legendary figure who refused to take up office. The reference to Confucius is presumably alluding to his willingness to accept disciples from near and far in his later years.

出也或身歾於他邦，或長幼而不歸。父母懷煢獨之思，室人抱東山之哀。親戚隔絕，閨門分離。無罪無辜，而亡命是效。古者行役，過時不反，猶作詩刺怨。故《四月》之篇，稱『先祖匪人，胡寧忍予。』又況無君命而自為之者乎！以此論之，則交游乎外、久而不歸者，非仁人之情也。」

tions leave their homes, some end up dying in other counties, while others do not return even after their children have grown up. Their own parents are left consumed with loneliness and their wives embracing the grief expressed in the ode 'Eastern Mountains.'* Parents and relatives lose contact with them, and wives become separated from them. Even though these sons and husbands have committed no crimes, they still imitate the actions of those who run away and no longer have their names recorded on the population register. In antiquity, if those conscripted for corvée did not return by the due time, they would even compose odes to criticize and complain. Hence the ode 'The Fourth Month' says: 'My forefathers, are you not humane?[22] / How can you bear that I should suffer so?'[23] Then how much more callous are those who, without being commanded to do so by their rulers, take it upon themselves to make such journeys! Thus argued, those who travel beyond their homes to form social connections, staying away for long periods without returning, fail to conform to the emotions of the humane person."

* *Odes*, Mao no. 156. The third stanza of this ode, Karlgren, *Odes*, 102, mod., reads: "The heron cries on the anthill / The wife sighs in the chamber / She sprinkles and sweeps / And the holes in the walls are stopped up…. / It has been three years now / Since she last saw me."

《曆數第十三》

昔者聖王之造曆數也，察紀律之行，觀運機之動，原星辰之迭中，窮晷景之長短。於是營儀以准之，立表以測之，下漏以考之，布筭以迫之。

13. *Astronomical Systems*

The subject of this essay is the importance of establishing and maintaining an accurate astronomical system so as to preserve a harmonious correlation between the realm of man and the realm of heaven. Xu Gan explains how in antiquity correct observance of the heavens, and the formulation of astronomical systems based on those observations, allowed the sage kings to ensure that "yin and yang were harmonized" and that "disasters and pestilence did not occur." They were able to do this because the realms of heaven and man are interdependent, and by observing the "heavenly patterns" and conforming with those patterns, rulers could safeguard the affairs of man. He discusses the origin of the astronomical systems in the time of the sage kings, their demise in the Eastern Zhou, and the various new systems developed and adopted during the Han: the Zhuan Xu, the Grand Inception, the Three Sequences, the Quarter-remaindcr, and the Celestial Images.

Long ago, when the sage kings created astronomical systems, they studied the motions of the sun, moon, and stars, and watched the turning of the Northern Dipper; they traced the alternation of the paths of stars and constellations as they crossed the observer's meridian due south; and they understood the significance of the length of the shadows cast by the sun.[1] Thereupon they made instruments to improve the accuracy of their observations, erected gnomons to measure, set up clepsydras to check the timing of star transits, and laid out counters to plot the lengths of the sun's

然後元首齊乎上，中朔正乎下，寒暑順序，四時不忒。夫曆數者、先王以憲殺生之期，而詔作事之節也。使萬國之民不失其業者也。

昔少皞氏之衰也，九黎亂德，民神雜揉，不可方物。顓頊受之，乃命南正重司天以屬神，北正黎司地以屬民。使復舊常，毋相侵黷。其後三苗復

shadow.*² Only then would the beginning of the year be arranged at the outset, the medial *qi* periods and the nodal *qi* periods be placed in proper order for the rest of the year, the heat and the cold follow one another in orderly succession, and the four seasons proceed without irregularity.† Astronomical systems were used by the former kings to promulgate the different periods for the taking of life and the nurturing of life, and, by edict, to set the seasons for conducting various activities, so ensuring that the people of all states would not fail to pursue their occupations.³

In the past, "with the decline of Shaohao, the Nine Li tribes threw virtue into disorder, the affairs of the people and the spirits became entangled with one another, and things could not be distinguished. Zhuan Xu acceded to the throne and ordered Zhong, Chief of the South, to be responsible for affairs relating to heaven, and entrusted him with matters concerning making offerings to the spirits.‡ He also ordered Li, Chief of the North, to be responsible for affairs relating to earth, and entrusted him with matters relating to putting the people in order.§ This was to ensure that the spirits and the people would resume their former regular practices and cease violating one another's domain and being contemptuous of one another. After this, the Three Miao tribes once again acted

*Instruments: this is possibly a reference to the "celestial sphere" (*huntian yi* 渾天儀) type of armillary sphere. The gnomon was used to measure the shadow cast by the sun to determine the solstices and equinoxes, as well as for stellar observations.

†The medial *qi* periods (*zhong qi* 中氣) and the nodal *qi* periods (*jie qi* 節氣 or *shuo qi* 朔氣) were the two divisions into which each month was divided, yielding the twenty-four periods of the year. On medial qi and nodal qi see Cullen, *Astronomy and Mathematics*, 107; Ho, *Li, Qi, and Shu*, 154–55; Major, *Heaven and Earth*, 90.

‡Zhuan Xu was a legendary predynastic ruler and descendent of the Yellow Emperor. On myths associated with Zhuan Xu, see Allan, *Shape of the Turtle*, 67–68. Nothing else is known of Zhong.

§According to *Zuo Commentary*, Zhao 29, 53.10a, Li was the son of Zhuan Xu.

九黎之德。堯復育重黎之後。不忘舊者，使復典教[a]之。故《書》曰：「乃命羲和，欽若昊天曆象，日月星辰，敬授民[b]時。」於是陰陽調和，災厲不作。休徵時至，嘉生蕃育。民人樂康，鬼神降福。舜、禹受之，循而勿失也。

及夏德之衰，而羲和湎淫。廢時亂日。湯武革命，始作曆明時，敬順天數。故《周禮》太史之職：「正歲年以序事。頒之

[a] On the basis of the *Guoyu*, 18.2b and *Shiji*, 26.1257 texts, omitting the character *jiao* 教.
[b] *Documents*, 2.9a, reads *li* 厤 for *li* 曆 and *ren* 人 for *min* 民.

in the manner of the Nine Li tribes.* Yao once more nurtured the descendants of Zhong and Li. Those among them who had not forgotten the old skills were made to restore the canons and instruct the others in their use."[4] Hence the *Book of Documents* says: "Then Yao commanded Xi and He to comply reverently with august heaven and its successive phenomena, with the sun, the moon, and stellar markers, and thus respectfully to bestow the seasons on the people."†[5] Thereupon the *yin* and *yang* forces became harmonized, calamities and pestilence did not arise, propitious omens arrived at due intervals, fine crops grew profusely, the people were happy and well, and the ghosts and spirits bestowed blessings. When Shun and Yu received the mandate to rule, they continued the practice [of employing the descendents of Zhong and Li] and so made no errors.[6]

When the virtue of the Xia declined, "the Xi and He families became dissolute and debauched. They abandoned the seasonal demarcations and threw the system for determining auspicious and baleful days into disarray."[7] When Tang and Wu changed the mandate to rule and began to devise systems to clarify the seasonal demarcations, they respectfully followed the "heavenly numbers."‡[8] Thus in describing the duties of the Grand Scribe, the *Rites of Zhou* says that he rectified the discrepancies between the tropical year and the synodic year so as to put affairs in order.§[9] His system was then promul-

---

*According to *Guoyu*, 18.2b, Wei commentary, the Three Miao were descendents of the Nine Li, who were active during the time of Gao Xin 高辛 or Ancestor Ku 嚳 (supposedly the father of Yao, see *Shiji*, 1.14). See also Knoblock, *Xunzi*, 2:217–18.

†Xi and He were the descendents of Zhong and Li.

‡Tang and Wu were founders of the Shang and Zhou dynasties, respectively. "Heavenly numbers": Han thinkers attached numerological significance to meteorological and astrological observations as described in terms of *yin yang* 陰陽, *wu xing* 五行, and so on.

§A tropical year is the length of time it takes the earth to complete a full orbit of the sun, 365.24 days. A synodic month is the time it takes the moon to make a full orbit of the earth. Twelve such orbits yield a total of approxi-

於官府及都鄙。頒告朔於 [c] 邦國。」於是分、至、啟、閉之日，人君親登觀臺以望氣，而書雲物為備者也。

故周德既衰，百度墮替，而曆數失紀。故魯文公元年閏三月，《春秋》譏之。其《傳》曰：「非禮也。先王之正時也，履端於始，舉正於中，歸餘於終。履端於始，序則不愆。舉正於中，民則不惑。歸餘於終，事則不悖。」又哀公十二年十二月:「螽。季孫問諸仲尼。仲尼曰：『丘聞之也，火(復)〔伏〕[d] 而後蟄者畢。今火猶西流，司曆過也。』」言火未伏，明非

[c] *Documents* reads *yu* 于 for *yu* 於.
[d] Following the *Zuozhuan* text and the *Bozi* version of *Discourses* in emending *fu* 復 to *fu* 伏.

gated in the offices of officials and throughout the capital and outlying districts. The ruler would issue the calendar for the following year to his vassal states.[10] Thereupon, on the days of the equinoxes, solstices, and the commencement and conclusion of the four seasons, the ruler would personally ascend his observatory to watch the *qi* and record the indications of clouds and other such objects in preparation for things to come.[11]

Hence when the charismatic virtue of the Zhou had declined, the one hundred standards of measurement were abandoned, and the astronomical system lost its standard of reckoning. For this reason, in the first year of the reign of Duke Wen of Lu [626 BCE], there was an intercalated third month. The *Spring and Autumn Annals* criticized him, and its tradition says: "This goes against propriety. In regulating the seasonal ordinances, the former kings made the beginning [the winter solstice] the starting point, determined the correct periods for demarcating the seasonal divisions based on the equinoxes and solstices, and reserved the surplus days for the year's end. Because they made the beginning as the proper starting point, no error was made in the proper order of the seasons; because they determined the correct periods for demarcating the seasonal divisions, the people were not confused; and because they reserved the surplus days for the year's end, affairs proceeded without any error."[12] Furthermore, the tradition records that in the twelfth month of the twelfth year of Duke Ai [483 BCE], "There were locusts. Jisun asked Confucius about it and he replied, 'I have heard that only after the Fire Stars disappear do the locusts stop their hibernation.* Yet now the Fire Stars still appear traversing the western skies. The manager of the calendar must have made a mistake.'"[13] This says that the Fire Stars had still not disappeared; thus clearly this

---

mately 354 days or one synodic year. The discrepancy between the tropical and synodic year was reconciled by the use of intercalated months. On intercalated months, see Cullen, *Astronomy and Mathematics*, 20–23.

*The Fire Stars (also known as *shangxing* 商星 and *dahuo* 大火) are part of the constellation Scorpius, including the star Antares.

立冬之日。自是之後，戰國搆兵，更相吞滅，專以爭強、攻取為務。是以曆數廢而莫脩，浸用乖繆。

大漢之興，海內新定，先王之禮法尚多有所缺。故因秦之制，以十月為歲首，曆用顓頊。孝武皇帝恢復王度，率由舊章。招五經之儒，徵術數之士，使議定漢曆。及更用鄧平所治。元起太初。然後分、至、啟、閉，不失其節。弦、望、晦、朔，可得而驗。成、哀之間，劉歆用平術而廣之，以為三統曆。比之眾家，最為備悉。至孝章皇帝，年曆疎闊，不及天時。及更用四分曆舊法。元起庚辰。至靈帝四分曆，

was not the day of the winter solstice. After this, the warring states engaged in battle and, one after another, swallowed one another up. The sole aim was to become the strongest state and to conquer the other states. In this way, astronomical systems fell into disuse and so ceased to be maintained, becoming more and more disordered and inaccurate.

With the rise of the Great Han, when the realm had been only newly pacified, there were still many lacuna in the rites and laws of the former kings. Because of this, the Han continued to use the Qin system of taking the tenth month as the beginning of the year and adopted the Zhuan Xu astronomical system.[14] Emperor Wu [r. 141–87 BCE] restored the standards of the former kings and followed the old regulations. He summoned scholars of the Five Classics and men skilled in the specialist arts and numerology to deliberate upon and fix the Han calendrical system, and also changed to using the revised astronomical system developed by Deng Ping, [Luoxia Hong, and others].[15] The epoch was calculated from the beginning of the Grand Inception reign period [24 December 105 BCE].[16] Thereafter, the equinoxes, solstices and the commencement and conclusion of the four seasons maintained a regular sequence, and the first and last quarters of the moon, the full moon, the last day of the moon, and the new moon could be accurately predicted. Between the reigns of Emperor Cheng [r. 33– BCE] and Emperor Ai [r. 7–1 BCE], Liu Xin [46 BCE–23 CE]—elaborating upon Deng Ping's system—devised the Three Sequences astronomical system.[17] Of all the different experts' systems, his was the most complete. By the time of Emperor Zhang [r. 75–89], however, the system had lost its precision, falling behind natural time, so it was replaced by the old methods of the Quarter-remainder astronomical system.*[18] The calendrical epoch was made to commence from 161 BCE.[19] By the period of Emperor Ling's reign [168–89],

* "Natural time" refers to time as measured by such phenomena as the phases of the moon.

猶復後天 ■ ■^e 半日。於是會稽都尉劉洪更造乾象曆,以追日月星辰之行。考之天文,於今為密。會宮車晏駕,京師大亂。事不施行,惜哉!

上觀前化,下迄於今,帝王興作,未有奉^f 贊天時,以經人事者也。故孔子制《春秋》,書人事,而因以天時,以明二物相須而成也。故人君不在分、至、啟、閉,則不書其時月。蓋刺怠慢也。

夫曆數者、聖人之所以測靈耀之賾,而窮玄妙之情也。非天下之至精,孰能致思焉?今纇論數家舊法,綴之於篇,庶為後之達者,存損益之數云耳。

^e Following other versions of *Discourses* (the two exceptions being the *Xue Chen* and *Han Wei congshu* versions) in omitting the two lacunas.
^f Following *Jianan qizi ji* in emending *hua* 化 to *dai* 代 and inserting the character *bu* 不 after *you* 有 and before *feng* 奉.

however, the system was still a further half a day behind natural time. Thereupon Chief Commandant Liu Hong of Guiji [alt. Kuaiji] developed the Celestial Images astronomical system so as to trace the movements of the sun, moon, and the stellar markers.[20] To this day, the precision of his system is verified by observation of the heavenly bodies and their movements. With Emperor Ling's recent passing, the capital has been thrown into a state of utter chaos and, unfortunately, matters of astronomy have been abandoned!

The matter of astronomical systems from former generations until the present having thus been reviewed, it is evident that when emperors and kings rose to power, they all respectfully employed natural time to assist them in conducting the affairs of man. Thus when Confucius wrote the *Spring and Autumn Annals* and recorded the affairs of man, he did so with reference to natural time so as to make it clear that the realms of heaven and man are integrated through codependence. Accordingly, if a ruler failed to observe the equinoxes, solstices, and the commencement and conclusion of the four seasons, then Confucius would not record the particular season or the month in which such matters occurred. This was done so as to criticize such rulers for being lax and remiss.

Astronomical systems are the means by which the sage fathoms the profundity of the sun and moon and thereby thoroughly penetrates the genuine condition of the mysterious and the sublime. If the most precise astronomical systems are not employed, then who would be able to apply their thoughts to these matters? Today, in very general terms, I have discussed the old systems of several experts and put them together in this essay. I hope that it will provide a record of what has been omitted from and what has been preserved of the principles of astronomical systems, for the benefit of enlightened men of future times.

《夭壽第十四》

或問：「孔子稱『仁者壽』，而顏淵早夭；

14. Distinguishing Between Premature Death and Longevity

This essay is another example of Xu Gan's disputation (*bian*). It is a critical analysis of two contemporary responses to a particular problem that had troubled thinkers from at least the third century BCE: if there is a normative moral order affecting the interrelationship of the human realm with the realm of heaven, why is it that morally exemplary people do not always meet with good fortune and morally reprehensible people do not always meet with misfortune? One of the views he criticizes argues that longevity can be achieved only by securing a virtuous reputation bestowed by posterity; the other argues that the idea that the accumulation of good deeds leads to blessings, and that the practice of humaneness results in longevity, is merely a ploy to keep "benighted people" on the straight and narrow. Xu then sets forth his own position, citing classical textual authority for his claim that the wiser ancient kings had achieved longevity in their own times by benign and wise government, and that hence longevity can be achieved in one's lifetime by acting humanely; it is only in anomalous circumstances that this is otherwise. As is made evident in "Cultivating the Fundamental," these irregularities are occasioned by certain adverse types of "circumstances which lie beyond human control" (*ming*) and which, in turn, are the product of "departures from regularities" (*bian shu*).[1]

Someone asked, "Confucius said that 'those who practice humaneness will live long' yet Yan Yuan died young; and al-

『積善之家必有餘慶』，而比干、子胥身陷大禍。豈聖人之言不信而欺後人耶？」

故司空潁川荀爽論之，以為古人有言，死而不朽，謂「太上有立德，其次有立功，其次有立言。」其身歿矣，其道猶存。故謂之不朽。夫形體者、人之精魄也。德義令聞者、精魄之榮華也。君子愛其形體，故以成其德義也。夫形體固自朽弊消亡之物。壽與不壽，不過數十歲。德義立與不立，差數千歲。豈可同日言也哉！顏淵時有百年之人，今寧復知其姓名耶？《詩》云：「萬有千歲，眉壽無有害。」人豈有萬壽千歲者，皆令德之謂也。

though 'the house that heaps good upon good is sure to have an abundance of blessings,' yet the bodies of Bigan and Zixu both fell into circumstances of great misfortune.*[2] Could it be that the words of the sages are untrustworthy, thus deceiving those of later times?"

To this, the former minister of works, Xun Shuang [128–90 CE] of Yingzhou, argued that the ancients had a saying, "to die yet not to perish," which means "that which is of utmost importance is the establishment of a legacy of virtue, next the establishment of meritorious deeds, and then the establishment of one's words."[3] If the physical body dies but one's way is still preserved, then one can call it "immune to corruption." The body is a person's refined life-spirit and to achieve renown for one's virtue and rightness is the finest flowering of that refined life-spirit.[4] Because the gentleman cares for his body, therefore he employs it to perfect his virtue and rightness. One's body is definitely something that will decay, die, disperse, and disappear. As to its longevity or lack thereof, it will in no case be more than several decades. The successful or unsuccessful establishment of virtue and rightness, however, can make a difference of several thousand years to one's reputation. How could such matters be discussed in the same breath? Contemporaneous with Yan Yuan were men who lived to one hundred years of age, yet today are we likely to hear their names still mentioned? The *Book of Odes* says: "For a myriad and again a thousand years/You will have a vigorous old age and no harm."[5] Is it conceivable that some men are able to live for myriads or thousands of years? Rather, in all cases, these lines refer to the reputation derived from an out-

---

*Bigan: See "Wisdom and Deeds," note 24. Zixu is the style of Wu Yun 伍員, loyal minister of King Fuchai 夫差 (r. 495–473) of Wu. The king ignored Wu's warnings about the state of Yue and instead believed slanders about him perpetrated by the traitorous Grand Steward, Bo Pi 伯嚭, forcing Wu to commit suicide. See *Zuo Commentary*, 57.2b–5a; *Shiji*, fascicle (*juan*) 66, and Knoblock, *Xunzi*, 2:195–97.

由此觀之，「仁者壽」、豈不信哉！《傳》曰：「所好有甚
於生者，所惡有甚於死者。」比干、子胥皆重義輕死者也。
以其所輕，獲其所重。求仁得仁，可謂慶矣！槌鍾擊磬，所
以發其聲也。煮豒燒薰，所以揚其芬也。賢者之窮厄戮辱，
此槌擊之意也。其死亡陷溺，此燒煮之類也。

　　北海孫翶以為死生有命，非他人之所致也。若積善有慶，
行仁得壽，乃教化之義。誘人而納於善之理也。若曰積善不
得報，行仁者凶，則愚惑之民將走(千)〔于〕惡以反天常。故
曰：「民可使由之，不可使知之。」

　　「身體髮膚，受之父母，不敢毀傷。孝之至[a]也。」若夫求
名之徒，殘疾厥體，冒厄危戮，以徇其名，則曾參不為也。
子胥違君而適讎國，以雪其恥，

[a] *Xiaojing*, 1.3a, reads *shi* 始 for *zhi* 至.

standing virtue. Viewing it thus, do you really not have faith that those who practice humaneness will be long-lived? One tradition states, "There is something which men are fonder of than life and something which men loathe more than death."[6] Bigan and Zixu were men who valued rightness and made light of death. In making light of death, they attained that which they valued. To seek and obtain humaneness can be called a blessing.[7] "It is by hitting bells and striking chimes that one produces sounds with them, and it is by boiling sacrificial wine and burning incense that one spreads one's fragrances." The hardship and humiliation suffered by a man of worth matches the meaning of "hitting and striking"; his death and downfall accord with the category of "burning and boiling."

Sun Ao of Beihai maintained that life and death are determined by circumstances which lie beyond human control and cannot be affected by other people.[8] Views such as "accumulating goodness leads to blessings" and "practicing humaneness results in a long life" have an edifying and transforming import; they entice people and draw them toward patterns of goodness. If one were to say that the accumulation of goodness reaps no reward, or that the practice of humaneness will lead to misfortune, then stupid and confused people would act badly and so contravene heaven's constant standards. Therefore it is said, "The people can be made to follow it but they cannot be made to understand it."[9]

One's body hair and skin are received from one's parents and to dare not destroy or injure them is the supreme expression of filial piety.[10] If a fame seeker were to mutilate his body or harm it risking life and limb in the pursuit of fame, then this is something Zeng Shen would not have done.* [Wu] Zixu turned against his lord [King Ping of Chu, r. 528–516] and went over to the rival state [of Wu] to cleanse his shame and

---

* Zeng Shen was a paragon of filial piety and the traditional author of *Xiaojing* (Classic of Filial Piety).

與父報讎。悖人臣之禮，長畔弒之原。又不深見二主之異量。至於懸首不化。斯乃凶之大者，何慶之為？

　　榦以為二論皆非其理也。故作《辨夭壽》云：榦聞先民稱「所惡於知者為鑿[b]也。」不其然乎？是以君子之為論也，必原事類之宜而循理焉。故曰：「說成而不可間也；義立而不可亂也。」

　　若無[c]二難者，苟既違本而死，又不以其實。夫聖人之言、廣矣！大矣！變化云為，固不可以一概齊也。今將妄舉其目以明其非。

　　夫壽有三：有

[b] See *Mencius*, 4B.26. *Mencius* reads *qi* 其 before *zuo* 鑿.
[c] Following the *Longxi jingshe* and *Liangjing yibian* versions of *Discourses* in reading *fu* 夫 not *wu* 無.

to avenge his father's death.* He transgressed a minister's code of decorum and abetted acts which contributed to a climate of treason and regicide. Moreover, he failed to see deeply enough into the differences in quality of the two rulers he served.† Right up until his head fell he did not change. This was a matter of great misfortunate; what blessings were there?‡

I think that the lines of argument advanced by Xun and Sun in regard to this matter are wide of the mark, and so I have written "Distinguishing Between Premature Death and Longevity," which is as follows: I have heard that people in former times said, "That which people detest in 'wise men' is strained arguments."[11] And is that not right? Thus in the course of making an argument, the gentleman must follow the principle of the matter by tracing affairs to their appropriate categories. Hence it is said, "When the argument is completed, it cannot be faulted; when the meaning is established, it cannot be confused."

As for Bigan and Wu Zixu, even if they did die as a consequence of going against that which is fundamental, this has still not touched upon the actuality of the matter. A sage's words are far-reaching and great. "In his alterations and transformations, and in his words and deeds," he certainly cannot be measured by the common rod.[12] Now I will presume to raise some points from Xun and Sun's arguments to show that they are fallacious.

There are three types of longevity. There is the longevity

---

*His father, Wu She 伍奢, had been thrown into jail, where he died, because King Ping was offended by Wu She's forthright criticisms of him.
†That is, King Helu 闔廬 (r. 514–496) of Wu, and his son and successor, King Fuchai. King Fuchai believed slanders about him and forced him to kill himself.
‡Sun Ao's argument might be seen as self-defeating. He wishes to maintain that there is no necessary connection between a man's moral actions and the blessings he enjoys; yet in describing Wu Zixu, he portrays him as unfilial, traitorous, lacking in propriety, and a bad judge of character. By the principle of "as ye sow so shall ye reap," it is not surprising that Wu Zixu failed to receive any blessings.

王澤之壽，有聲聞之壽，有行仁之壽。《書》曰：「五福：
一曰『壽。』」此王澤之壽也。《詩》云：「其德不爽，壽考
不忘。」此聲聞之壽也。孔子曰：「仁者壽。」此行仁之壽
也。孔子云爾者，以仁者壽利養萬物，萬物亦受利矣。故必
壽也。

　　荀氏以死而不朽為壽。則《書》何故曰：「在昔[d]殷王中
宗，嚴恭寅畏。天命自度。治民祇懼。不敢荒寧。肆中宗之
享國[e]七十有五年。其在高宗，寔[f]舊勞於外，爰暨小人。作
其即位，乃或亮陰，三年不言。〔其〕惟〔不言〕[g]，言乃雍。
不敢荒寧。嘉靖殷國[h]。至於小大，無時或

[d] *Documents* reads *xi zai* 昔在.
[e] The *Documents* text and the *Longxi jingshe* and *Liangjing yibian* versions of
Discourses read *bang* 邦 not *guo* 國.
[f] *Documents* reads *shi* 時 not 寔.
[g] Following *Documents* in emending the *Discourses* text to read *qi wei bu yan*
其惟不言.
[h] The *Documents* text and the *Longxi jingshe* and *Liangjing yibian* versions of
Discourses read *bang* 邦 not *guo* 國.

derived from the king's beneficence, the longevity derived from reputation, and the longevity derived from practicing humaneness. The *Book of Documents* says, "There are five blessings. The first is called longevity."[13] This is the longevity derived from the king's beneficence. The *Book of Odes* says: "His virtue is faultless / May he live long and not be forgotten."[14] This is the longevity derived from reputation. Confucius said, "Those who practice humaneness are long-lived."[15] This is the longevity derived from practicing humaneness. Confucius said that because the longevity of the humane is beneficial to the nourishment of the myriad creatures, so the myriad creatures also benefit from the practice of humaneness. For this reason the humane will certainly be long-lived.

Mr. Xun maintained that "to die but not to perish" was longevity. Why then does the *Book of Documents* record the following?

In the past, Zhongzong, king of Yin, was solemn, humble, reverential, and awesome.* He measured himself with reference to the appointment of heaven and cherished a reverent apprehension in governing the people, not daring to indulge excessively in pleasures. It was thus that Zhongzong enjoyed the throne for seventy-five years.

Coming to the time of Gaozong, he worked for a long time outside together with the common people.† When he came to the throne, it is said that while he was in the mourning shed, he did not speak for three years. When he did speak, people listened in harmony to his words. He did not dare to indulge excessively in pleasures, and made the Yin domain beautiful and peaceful, until there was no one, young or old, who ever harbored

---

*Zhongzong was the temple name of Da Wu 大戊, fifth ruler of the Shang dynasty.
†Gaozong was the temple name of Wu Ding 武丁, twelfth ruler of the Shang-Yin dynasty.

怨。肆高宗之享國五十有九年。其在祖甲，不義惟王，舊為小人。作其即位，爰知小人之依，能保惠庶民，不侮[i]鰥寡。肆祖甲之享國三十有三年。自時厥後立王，生則逸[j]！不知稼穡之難艱[k]。不知小人之勞苦[l]。惟耽樂是[m]從。自時厥後，亦罔或克壽。或十年，或七八年，或五六年，或三四[n]年者。」周公不知夭壽之意乎？故言聲聞之壽者，不可同於聲聞[o]。是以達人必參之也。

孫氏專以王教之義也。惡愚惑之民將反天常。孔子何故曰：「有殺身以成仁，無求生以害仁。」又曰：「自古皆有死。

i Documents reads hui yu shu min bu gan wu 惠于庶民不敢侮.
j The phrase sheng ze yi 生則逸 is reduplicated in Documents.
k The two characters nan jian 難艱 are reversed in Documents.
l Documents reads bu wen xiao ren zhi lao 不聞小人之勞.
m The Documents text and the Longxi jingshe and Liangjing yibian versions of Discourses read zhi 之 not shi 是.
n Documents reads si san 四三.
o Following the Longxi jingshe and Liangjing yibian versions of Discourses in reading wang ze 王澤 not sheng wen 聲聞.

ill feelings toward him. It was thus that Gaozong enjoyed the throne for fifty-nine years.

In the case of Zujia, he did not think it right that he
should be king, and for a long time he remained one of
the common people.* When he came to the throne, he
understood the pain of the people, and was able to protect and love them, and did not treat the widower and
widow with contempt. Thus it was that Zujia enjoyed
the throne for thirty-three years.

The kings who were subsequently established on the
throne enjoyed ease all their lifetimes. They did not
understand the arduous toil of sowing and reaping, nor
did they know of the hard labors of the common people.
They sought only indulgence in pleasure. After this, not
one of the kings was long-lived. Some reigned for ten
years, some for seven or eight, some for five or six, and
some only for three or four.[16]

Did the duke of Zhou not understand the meaning of premature death and longevity?† Hence in speaking about the longevity derived from reputation, one cannot equate it with the
longevity derived from the king's beneficence.‡ This being so,
men of penetrating understanding should take note of it.

Mr. Sun has focused on the significance of the kingly teachings. He detests those "stupid and confused people" who would
contravene heaven's constant standards. Why, on the one hand,
did Confucius say, "One may sacrifice one's life in order to
make whole one's humaneness; one may not seek to stay alive
at the expense of harming one's humaneness," yet, on the other
hand, say, "From antiquity, death has been the lot of all men;

---

*Zujia was the son of Wu Ding.
†The above passage comes from a speech attributed to the duke of Zhou.
‡Here Xu Gan has countered Xun Shuang's claim that longevity is simply
a long-lasting posthumous reputation by arguing that the wiser ancient
kings achieved longevity in their own lifetimes by benign and wise government, which in turn afforded them long terms as rulers.

民無信不立。」欲使知去食而必死也。昔者仲尼乃欲民不仁不信乎？夫聖人之教，乃為明允君子。豈徒為愚惑之民哉！愚惑之民威以斧鉞之戮，懲以刀墨之刑，遷之他邑，而流於裔土，猶或不悛。況以言乎？故曰：「惟上智與下愚不移。」然則荀、孫之義，皆失其情，亦可知也。

昔者帝嚳已前尚矣。唐虞三代厥事可得略乎？聞自堯至於武王，自稷至於周召，皆仁人也。君臣之數不為少矣。考其年壽

if, however, the people have no faith in their government, then it will not stay standing"?[17] He did so because he wished to make it known that if food were taken away, death would be certain. In the past, was it then really the case that Confucius did not wish the people to practice humaneness and good faith?* A sage's teachings are for enlightened and sincere gentlemen. They would hardly be only for stupid and confused people. Stupid and confused people may be threatened with decapitation by axe, punished with tattooing by knife on the forehead, forcibly moved to other districts, or banished to the frontiers, and still some will not change! How much more so will this be the case if words, rather than punishments, are used? Thus it is said, "It is only the most wise and the most stupid who do not change."[18] This being the case, we can also understand how the import of Xun and Sun's arguments fail to relate to the actual situation.

Ancestor Ku lived such a long time ago that there is little known about him.[19] General accounts can be heard of the affairs of Tang and Yu and the Three Dynasties; and I have heard that from Yao to King Wu, Ji to Zhou and Shao, all were humane men.† The number of rulers and their ministers in this period was many, but if their lifespans are examined,

---

*That is, on the one hand, there are certain people who are prepared to give up their lives so that they will not compromise their virtue, yet, on the other hand, there are people whose primary concern is to stay alive. Xu Gan is arguing that while Confucius did not want stupid and confused people to be lacking in these virtues, yet, as far as these people were concerned, food was of primary concern, not the virtues. Hence the kingly teachings were ultimately of little relevance to the common people, and their edifying import would have held little sway over their actions.
† Ku was a traditional predynastic sage ruler and the progenitor of the Shang. Tang and Yu: the traditional predynastic sage rulers Yao and Shun. The Three Dynasties were the Xia, Shang, and Zhou. Ji was Hou Ji 后稷 (Lord Millet), who, according to traditional accounts, was the minister of agriculture under Shun. The dukes of both Zhou and Shao were known as virtuous administrators. For further details, see *Shiji*, fascicle 4. Xu Gan cannot mean literally all rulers, as this would include such rulers as the tyrants Jie and Zhou.

不為夭矣。斯非仁者壽之驗耶！又七十子豈殘酷者哉？顧其仁有優劣耳，其夭者惟顏回。據一顏回而多疑其餘，無異以一鉤之金，權於一車之羽，云金輕於羽也。天道迂闊，闇昧難明。聖人取大略以為成法。亦安能委曲不失，毫芒無差跌乎！且夫信無過於四時，而春或不華，夏或隕霜，秋或雨雪，冬或無冰。豈復以為難哉！

所謂禍者、己欲違之而反觸之者也。比干、子胥已知其必然而樂為焉！天何罪焉！天雖欲福仁，亦不能以手臂引人而亡[p]之。非所謂無慶也。苟令以此設難，而解以槌擊燒薰。於事無施。

[p] Following *Zhuzi pingyi bulu*, 80, in emending *wang* 亡 to *yu* 與 (与).

it can be seen that they did not die prematurely. Is this not proof that those who practice humaneness are long-lived? Furthermore, is it possible that among Confucius's disciples there were those who were cruel and violent? If, however, one looks at their practice of humaneness, there were those who were outstanding and those who were inferior. Of those who died prematurely, there was only Yan Hui. Yet to take this instance of Yan Hui's premature death to suspect most of the other disciples is no different from weighing a cartload of feathers against a hook of gold, and then saying that the gold is lighter than the feathers!* The way of heaven, however, is distant, vast, dim, and obscure. Because sages take the overall situation to be the established norm, how could they possibly avoid some slight inaccuracies when trying to accommodate things to conform to the overall situation? Furthermore, in matters of trustworthiness, nothing surpasses the four seasons, yet in spring sometimes the flowers do not blossom; in summer sometimes frosts fall; in autumn sometimes it snows; and in winter sometimes there is no ice. Are you really suggesting that we should return to the argument by employing such anomalies as refutations?

That which is called calamity occurs when one encounters something contrary to one's desire. Bigan and Zixu already knew that their deaths were inevitable and yet were still happy to do what they did. What fault rested with heaven? Although heaven desires that men enjoy good fortunes, it still cannot take them by the hand and make them so. The fates of Bigan and Zixu, accordingly, are not what is known of as "a lack of blessings." Xun employed Bigan and Zixu as counter-examples to the thesis that the house that heaps good upon good is sure to have an abundance of blessings, and then proceeded to employ the metaphors of hitting, striking, and burning incense to resolve the difficulty. Yet this did not help

---

*That is, to suspect that the reason they were able to live full lives had nothing to do with their being humane.

孫氏譏比干、子胥，亦非其理也。殷有三仁，比干居一。何必啟手然後為德。子胥雖有讎君之過，猶有觀心知仁。懸首不化。固臣之節也。

　　且夫賢人之道者，同歸而殊途，一致而百慮。或見危而授命，或望善而遲舉，或被髮而狂歌，或三黜而不去，或辭聘而山棲，或忍辱而俯就。豈得責以聖人也哉？於戲！通節之士，實關斯事，其審之云耳。

clarify matters at all. In ridiculing Bigan and Zixu, Mr. Sun also advanced an inappropriate line of argument. In the Yin dynasty, there were three humane men. Bigan was one of them. Why should he have needed to "uncover his hands" and only then be considered to be a man of virtue?* Although Zixu can be faulted for having made his ruler [King Ping of Chu] his enemy, nevertheless he still knew whether his actions were humane by contemplating his own heart. Right up until he committed suicide he did not change. This is certainly the integrity of a minister.

The ways pursued by worthy men all return to a common point, even though they follow different routes; and they all arrive at a common destination, even though a multitude of considerations comes into play.[20] Some, "seeing danger, were willing to lay down their lives."[21] Some looked for the good and traveled far to find it.[22] Some let their hair down, madly singing.[23] Some were dismissed thrice from office yet still did not leave their home country.[24] Some refused calls to office, retiring instead to the mountains.[25] Some bore humiliation and accepted demeaning positions.[26] As sages, can they be reproached? Oh, gentlemen of understanding and integrity, there is a common actuality which links all these matters; I urge you to examine it.

---

*"Uncover his hands": not to have brought harm to his body. This is a reference to *Analects*, 8.3; Lau, *Analects*, 92, and Legge, *Chinese Classics*, 1:208, mod: "When he was seriously ill, Zengzi summoned his disciples and said, 'Uncover my feet, uncover my hands. The *Book of Odes* [Mao no. 195] says: "In fear and trembling/As if approaching a deep abyss/As if walking on thin ice." Now and hereafter, I know I will be spared, my young friends.'" Zheng Xuan, cited in Huang Kan's *Lunyu yishu*, 8.23a, comments that Zengzi maintained that because his body had been received from his parents, he should never dare to harm or injure it. He thus instructed his disciples to remove the bed coverings and see what good condition his hands and feet were in.

《務本第十五》

人君之大患也，莫大於詳於小事，而略於大道；察其[a]近物，
而闇於遠圖。故自古及今，未有如此而不亂也，未有如此而
不亡也。夫詳於小事，而察於近物者，謂耳聽乎絲竹歌謠之
和，目視乎琱琢采色之章，口給乎辯慧切對之辭，心通乎短
言小說之文，手習乎射御書

---

[a] Following *Qunshu zhiyao*, 46.16b, in emending *qi* 其 to *yu* 於.

15. *Attend to the Fundamentals*

This essay is the first of several in the final quarter of *Discourses* which read as if they might originally have been memorials written to a ruler (presumably Cao Cao). By "fundamental," Xu Gan means the formulation of long-term strategies. Thus, rather than letting himself be distracted by trifling matters which may be "flavorsome and sweet," the ruler should formulate strategies based on the way, even though the way is "plain and tasteless." Unlike the importance that Xu Gan attaches to the arts in several other essays, here the arts are characterized as being favored by "those of mediocre talents," and their accomplishment is not something to which a ruler of men should devote his energies. Citing the example of several historical rulers, Xu Gan argues that practical wisdom in matters of statecraft is of far greater importance to a ruler than being adept in the arts or even observant of ritual propriety.

Of all the major troubles that can befall a ruler, none is greater than to be exacting in trifling matters and yet overlook the great way; to investigate into things close at hand yet to be unapprised of long-term strategies. Hence from antiquity until the present, whenever such a situation has occurred, chaos and destruction have ensued. "To be exacting in trifling matters" and "to investigate into things close at hand" refer to the following: to listen to the harmony of music and song with one's ears; to look at engraved and colorful designs with one's eyes; to be glib of tongue in disputatious cunning and words of sharp retort; to know by heart maxims and the words of the philosophers; to be dexterous at archery, charioteering, writ-

數之巧，體騖乎俯仰折旋之容。凡此〔數〕[b]者，觀之足以盡人之心，學之足以動人之志。且先王之末教也，非有小才小智，則亦不能為也。是故能為之者，莫不自悅乎其事，而無取於人。以人皆不能故也。

夫〔君〕[c]居南面之尊，秉生殺之權者，其勢固足以勝人也。而加〔之〕[d]以勝人之能，懷是己之心，誰敢犯之者乎？以匹夫行之，猶莫之敢規也。而況〔於〕[e]人君哉？故罪惡若山而己不見也。謗聲若雷而己不聞也。豈不甚矣乎！

夫小事者味甘，而大道者醇淡。〔而〕[f]近物者易驗，而遠數者難效。非大明君子則不能兼通者也。故皆惑於所甘，而不能至乎所淡；眩於所易，而不能反[g]於所難。是以

[b] Following *Qunshu zhiyao*, 46.16b, in reading *shu* 數.
[c] Not following the *Concordance to Zhonglun* in introducing this character into the text.
[d] Not following the *Concordance to Zhonglun* in introducing this character into the text.
[e] Not following the *Concordance to Zhonglun* in introducing this character into the text.
[f] Not following the *Concordance to Zhonglun* in introducing this character into the text.
[g] Following *Qunshu zhiyao*, 46.17b, in emending *fan* 反 to *ji* 及.

ing, and mathematical reckoning; and for one's body to be well trained in lowering and raising the head and turning to either side.* Just to observe these activities is enough to exhaust a person's mind, while to practice them is enough to waver a person's aims. They are, moreover, the peripheral teachings of the former kings, so unless one possesses a modicum of talent and wisdom, one will surely be incapable of accomplishing these skills. It is for this reason that those who are capable of accomplishing these skills are all so pleased with their accomplishments that they refuse to seek the advice of others on the grounds that others are all incapable of accomplishing these skills.

He who occupies the exalted position of the throne, and who holds the balance on which hang matters of life and death, certainly has sufficient power to overcome others. Thus who would dare to oppose one who, in addition to his ability to overpower others, believes that what he does is correct? Even if it were an ordinary person who wielded such power, there would still be no one who would dare to admonish him. How much more so, then, would this be the case if the person wielding this power were a ruler of men? Thus despite the fact that his criminal wrongdoings are as great as a mountain, he himself fails to see them; although public criticism of him rings out like thunder, he himself fails to hear it. Is this not intolerable?

Trifling matters are flavorsome and sweet, but the great way is plain and tasteless. It is easy to show proof of matters which lie close at hand, but it is difficult to verify the outcome of long-term strategies. Unless one is a gentleman of great clarity of vision, it is impossible to achieve both goals. Hence being consistently misled by what is sweet, one is unable to attain an appreciation of what is tasteless; dazzled by that which is easy, one is unable to reach that which is difficult. This is why there

* Movements associated with the deportment appropriate to ceremonial.

治君世寡，而亂君世多也。故人君之所務者，其在大道遠數乎！

大道遠數者，為仁足以覆燾群生，惠足以撫養百姓，明足以照見四方，智足以統理萬物，權足以（變應）[h]〔應變〕無端，義足以阜生財用，威足以禁遏姦非，武足以平定禍亂。詳於聽受，而審於官人。達於興廢之原[i]，通於安危之分。如此、則君道畢矣。

夫人君非無治為也，失所先後故也。道有本末，事有輕重。聖人之異乎人者，無他焉。蓋如此而已矣。魯桓公容貌美麗，且多技藝。然而無君才大智，不能以禮防正其母，使與齊侯淫亂不絕，驅馳道路。故《詩》刺之曰：「猗嗟名兮，美目清兮。儀既成兮，終日射侯，

[h] Following *Qunshu zhiyao*, 46.17b, in reversing *bian ying* 變應 to *ying bian* 應變.
[i] Following *Qunshu zhiyao*, 46.17b, in emending *yuan* 原 to *yuan* 源.

have been very few ages when there were capable rulers and many ages when there were incompetent ones. Thus it is to the formulation of long-term strategies, based on the great way, that the ruler applies himself.

Long-term strategies based on the great way are as follows: humaneness sufficient to blanket all living things; grace sufficient to nourish all the people; a clarity of vision sufficient to illuminate the four quarters; wisdom sufficient to draw together and order the myriad things; responsiveness to expedients sufficient to respond to transformations endlessly; rightness sufficient to foster the production of wealth and goods; awe sufficient to deter treachery and wrongdoing; and martial strength sufficient to quell trouble and chaos. By being circumspect in heeding advice and being judicious in placing officials, one will come to comprehend the wellspring of rise and decline and to understand the preordained factors leading to stability and instability. If one can be like this, then the way of the ruler will be complete.

It is not due to a lack of political measures that rulers have failed, but rather because they have not maintained the appropriate priorities. The way has its fundamental and peripheral aspects, just as affairs have their trifling and important aspects.[1] Is it not precisely in his ability to recognize this that the sage differs from other people? This is probably the sum of it. Although Duke Huan of Lu [r. 711–697] was both handsome and very skilled in the arts, he lacked great talent and immense wisdom. He was incapable of using propriety either to take precautions against or to rectify his mother's behavior. This resulted in her unending debauchery with the Marquis of Qi and her hastening down the road to meet him.[2] Therefore the *Book of Odes* satirized him, saying:

> Lo! How renowned!
> His beautiful eyes so clear!
> His manner so complete!
> Shooting all day at the target,

不出正兮。展我甥兮。」

　下及昭公，亦善有容。儀之習以豫。其朝晉也。自郊勞至
於贈賄，禮無違者。然而不恤國政。政在大夫，弗能取也。
子家羈賢，而不能用也。奸大國之明禁[j]，凌虐小國，利人之
難，而不知其私。公室四分，民食其他。思莫在於公，不圖
其終。卒有出奔之禍。《春秋》書而絕之曰：「公孫於齊，
次於[k]陽州。」故《春秋外傳》曰：「國君者[l]服寵以為美，
安民以為樂，聽德以為聰，致遠以為明。」又《詩》陳文王
之德，曰：

[j] Following *Zuo Commentary*, Zhao 5, 43.7a, in reading *meng* 盟 instead of
ming jin 明禁.

[k] The *Spring and Autumn Annals* text reads *yu* 于 not *yu* 於 (twice) in this
sentence.

[l] The *Guoyu* text reads *guo jun* 國君 for *guo jun zhe* 國君者.

And never hitting outside the bull's-eye!
Indeed he is our ruler's nephew.*[3]

Coming to the time of Duke Zhao [r. 541–510] of Lu, he, too, was of fine appearance and frequently observed the practices of ritual decorum. Thus when he visited Jin to pay his respects at court, whether it was extending his greetings to people living in the outlying areas of the capital or bestowing gifts, everything was done in accord with propriety. Despite this, he showed no concern for the affairs of his own state of Lu; indeed, political control was in the hands of the senior ministers and could not be taken from them. Zijia Ji was a worthy man, yet Duke Zhao was unable to employ him.† He violated a treaty with the large state of Jin and bullied and cruelly treated the small state of Ju. While benefiting from the difficulties of others, he remained ignorant of his own problems. The state was rent with disunity and the people had to rely on the three great families of Lu for their sustenance.‡ He gave no consideration to the public good and made no plans for future eventualities. Finally, he had to face the calamity of fleeing his own country.[4] The *Spring and Autumn Annals* dismissed him in the following terms: "The Duke 'removed himself' to Qi, stopping at Yangzhou."[5] Thus the *Outer Commentary of the Spring and Autumn Annals* says: "Rulers deem it good to bestow favors on the worthy, a joy to make their people content, smart to heed the virtuous, and clear-sighted to attract people from distant places."[6] Furthermore, in describing King Wen's virtue, the *Book of Odes* says:

---

* Duke Huan is a mistake for Duke Zhuang 莊 (r. 693–662). The reference to "hastening down the road" refers to Wenjiang's 文姜 (style of the wife of Duke Huan of Qi) incestuous assignations with her elder brother, Marquis Xiang 襄 of Qi.
† Zijia Ji was a senior minister of Lu.
‡ The great families of Lu were the Jisun 季孫, Shusun 叔孫, and Mengsun 孟孫.

「惟此文王[m]，帝度其心，貊[n]其德音。其德克明，克明克類。克長克君，王此大邦。克順克比，比于文王。其德靡悔。既受帝祉，施于孫子。」「心能制義曰『度』。德政應和曰『貊』。照臨四方曰『明』。施勤[o]無私曰『類』。教誨不倦曰『長』。賞慶刑威曰『君』。慈和徧服曰『順』。擇善而從曰『比』。經緯天地曰『文』。」如此則為九德之美。何技藝之尚哉？

今伻人君視如離婁，聰如師曠、御如王良，射如夷羿，書如史籀，計如隷首，走追駟馬，力折門鍵。

---

[m] The modern *Odes* text reads *wei ci wang ji* 維此王季. The *Discourses* text follows the reading of this ode as cited in the *Zuo Commentary*.

[n] The *Zuo Commentary* version of this ode reads *mo* 莫 for *mo* 貊.

[o] The *Zuo Commentary* text reads *zheng* 正 for *zheng* 政; *mo* 莫 for *mo* 貊; *lin* 臨 for *jian* 監; *qin shi* 勤施 for *shi qin* 施勤. I have followed the readings in the *Discourses* text, except in the case of the *Zuo Commentary* reading of *qin shi*.

Now this King Wen,
Di enabled him to take measure of his heart,
And to maintain with tranquility the good name of
    his virtuous rule.
His virtue enabled him to become clear-sighted;
Having been able to become clear-sighted he was able
    to be an exemplar.
Being able to lead, he was able to be a ruler
And rule this great country.
Being able to effect compliance,
He was able to emulate [good men].
The people then emulated King Wen.
His virtue left nothing with which to be dissatisfied,
And he received Di's blessing,
And it extended to his sons and grandsons.[7]

"The phrase 'to measure' describes the heart's ability to work out what is right. When virtuous government is responded to harmoniously, this is called 'to maintain with tranquility.' To illuminate and watch over the four quarters is called 'being clear-sighted.' To exert oneself strenuously without partiality is called 'being an exemplar.' To instruct without tiring is called 'leading.'[8] To bring blessings through rewards and to inspire awe through punishments is called 'being a ruler.' To bring about pervasive submission through compassion and harmony is called 'effecting compliance.' To choose good men and follow them is called 'emulating [good men].'[9] Employing heaven and earth as the warp and woof is called 'making a pattern.'"[10] Such is the excellence of the nine virtues. What skill or art is held in such admirable regard as these nine virtues?

Now suppose that a ruler were able to see as well as Li Lou, to hear as well as Shi Kuang, to drive a chariot as well as Wang Liang, to excel in archery as well as Yi Yi, to be as accomplished in writing as Secretary Zhou, and to calculate as well as Li Shou, then his physical prowess would be ex-

有此六者，可謂善於有司之職矣。何益於治乎？無此六者，可謂乏於有司之職矣。何增於亂乎？必以廢仁義、妨道德〔矣〕[p]。何則小器弗能兼容？治亂既不繫於此，而中才之人〔所〕好[q]也。昔路豐[r]舒、晉知〔伯瑤〕(其)〔之〕亡也[s]。皆怙其三才，恃其五賢，而以不仁之故也。

[p] Not following the *Concordance to Zhonglun* in introducing this character into the text.

[q] Following *Qunshu zhiyao*, 46.18a, in emending *ren hao* 人好 to *ren suo hao* 人所好.

[r] Following the *Zuo Commentary* readings of *Lu* 潞 for *Lu* 路, and *Feng* 酆 for *Feng* 豐.

[s] Following *Qunshu zhiyao*, 46.18a, in reading *Jin Zhi Boyao zhi wang* 晉智伯瑤之亡 instead of *Jin Zhi qi wang* 晉知其亡. Not following the *Concordance to Zhonglun* in retaining the character *ye* 也.

traordinary.*[11] To possess these six skills may be deemed to be accomplished in the requirements of office for a petty officer; yet what benefits do they bring to good order? Not to possess them may be said to be lacking in the requirements of office for a petty officer; yet how does it increase chaos? To increase chaos necessarily requires abandoning humaneness and rightness and obstructing the way and virtue. So why is it that those of narrow capacities are incapable of combining all that it would take to bring about such an increase of chaos? It is because order and chaos are not connected with the cultivation of these skills and it is these skills alone that those of mediocre talents favor. In the past, Feng Shu of the Lu and Zhi Boyao of Jin died because they relied on their three talents and five worthy qualities, while failing to act humanely.†[12]

---

* Li Lou, said to have been a contemporary of the predynastic Yellow Emperor, was gifted with exceptional eyesight. See *Mencius*, 4A.1; Zhao commentary, *Mengzi zhengyi*, 1:475. As the following five persons cited by Xu Gan are all skilled in one of the six arts, it would seem that he is making a tenuous connection between a discerning eye and the first of the six arts, the five rituals. Shi Kuang was the music master of Duke Ping of Jin (r. 557–537). See Zhao commentary, *Mengzi zhengyi*, 1:475. Wang Liang was a great charioteer, see *Mencius*, 3B.1; *Han Feizi*, 11.4a. Yi Yi was a great archer; see *Analects*, 14.5; *Xunzi*, 4.14b–15a. Zhou was the secretary of King Xuan of Zhou (r. 827/25–782). In the *Hanshu* "Treatise on Bibliography," a work in fifteen chapters, in *dazhuan* 大篆 (great seal) script, is attributed to him. Li Shou was said to have been a contemporary of the Yellow Emperor and to have invented methods of mathematical computation. See *Shiben* 世本, quoted at *Shiji*, 26.1256, n.2.

† The Lu were a branch of the Di 狄 tribes. Feng Shu, the chief minister of the Lu, having usurped power, injured the ruler of the Lu and killed his wife. The wife's younger brother, the marquis of Jin, wished to send a punitive expedition against Feng Shu but was advised by his ministers not to do so because Feng Shu was said to possess three outstanding talents. The marquis of Jin, however, argued that Feng Shu's faults outweighed his talents, and so he attacked the Lu and killed his enemy. The three talents are nowhere listed. See *Zuo Commentary*, Xuan 15, 15.9b–10a. The five qualities are listed in *Guoyu* as growing a profusion of beautiful temple-hair, might in archery and charioteering, a mastery of the arts, skill in literary phrasing and disputation, and resoluteness in determination.

故人君多技藝，好小智，而不通於大倫[t]者，適足以距[u]諫者
之説而鉗忠直之口也。（秖）〔祇〕足以追亡國之迹，而背安家
之軌也。不其然耶？不其然耶？

[t] Following *Qunshu zhiyao*, 16.18a, in reading *da dao* 大道 instead of *da lun* 大倫.
[u] Following *Qunshu zhiyao*, 16.18a, in emending *ju* 距 to *ju* 拒.

Hence if a ruler of men is accomplished in many skills and arts and is fond of petty wisdom, but remains ignorant of the great way, then this is just what it will take to dismiss the exhortations of those who would remonstrate with him, thereby forcing the loyal and upright to silence. By the same token, this is also just what it will take to lead to the demise of his ducal state and to the loss of stability in the ministerial domains within his state. Is this not so? Is this not so?

《審大臣第十六》

16. Examining the Selection of High Officials

The first part of this essay argues the importance of the ruler's personally selecting candidates for high office, reputation alone being unreliable. The point is illustrated with the examples of several well-known historical and prehistorical figures. Xu Gan then comments how, in his own times, the ruler (Emperor Ling?) no longer personally investigated the worth of individual candidates for high office, relying exclusively on the evidence of individual reputations; and because the common people ignored the great worthies who preferred to live in obscurity rather than compete for reputation and office, these worthies did not come to the attention of the ruler. In the second half of the essay, which takes the form of a dialogue, Xu Gan comments on this state of affairs, remarking that "those who are not praised by the common people of the day are not necessarily those who do not deserve a reputation, while those who are praised are not necessarily those who do deserve a reputation," and proceeds to blame such anomalies on the disturbing effects of chaos (*luan*) and "departures from regularities." In what appears to be a personal lament, he also complains about the consequences of not having been born in the right times, such as the inability to secure an appropriate reputation and being controlled by men of no real worth. He concludes the essay on the following plaintive note: "Thus long has been the history of names not matching actualities. In what age has it not been so? Only when the world has the way will this state of affairs come to an end!"

帝者昧旦而視朝廷。南面而聽天下。將與誰為之？豈非群公卿士歟！故大臣不可以不得其人也。大臣者、君之股肱、耳目也。所以視聽也，所以行事也。先王知其如是也，故博求聰明睿哲君子，措諸上位，〔使〕[a] 執邦之政令焉。執政〔聰明叡哲〕[b]，則其事舉。其事舉，則百僚〔莫不〕任其職。百僚〔莫不〕任其職，則庶事莫不致其治。庶事〔莫不〕[c] 致其治，則九牧之民莫不得其所。故《書》曰：「元首明哉。股肱良哉。庶事康哉。」

故大臣者、治萬邦之重器也。不可以眾譽著也。人主所宜親察也。眾譽者可以聞斯人而已。故堯之聞舜也，以眾譽。及其任之者，則以心之所自見。又有不因眾譽而獲大賢。

[a] Following *Qunshu zhiyao*, 46.18b, in inserting *shi* 使.
[b] Following *Qunshu zhiyao*, 46.18b, in inserting *cong ming rui zhe* 聰明叡哲.
[c] Following the *Concordance to Zhonglun* in adding the three occurrences of *mo bu* 莫不.

At dawn, the emperors hold court. Facing south they heed the affairs of the realm. Together with whom will they manage these affairs? Who else but the assembly of the three dukes, the nine ministers and the low ranking officials? Hence it is imperative that the right men are appointed as senior officials. Senior officials are a ruler's arms, legs, ears, and eyes—the means by which he sees, hears, and has his business carried out. Being aware of this, the former kings therefore sought far and wide for gentlemen who were perceptive and wise, placing them in top positions and granting them responsibilities for upholding the decrees of the state's government. If those who hold positions of authority in government are intelligent and wise, then all of the country's affairs will be carried out. If the country's affairs are acted on, the various other officials will fulfill the duties of their offices. If the various officials fulfill the duties of their offices, then all of the various affairs will be brought to order. If the various affairs are brought to order, the people of the nine regional representatives will all have their proper places.* Thus the *Book of Documents* says: "When the head is intelligent and the limbs are good, then the various affairs will be happily settled."[1]

Thus a senior official is an important instrument for governing the myriad states.† One cannot select him on the basis of an outstanding popular reputation; rather, the ruler should personally investigate him. A popular reputation enables one to know about somebody only by hearsay. Thus although Yao learned of Shun by his popular reputation, when it came to employing him he made up his mind on the basis of what he saw for himself. There are other examples in which the service of worthies has been secured despite the lack of a popular repu-

*The "regional representatives" were representatives of the "nine regions," thus the "people of the nine regional representatives" were the people of China. On the nine regions, see "Establishing Models and Exemplars," note 15.
†Xu Gan had in mind a government modeled on that of the early Zhou rulers, who held dominion over a confederacy of vassal states.

其文王乎！畋於渭水邊，道遇姜太公。皤然皓首，方秉竿而釣。文王召而與之言，則帝王之佐也。乃載之歸，以為太師。姜太公當此時貧且賤矣。年又老矣。非有貴顯之舉也。其言誠當乎賢君之心，其術誠合乎致平之道。文王之識也，灼然若披雲而見日，霍然若開霧而觀天。斯豈假之於眾人哉！

非惟聖然也，霸者亦有之。昔齊桓公夙出，甯戚方為旅人。宿乎大車之下。擊牛角而歌。歌聲悲激。其辭有疾於世。桓公知其非常人也。召而與之言，乃立功之士也。於是舉而用之，使知國政。

凡明君之用人也，未有不悟乎己心，而徒因眾譽也。用人而因眾譽焉，斯不欲為治也。將以為名也。然則見之不自知，而以眾譽為驗也，此所謂效眾譽也。非所謂效得

tation. Such an example is that of King Wen. While hunting beside the Wei River, King Wen met Jiang Taigong on the way, by chance. The white-haired old man was holding a rod fishing. King Wen called him over and talked to him, discovering him to be a man who could assist a sovereign. He took him back with him in his carriage, making him grand preceptor.*[2] When this happened, Jiang Taigong was poor, of humble background, and, moreover, old. He was not promoted to office by the noble or the famous; rather, his words were in genuine accord with the heart of the worthy ruler, and his methods were in genuine accord with the way for bringing about stability of rule. Wen Wang's powers of recognition were brilliant, just like the clouds opening up to reveal the sun, and swift, just like mist dispersing to show the sky. Could this have been brought about by relying on popular opinion?

Yet this applied not only to sage rulers but also to overlords. Duke Huan of Qi went out early one morning. Ning Qi, a traveler, was beating an ox horn and singing a sad and moving song as he spent the night under his large cart.[3] The lyrics of the song were critical of the times. Realizing that Ning Qi was no ordinary person, Duke Huan called him over and spoke with him, discovering him to be a fellow who could contribute real achievements.[4] Thereupon he promoted him to office, placing him in charge of the country's government.[5]

In employing men, no clear-sighted ruler would ignore his own feelings and rely solely on popular reputation. If, in employing men, a ruler were to rely on accounts of popular reputation conveyed to him, this would be not because he wanted to bring about order but rather because he wanted to make a name for himself. Accordingly, if he looks at a candidate on the evidence of popular reputation and not on the basis of his own understanding, this is what is called "confirming popular reputation" and not what is known as "confirming the em-

*On Tai Gongwang, see "Titles and Emoluments," note 9 and page 127, note †.

賢能也。苟以眾譽為賢能，則伯鯀無羽山之難，而唐虞無九載之費矣。聖人知眾譽之或是或非，故其用人也，則亦或因或獨。不以一驗為也。況乎舉非四嶽也。

世非有唐虞也。大道寢矣，邪說行矣。臣已詐矣，民已惑矣。非有獨見之明，專任眾人之譽，不以己察，不以事考，亦何由獲大賢哉？且大賢在陋巷也，固非流俗之所識也。何則？大賢為行也，哀然不自〔見〕[d]，偏然若無能。不與時爭是非，不與俗辯曲直。不矜名，不辭謗，不求譽。其味至淡，其觀

---

[d] Following the *Longxi jingshe* and *Liangjing yibian* versions of *Discourses* in reading *man* 滿 not *jian* 見.

ployment of the worthy and able." If popular reputation can be taken as evidence of worth and ability then Baron Gun would not have been exiled to Mount Yu, and neither would Tang and Yu [Yao and Shun] have wasted nine years. Because the sage knows that a popular reputation is sometimes right and sometimes wrong, therefore in employing men, he sometimes accepts it and sometimes he makes his own judgment. He does not, however, make a decision on the basis of a single piece of evidence. How much more should this be so when those who are promoting men to office are not the four chiefs of the feudal lords?*

In this age, there is no Tang or Yu; the great way is dormant and heresies are spread. Ministers deceive themselves and people confuse themselves. If the ruler does not have the perspicacity of singular judgment but exclusively heeds popular reputations; if he does not investigate candidates himself, nor examines into matters, then he would surely have no way to obtain the services of great worthies. Great worthies living in mean dwellings will certainly not be recognized by the common people.[6] Why is this the case? In his actions and behavior the great worthy is self-effacing and without conceit. His down-and-out appearance makes him seem like a person of no ability. He does not quarrel with his contemporaries about matters of right and wrong, nor dispute with ordinary people over distinctions about what is bent and what is straight. He does not attach importance to reputation, nor does he deny slanderous allegations made against him. He does not seek reputation. He has the air of being utterly plain and the appear-

*Baron Gun was the father of Great Yu. See *Documents*, 2.19b–20a, 3.14b: "The emperor said, 'Oh, four chiefs of the feudal lords, the swelling floods have caused widespread disaster.… Is there a capable man who can bring them under control?' They all said, 'Yes, there is Gun.' The emperor said, 'Alas, no! He disobeys orders and harms good men.' The four chiefs said, 'Select him. See if he is suitable and then decide if he should be appointed to the post.' The emperor said to Gun, 'Go and take care!' After nine years he had still made no progress.… He was exiled to Mount Yu." The emperor in this passage is Yao, while it was Shun who exiled Gun to Mount Yu.

至拙。夫如是，則何以異乎人哉？其異乎人者，謂心統乎群理而不繆，智周乎萬物而不過，變故暴至而不惑，真偽叢萃而不迷。故其得志則邦家治以和，社稷安以固，兆民受其慶，群生賴其澤。八極之內同為一。斯誠非流俗之所豫知也。

「不然，安得赫赫之譽哉？」

「其赫赫之譽者，皆形乎流俗之觀，而曲同乎流俗之聽也。君子固不然矣。昔管夷吾嘗三戰而皆北。人皆謂之無勇。與之分財，取多。人皆謂之不廉。不死子(糾)〔糾〕之難，人皆謂之背義。若時無鮑叔之舉、霸君之聽，休功不立於世，盛名不垂於後，

ance of being utterly clumsy. This being the case, then how can he be differentiated from other people? Where he differs from other people is that by bringing the myriad principles together in his mind, he makes no error, and by encompassing the myriad creatures with his wisdom, he makes no mistakes. When unexpected events suddenly occur, he does not become flustered, and when truth and falsehood proliferate side by side he does not become confused.[7] Thus, should he secure office, the country will be harmoniously ordered, and the nation will be securely stabilized; the multitudes of people will receive his blessings, the myriad living things will rely on his beneficence for their sustenance, and all within the eight directions will be as one.* Truly, this is not something of which the common people could have foreknowledge.

[An interlocutor may object, saying] "That is not right!†[8] Otherwise, how could there possibly be glowing reputations?"‡

[I would reply,] "All glowing reputations are formed on the basis of the observations of the common people, and so they are twisted to conform with what the common people want to hear. The gentleman is certainly not like this. In the past, Guan Yiwu [Guan Zhong] was thrice defeated in battle, and people all said that he was cowardly.[9] When he divided up goods with Bao Shu, he took the most, and people all said that he was corrupt.§[10] Furthermore, he did not share the misfortune of Zijiu's death.|| People all said that he had turned his back on doing what is right and if, at the right times, he had not had Bao Shu's recommendation and the overlord's ear, he would not have been able to make great achievements in his own age, nor would his grand reputation have been passed on to pos-

---

* North, south, east, west, northeast, northwest, southeast, and southwest.
† That is, it is not the case that only "nameless" worthies can make real achievements.
‡ That is, there must be some basis on which popular reputations stand.
§ Bao Shu was a long-time friend of Guan Zhong's.
|| See *Analects*, 14.17 and "Wisdom and Deeds," page 111, note §.

則長為賤丈夫矣。

「魯人見仲尼之好讓而不爭也。亦謂之無能。為之謠曰：『素鞞羔裘，求之無尤。(黑)ᵉ〔羔〕裘ᶠ素鞞，求之無戾。』夫以聖人之德，昭明顯融，高宏博厚，宜其易知也。且猶若此，而況賢者乎？以斯論之，則時俗之所不譽者，未必為非也。其所譽者，未必是也。故《詩》曰：『山有扶蘇，隰有荷華。不見子都，乃見狂且。』言所謂好者非好，醜者非醜。

ᵉ Notes following the *Concordance to Zhonglun* in emending *hei* 黑 to *gao* 羔.
ᶠ Emending the two occurrences of *qiu* 裘 to *tou* 投 on the basis of a similar ditty found in *Lüshi chunqiu*, 16.11a–11b.

terity, and so he would have been forever known as a despicable man.[11]

"When the people of Lu saw that Confucius did not vie with others because of his fondness for deference, they also said that he was incompetent and made up a song about him:

> The one who wears undyed ceremonial knee covers
>   and a lambskin coat,
> There is no crime in getting rid of him.
> The one who wears a black coat and undyed
>   ceremonial knee covers,
> There is no crime in getting rid of him.*

Given that Confucius's virtue was brilliant, bright, resplendent, shining, lofty, vast, extensive, and sincere, it should have been easy to recognize him as a sage. If even a sage should be thus treated, then how much more so will it be the case for worthies? Argued in this way, then those who are not praised by the common people of the day are not necessarily those who do not deserve a reputation, while those who are praised are not necessarily those who do deserve a reputation. Thus the *Book of Odes* says:

> On the mountains are the mulberry trees:
> In the marsh are the lotus flowers.
> I do not see Zidu,
> But I see this crazy man.†[12]

This is saying that that which is called beautiful was, in fact, not beautiful and that which is ugly was, in fact, not ugly.

---

*The *Lüshi chunqiu*, 16.11a–b, version of the ditty relates that people of Lu made fun of Confucius when he first took up an appointment in Lu, but within three years his meritorious achievements were already obvious. See also *Kongcongzi*, 5.9a.

†According to the Mao preface, this ode satirized Duke Zhao of Zheng (r. 696–695) for regarding as beautiful that which was not beautiful. See *Odes*, 4/3.8a. On Zidu, see "Valuing Proof," page 69, note *.

亦由亂之所致也。治世則不然矣。

　「叔世之君生乎亂。求大臣、置宰相而信流俗之說，故不免乎國風之譏也。而欲與之興天和，致時雍，遏禍亂，弭妖災。無異策穿蹄之乘，而登太行之險，亦必顛躓矣。故《書》曰：『肱股墮哉，萬事隳[g]哉。』此之謂也。」

　「然則君子不為時俗之所稱？」

　曰：「孝悌忠信之稱也，則有之矣。治國致平之稱，則未之有也。其稱也無以加乎習訓詁之儒也。夫治國致平之術，不兩[h]得其人，則不能相通[i]也。其人又寡矣。寡不稱眾，將誰使辨之。

　「故君子不遇其時，則不如流俗之士聲名章徹也。非徒如此，又為流俗之士所裁制焉。高下之分，

[g] Standard versions of the modern *Documents* text read *duo* 惰 for *duo* 墮 and *duo* 噎 for *duo* 隳.

[h] Following the *Longxi jingshe* and *Liangjing yibian* versions of *Discourses* in reading *duo* 多 not *liang* 兩.

[i] Emending *tong* 通 to *yu* 遇 on the grounds of coherence.

This, too, was brought about by chaos. In times of order things are not so.

"The last ruler of a dynasty is born out of chaos. In seeking senior ministers and establishing a grand counselor, he listens to common rumors and so does not escape the ridicule of 'Airs of the States'; yet with the aid of his ministers he hopes to reestablish harmony with heaven, cause the seasons to proceed in their proper sequence, put an end to disaster and chaos, and dispel demons and pestilence.* This is no different from flogging a worn-out horse to pull a carriage up the precipices of the Taihang mountains—it is bound to trip over. Therefore, the *Book of Documents* says: '[When the ruler concerns himself with petty matters at the expense of important concerns,] his ministers will grow idle and affairs will all go to ruin.'[13] This is what is meant."

"This being so, are gentlemen then not commended by the common people of the day?"

I replied, "While there are commendations for being 'filially pious,' 'deferring to one's elders,' 'doing one's best on behalf of others,' and 'living up to one's word,' yet there are none for 'being able to bring stability to the country.'† This is because such commendations could not be applied to those scholars who practice textual glossing. If the services of more men who are skilled in the methods of bringing stability to the country are not secured, then there is no way that the ruler and they can meet. Moreover, there are already very few such men as it is. If the few are outnumbered by the many, then who will be able to distinguish them from among the many?

Hence if the gentleman is not born in the right times, not only will he fail to match the splendid and widespread reputation of vulgar persons, but, moreover, he will be controlled by them, for the distinctions of high and low and the worth

---

* The "Airs of the States" is the first division of the *Odes*. Mao no. 84 (see above, "On the mountains are the mulberry trees … ") is in this division.
† Here he is presumably referring to the reign of Emperor Ling (168–89 CE), rather than the puppet Emperor Xian (r. 189–220).

貴賤之賈，一由彼口。是以沒齒窮年，不免於匹夫。昔荀卿生乎戰國之際，而有叡哲之才。祖述堯舜，憲章文武，宗師仲尼，明撥亂之道。然而列國之君以為迂闊，不達時變。終莫之肯用也。至於游說之士，謂其邪術，率其徒黨，而名震乎諸侯。所如之國，靡不盡禮郊迎，擁篲先驅。受爵賞為上客者，不可勝數也。故名、實之不相當也，其所從來尚矣。何世無之？天下有道，然後斯物廢矣。」

of noble and base are all determined by what they say. Because of this, then, till the end of his days he will not be able to rise above the rank of a common person. In the past, Xun Qing [c. 335–c. 238], who lived in the Warring States period, was a man of brilliant intellectual talents. He transmitted the ancient traditions of Yao and Shun, modeled himself on the teachings of Wen and Wu, took Confucius as his teacher, and made clear the way to bringing chaos to order. Despite this, the rulers of the various states thought him impractical and out of step with the changing times; to the end of his days none of them was prepared to employ him.[14] As to those men of education and local standing engaged as itinerant persuaders, because they spouted their pernicious methods and led their groups of followers, their reputations resounded in the courts of the feudal lords. There was no state that they went to where the ruler failed to carry out ceremonial to the fullest by proceeding in person to the outskirts of the city to meet them, to sweep the ground in front of them, and to lead them into the city.[15] Countless numbers received titles and were treated as top-ranking guests. Thus long has been the history of names not matching actualities. In what age has it not been so? Only when the world has the way will this state of affairs come to an end!

《慎所從第十七》

夫人之所常稱曰：「明君舍己而從人，故其國治以安；闇君違人而專己，故其國亂以危。」乃一隅之

17. Be Careful of the Advice One Follows

The theme of this essay is the importance of the ruler's being able to discern which advice to follow: good advice will lead to order, while bad advice will lead to disorder. The ruler is able to secure good advice only if he personally makes senior appointments. Xu Gan cites several historical examples to illustrate the consequences of (1) following bad advice, (2) selecting the best strategies from among many, and (3) rejecting sound strategies. The final portion of the essay takes the form of a dialogue, in which Xu Gan cites the example of the famous general Xiang Yu to argue that fighting battles is not the way to take dominion of the empire. Instead, he proposes the two principles of wisdom and humaneness, because "If a ruler is humane, then all the other states will come to incline toward him, and if he is wise, then outstanding men will rally to him." These views are inconsistent with those he develops in "Wisdom and Deeds," suggesting that the two essays were written at different periods in Xu Gan's life. One might speculate that Xu Gan cited the example of Xiang Yu to curb Cao Cao's use of military force.

There is something which people often say: "The clear-sighted ruler gives up his own views and defers to the views of others.[1] Accordingly, his country is in good order and stable. The dim-sighted ruler, however, heeds only his own views and turns his back on the views of others. For this reason, his country is disordered and unstable." Now this is a one-sided doc-

偏說也，非大道之至論也。

　凡安危之勢，治亂之分，在乎知所從，不在乎必從人也。人君莫不有從人，然或危而不安者，失所從也；莫不有違人，然或治而不亂者，得所違也。若夫明君之所親任也，皆貞良聰智。其言也，皆德義忠信。故從之則安，不從則危。闇君之所親任也，皆佞邪愚惑。其言也，皆姦回（諂）〔諂〕諛。從之安得治，不從之安得亂乎？

　昔齊桓公從管仲而安。二世從趙高而危。帝舜違四凶而治。殷紂

trine which looks at only one perspective. It is not one of the supreme tenets of the great way.

In all cases, it is knowing what advice should be followed which determines the propensity toward stability or instability, and the difference between order and disorder; it does not depend upon necessarily following the advice of others. All rulers have occasion to follow the advice of others. So if some advice leads to instability rather than stability, this is because the ruler made the wrong choice in what he followed. All rulers have occasion to reject the advice of others. So if some advice promotes order and not disorder, this is because the ruler made the right choice in what he rejected. Those whom the clear-sighted ruler personally appoints to office are all incorruptible, good, intelligent, and wise, and their words all full of virtue, rightness, loyalty, and good faith. Hence to follow their advice will lead to stability and not to do so will lead to instability. Those whom the dim-sighted ruler personally appoints to office are all fawning, aberrant, stupid, and confused, and their words are all crafty, treacherous, obsequious, and flattering. If their advice were followed, how could it lead to order? And if it were not followed, how could it lead to disorder?

In the past, Duke Huan of Qi followed the advice of Guan Zhong and brought about stability.* The Second Emperor of Qin [r. 210–207] followed the advice of Zhao Gao and brought about instability.† Emperor Shun rejected the advice of the four wicked ones and brought about order.‡ Zhou of the Yin

---

* See "Wisdom and Deeds," note 10.
† Zhao Gao was chamberlain for attendants under the Second Emperor of Qin. He later forced the Second Emperor to commit suicide. Within two years the dynasty fell. See *Shiji*, 87.2552–63.
‡ The term, "four wicked ones" (*si xiong*), seems to be used by Xu Gan to refer to the four officials described in *Documents*, 3.14a–b (also *Mencius*, 5A.3): "Shun exiled Gong Gong 共工 to Youzhou, banished Huan Dou 驩兜 to Mount Chong, drove San Miao 三苗 into Sanwei, and banished Gun 鯀 to Mount Yu." However, in *Zuo Commentary*, Wen 18, 20.19b–20a, *si xiong* is used to refer to the following four clan leaders: "After

違三仁而亂。故不知所從而好從人，不知所違而好違人，其
敗一也。孔子曰：「知不可由，斯知所由矣。」

　　夫言或似是而非實，或似美而敗事，或似順而違道。此三
者、非至明之君不能察也。燕昭王使樂毅伐齊。取七十餘
城。莒與即墨未拔。昭王卒。惠王為太子，時與毅不平。即
墨守者田單，縱反間於燕，使宣言曰：「王已死，城之不拔
者三[a]耳。樂毅與新王有隙。懼誅而不敢歸。外以伐齊為名，
實欲因齊人未附。故且緩即墨以待其事。齊人所懼，惟恐他
將之來。即墨殘矣。」惠王以為然，使騎劫代之。大為田單
所破。

[a] Emending *san* 三 to *er* 二 on the grounds that it contradicts the earlier
passage which lists only Ju and Jimo. According to *Shiji*, 80.2429, 82.2453,
there were only two cities. *Shiji*, 34.1558, and the Bao Biao commentary to
Zhanguoce, 9.35b, however, list three cities.

dynasty failed to follow the advice of the three humane ones and brought about disorder.* Thus not knowing which course should be followed and so preferring to follow the advice of others, and not knowing which course should not be followed and so preferring not to follow the advice of others, equally result in failure. Confucius said: "Knowing what may not be followed is to know what to follow."[2]

Some words appear to be right yet are not so in actuality. Some appear to be excellent yet destroy one's affairs. Some appear as if one may act in accord with them, yet are contrary to the way. Only the most clear-sighted ruler is able to discern the difference in each of these three cases. King Zhao of Yan [311–279] sent Yue Yi on a punitive expedition against Qi and he took more than seventy cities. Only Ju and Jimo were still holding out when King Zhao died. When King Zhao's successor, King Hui [278–272], was still crown prince, he and Yue Yi had often failed to see eye to eye. The defender of Jimo, Tian Dan, spread dissension in Yan, starting a rumor that "the king is already dead. There are only two cities which Yue Yi has failed to capture. Because of the rift between Yue Yi and the new king, Yue Yi dares not return for fear of being killed. While in name he puts on a facade of attacking Qi, in actuality he wants to side with the people of Qi, but as yet has not joined them. It is for this reason that in the interim he is proceeding very slowly with his attack on Jimo, as he waits for the conclusion of his negotiations with Qi. The only thing that the people of Qi are worried about is that another general might come. Jimo would then be finished." King Hui believed the rumor and ordered Qi Jie to replace Yue Yi. Qi Jie was badly defeated

---

Shun began to serve Yao, he made four gates in the city wall and banished the four clan leaders—Hundun 渾敦, Qiongqi 窮奇, Taowu 檮杌 and Taotie 饕餮 —to the four distant border regions, where they were to stop evil demons from entering. Because of this, when Yao died, the empire was united." It may even be that Xu Gan equated the two groups.
* See "Wisdom and Deeds," note 24.

此則似是而非實者也。

　　燕相子之有寵於王。欲專國政。人為之言於燕王噲曰：「人謂堯賢者，以其讓天下於許由也。許由不受，有讓天下之名，而實不失天下。今王以國讓於相子之，子之必不敢受。是堯與王同行也。」燕噲從之。其國大亂。此則似美而敗事者也。

　　齊景公欲廢太子陽生，而立庶子荼。謂大夫陳乞曰：「吾欲立荼，如何？」乞曰：「所樂乎為君者。欲立則立之，不欲立則不立。君欲立之，則臣請立之。」於是立荼。此則似順而違道者也。

　　且夫言畫施於當時，事效在於後日。後日遲至，而當時速決也。故今巧者常勝，拙者常負，其勢然也。此謂中主之聽也。至於闇君，則不察辭之巧拙也。二策並陳，而從〔致〕[b]己之欲者。明君不[c]察辭之巧拙也。二策並陳，而從

[b] Following Qian's *Reading Notes* in inserting *zhi* 致.
[c] Omitting the character *bu* 不 on the grounds of coherence.

by Tian Dan.*[3] This is an example of words appearing to be the case, yet not being so in actuality.

Minister Zizhi was greatly favored by King Kuai of Yan [r. 320–312]. Because he wanted to take control of the government of the state, he had someone say to King Kuai, "People say that Yao was a worthy because he yielded his throne to Xu You. Xu You, however, declined to accept, and so Yao earned the name of having yielded his throne, while in actuality he did not lose his kingdom.[4] Now if you were to yield the throne to Minister Zizhi, he would certainly not dare to accept, and so your virtuous action would be the same as that of Yao." King Kuai of Yan followed his advice and his state was thrown into complete chaos.[5] This is an example of words which appear to be excellent destroying one's affairs.

Duke Jing of Qi [r. 547–490] wanted to depose the crown prince, Yang Sheng, and appoint Tu, the son of a concubine, in his place. He said to counselor Chen Qi, "I wish to appoint Tu as my successor. What do you think?" Chen Qi replied, "Do whatever pleases you, sir. If you wish to appoint him, then do so; if not, then do not. If you wish to appoint him, then I beg leave to carry out your wishes." And so, Tu was appointed.[6] While this would seem to be acting in accord with the duke's wishes, in actuality it works against the way.

Furthermore, words and plans put into practice today will come to be realized in affairs at a later date. Although this later date may be slow to come, what it has in store will have been already swiftly determined today. Thus those who offer clever advice today will usually be successful, while those who offer stupid advice today will usually be unsuccessful. The potential in each case makes it so. This is called "securing the ruler's attention." The clear-sighted ruler investigates whether advice is clever or stupid. When two strategies are presented, he opts

*Yue Yi was a native of Wei, who became a general for Yan heading an alliance of troops from Yan, Wei, Zhao, and Qin to defeat the Qi army in 284 BCE. See also "Destruction of the State."

其致己之福者[d]。故高祖、光武，能收群策之所長，棄群策之所短，以得四海之內，而立皇帝之號也。吳王夫差、楚懷、〔王襄[e]〕〔襄王〕，棄伍員、屈平之良謀，收宰噽上官之諛言，以失江漢之地，而喪宗廟之主。此二帝三王者，亦有從人，亦有違人。然而成敗殊馳、興廢異門者。見策與不見策耳。

<hr>

[d] Reversing the order of these two sentences on the grounds of rhetorical coherence as suggested by the *zhi yu* 至於 … *ze* 則 construction. The revised order thus reads: 明君 … 致己之福者 … 至於闇君則 … 致之欲者.
[e] Adding an additional character *wang* 王 after *Xiang* 襄; not following the emendment proposed in the *Concordance to Zhonglun*.

for the one that will bring him good fortune. As to the dull-sighted ruler, he does not investigate whether advice is clever or stupid; when two strategies are presented, he opts for the one that accords with his own desires. It is for this reason that emperors Gaozu [r. 206–194] and Guangwu [r. 25–58] were able to select the best strategies from amongst many and reject the inferior ones, thereby securing rule over the whole empire and establishing themselves as emperors.* King Fuchai of Wu [r. 495–473] and King Huai [r. 328–299] and King Xiang [r. 298–263] of Chu rejected the sound strategies of Wu Yun and Qu Ping, while accepting the flatteries of Grand Steward Pi and senior official Jin Shang. As a result, Chu lost the territories along the Han and Yangtze rivers and Wu lost the tablets of its ancestral hall.† These two emperors and three kings were alike in that sometimes they followed the advice of others and sometimes they rejected the advice of others; yet victory and defeat each ride in different carriages; and success and failure each take different doors. It all comes down to a question of appreciating the merits of a strategy. To

---

* Gaozu and Guangwu were the first emperors of the Western and Eastern Han, respectively.
† See "Distinguishing Between Premature Death and Longevity," page 193, note †, on Wu Yun, King Fuchai, and Bo Pi. Being slandered by Bo Pi, King Fu Chai ignored the advice of Wu Yun that he should completely destroy the state of Yue. As Wu had predicted, King Goujian 勾踐 of Yue (r. 496–465) eventually rose to strength again, destroying Wu and killing Fuchai. The phrase "lost the tablets of its ancestral hall" refers to this destruction. See *Zuo Commentary*, 60.17a. Because King Huai of Chu believed the lies that the senior official, Jin Shang, had told about Qu Yuan 屈原 (Qu Ping), he chose to ignore Qu Yuan's advice not to travel to meet King Zhao of Qin because it might be a trap. This led to his being taken prisoner, eventually dying in Qin, and also to the loss of the territory north of the Han River, as well as the Shangyong 上庸 region (see *Shiji*, 40.1735) in 280, and the Wu 吳 commandery and part of the Jiangnan 江南 region in 277 (see *Shiji*, 5.213). His son and successor, King Xiang, learning of Qu Yuan's criticism of the king's failure to heed his advice, expelled him from Chu. Qu Yuan subsequently committed suicide. See *Shiji*, fascicle 84.

不知從人甚易，而見策甚難。夷考其驗，斯為甚矣。

問曰：「夫人莫不好生而惡死，好樂而惡憂。然觀其舉措也，或去生而就死，或去樂而就憂。將好惡與人異乎？」

曰：「非好惡與人異也，乃所以求生與求樂者失其道也。譬如迷者欲南而反北也。今略舉一驗以言之。昔項羽既敗，為漢兵所追，乃謂其餘騎曰：『吾起兵至今八年，身經七十餘戰。所擊者服，遂霸天下。今而困於此，此天亡我，非戰之罪也。』斯皆存亡所由欲南反北者也。

「夫攻戰、王者之末事也。非所以取天下也。王者之取天下也，有大本，有仁智之謂也。仁則萬國懷之，智則英雄歸之。御萬國，摠英雄，以臨四海。其誰與爭？若夫攻城必拔，野戰必克，

follow someone's strategy in ignorance is very simple, but to appreciate the merit of a strategy is exceedingly difficult. The most important thing lies in determining the efficacy of a strategy.

Someone asked, "All people are fond of life and detest death, just as they are fond of joy and detest sorrow. Yet on examining the actions of some, it would appear that they reject life and happiness and draw near to death and sorrow. Could it be that their likes and dislikes are different from those of other people?"

I replied, "No, it is not the case that their likes and dislikes are different from those of other people; rather, they have lost the way by which to seek life and happiness. They may be likened to someone lost who wants to go south but travels north. Now I will outline an example to explain what I mean. In the past, when Xiang Yu had already been defeated and was being pursued by Han soldiers, he said to his remaining horsemen: 'It has been eight years since I first raised troops. In that time I have participated in more than seventy battles. Everyone I attacked submitted, and so I came to assert my domination of the realm. But now I am besieged here! This is because heaven would destroy me, and not because of any faults made in battle.'[7] These words are an example of regarding warfare as the sole means for determining all matters of survival and defeat. They are like the words of the lost soul who wants to go south but travels north.[8]

"Fighting battles is a matter of peripheral importance to a king. It is not the means by which to take dominion of the realm. In taking dominion of the realm, the king has two great fundamentals at his disposal: they are called possessing humaneness and wisdom. If a ruler is humane, then all the other states will come to incline toward him, and if he is wise, then outstanding men will rally to him. He presides over the whole empire through his command of all the states and his assembly of outstanding men—who would vie with him? As for making sure that a city is taken when attacked, and that victory is certain when fighting a battle in the open, these mat-

將帥之事也。羽以小人之器，闇於帝王之教，謂取天下一由攻戰。矜勇有力。詐虐無親。貪嗇專利。功勤不賞，有一范增。既不能用，又從而疑之。至令憤氣傷心，疽發而死。豪傑背叛，謀士違離。以至困窮，身為之虜。然猶不知所以失之，反瞋目潰圍，斬將取旗，以明非戰之罪。何其謬之甚歟？高祖數其十罪。蓋其大略耳。若夫纖介之失，世所不聞，其可數哉？

「且亂君之未亡也，人不敢諫。及其亡也，人莫能窮。是以至死而不寤，亦何足怪哉？」

ters are the business of generals and military leaders. Being the sort of man who had the mettle of a petty individual, and who was ignorant of the teachings of the sage emperors and kings, Xiang Yu said that the only way to take dominion of the empire was to fight battles. He vigorously boasted of his bravery and cold-heartedly deceived and oppressed everyone. Greedy and parsimonious, he did not share his gains, nor did he reward achievement and hard work. There was a man called Fan Zeng, for whom Xiang Yu had no further use and whom he came to suspect. This caused Fan Zeng to be so angry and hurt that he developed an abscess and died.[9] The leading men turned against Xiang Yu, and his strategists left him. Finally, besieged and with nowhere else to turn, he was taken prisoner. Yet Xiang Yu still did not understand how he had come to fail to take dominion of the realm. On the contrary, with glaring eyes he broke through the encirclement of his besiegers, beheading some of their generals and taking their flags, so as to prove that it was not his fighting that was at fault.[10] How absurd! Han Gaozu [Liu Bang] enumerated Xiang Yu's ten faults.[11] Yet this was only a general outline. Could his minor faults that went unrecorded ever be enumerated?

"Until a reckless ruler has been destroyed, people are unable to know the full extent of his failings. Once he has been destroyed, however, there is no end to people's criticisms. Is it any wonder that such rulers remain unawakened right up until when they die?"

《亡國第十八》

凡亡國之君，其朝未嘗無致治之臣也，其府未嘗無先王之書也。然而不免乎亡者，何也？其賢不用，其法不行也。苟書法而不

18. Destruction of the State

Xu Gan attributes the demise of states in former times to two failures of rulers: to employ worthy men and to put into practice the models that had been laid down. Once again, he draws on the examples of legendary and historical figures to illustrate his claims. He then develops an extended criticism of rulers who go to great lengths to employ worthy men but, having secured their services, fail to consult them. Wang Mang (r. 9–23 CE), in particular, is criticized for forcing worthy men into his service and making it impossible for them either to proffer advice or to retire from service. In contrast to such figures as Wang Mang and the rulers of the late Han are the clear-sighted rulers who gain the hearts and minds of the worthies who enter their service. It is a ruler's right actions which determine whether worthy men will, of their own accord, come to serve him. The essay concludes with Xu Gan's lament that in his own times rulers are not like this, and so they are unable to obtain the services of worthy men who prefer to feign madness rather than to compromise their principles. Again, one might speculate that the intended audience of this essay was Cao Cao.

In all cases, the rulers of states which have been destroyed have lacked neither ministers in their courts who were capable of bringing about order nor the books of the former kings in their archives. Despite this, they were unable to avoid the destruction of their states. Why? Because they did not employ worthies, nor did they implement the laws bequeathed by the former kings. If these models are written down but are not, in

行其事，爵賢而不用其道，則法無異乎路說，而賢無異乎木主也。

昔桀奔南巢，紂踣於京，厲流於彘，幽滅於戲。當是時也，三后之典尚在，〔而〕[a] 良謀之臣猶存也。下及春秋之世，楚有伍舉、左史倚相、右尹子革、白公子張[b]，而靈王喪師。衛有太叔儀、公子鱄、蘧伯玉、史鰌，而獻公出奔。晉有趙宣(子)〔孟〕[c]、范武子、太史董狐，而靈公被殺。魯有子家羈、叔孫婼，而昭公野死。齊有晏平仲、南史氏，而莊公不免〔弒〕[d]。虞、虢有宮之奇、舟之僑，而二公絕祀。由是觀之，苟不用賢，雖有

[a] Not following the *Concordance to Zhonglun* in introducing this character into the text.
[b] Following *Qunshu zhiyao*, 46.19a, in omitting *bai gong zi zhang* 白公子張.
[c] Not following the *Concordance to Zhonglun* in emending *zi* 子 to *meng* 孟.
[d] Following *Qunshu zhiyao*, 46.19b, in inserting *shi* 弒.

fact, implemented, and if worthies are given titles but their ways are not employed, then the models will be no different from casual talk and the worthies no different from the wooden tablets on the ancestral altar.

In the past, Jie fled to Nanchao; Zhou fell into the fire in the capital; Li was banished to Zhi; and You [r. 781–771] was destroyed at Xi.* During the periods of their rule, the books of Yao, Shun, and Yu still existed and they still had ministers who were fine strategists. By the Spring and Autumn period, Chu had Wu Ju, the scribe of the left, Yi Xiang, and the director of the right, Zige, yet King Ling's [r. 541–529] army was still defeated.[1] Wei had Taishu Yi, Prince Zhuan, Qu Boyu, and Shi Qiu, yet Duke Xian [r. 546–544] fled from his own state.[2] Jin had Zhao Xuanzi, Fan Wuzi, and the grand scribe, Dong Hu, yet Duke Ling [r. 620–607] was murdered.[3] Lu had Zijia Ji and Shusun Ruo, yet Duke Zhao [r. 541–510] died in the countryside.[4] Qi had Yan Pingzhong and the scribe of the south, yet Duke Zhuang [r. 553–548] still did not avoid being assassinated.[5] The states of Yu and Guo had Gong Zhiqi and Zhou Zhiqiao, yet the sacrifices of both states were discontinued.† In view of this, it would appear that even though a ruler has worthies in his employ, if he does not act on their

---

* Jie is the traditional wicked last ruler of the Xia. On his flight to Nanchao, see *Shiji*, 2.89, *Zhengyi Commentary*. Zhou was the wicked last ruler of the Yin. See *Shiji*, 3.108. Li was a cruel Zhou king (r. 857/53–842/28) who was banished to Zhi by his own people. See *Zuo Commentary*, Zhao 26.7a–b; *Shiji*, 4.141–42. You was the inept successor to Li. He was killed by "barbarians" and his demise marks the end of the Western Zhou. See *Shiji*, 4.148–49.

† In 658 BCE, Jin presented some precious gifts to the state of Yu, requesting them to let the Jin army pass through Yu to attack the small state of Guo. The Yu ruler agreed. After three years, the same request was made again, but this time Gong Zhiqi said to the Yu ruler that it would be foolish to do so, for Guo acted as a buttress defense for Yu. See *Zuo Commentary*, Xi 2, 12.5b–6b; Xi 5, 12.22a. The ruler of Yu ignored Gong Zhiqi's advice, and Jin proceeded to destroy both Guo and Yu. Before fleeing to Jin, Zhou Zhiqiao had warned the duke of Guo of imminent disaster. See *Zuo Commentary*, Min 2, 11.6b–7a.

無益也。然此數國者，皆先君舊臣世祿之士，非遠求也。

乃有遠求而不用之者。昔齊桓公立稷下之官，設大夫之號，招致賢人而尊寵之。自孟軻之徒皆遊於齊。楚春申君亦好賓客。敬待豪傑。四方並集。食客盈館。且聘荀卿，置諸蘭陵。然齊不益強，黃歇遇難。不用故也。

夫遠求賢而不用之，何哉？賢者之為物也，非若美嬪麗妾之可觀於目也。非若端冕帶裳之可加於身也。非若嘉肴庶羞之可實於口也。將以言策，策不用，雖多亦奚以為！若欲備百僚之名，而不問道德之實，則莫若鑄金為人，而列於朝也。且無食祿之費矣。

counsels, there is no advantage in their being there. In all of these states there were men of standing whose forebears had served as ministers under former rulers, generation after generation; they did not go far to find such men.

There were, however, rulers who sought candidates from far away only to ignore their counsels. In the past, Duke Huan [r. 363–357?] of Qi established the office of Jixia, set up the title of counselor, and summoned worthy men whom he honored and favored.[6] Because of this, Mencius's followers sojourned in Qi.[7] Huang Xie, Lord Chunshen [r. ?–238] of Chu, was also fond of retainers. He received with courtesy the outstanding men who flocked to Chu from all over the country, and his buildings were full of retainers on his payroll. Moreover, he employed Xun Qing and posted him at Lanling.[8] But Qi did not become stronger, and Huang Xie encountered tragedy. This was because Huang ignored the counsels of his ministers.[*]

What is the point of seeking worthy men from faraway places and yet not using them? As objects, worthies cannot please the eye as can beautiful concubines; they cannot adorn the body as can hats, belts, or lower garments; and they cannot satisfy the palate as can fine viands and assorted delicacies. If a worthy's function is to advise the ruler of particular strategies, but the strategies are not adopted, then even if the ruler engages the services of more of them, what would be the point of doing so? If a ruler wants a reputation for having a full corps of officials in his service, yet does not concern himself with the actuality of the way and virtue, then he would be better off casting men out of bronze and arranging them in his court. Moreover, there would be no expenses for food and

---

[*]Huang Xie, Lord Chunshen, served as prime minister of Chu for more than twenty years. He was celebrated as a host and at times had more than three thousand retainers. Because he had failed to heed the warning of the minister, Zhu Ying, about the coup being plotted by Li Yuan, he and his family were murdered by Li Yuan's troops. See *Shiji*, 78.2397–98. Within fifteen years of Huang's death, Chu fell to Qi.

然彼亦知有馬必待乘之而後致遠，有醫必待行^{e}之而後愈疾。至於有賢，則不知必待用之而後興治者，何哉？賢者難知歟？何以遠求之？易知歟？何以不能用也？豈為寡不足用、欲先益之歟？此又惑之甚也。

賢者稱於人也，非以力也。力者必須多，而知者不待眾也。故王〔卒〕^{f}七萬，而輔佐六卿也。故舜有臣五人而天下治。周有亂臣十人而四海服。此非用寡之驗歟！且六國之君，雖不用賢，及其致人也，猶脩禮盡意，不敢侮慢也。至於王莽，既不能用，及其致〔之〕^{g}也，尚不能言。莽之為人也，內實姦邪，外慕古義。亦聘求名儒，徵命術士。政

^{e} Following *Qunshu zhiyao*, 46.19b, in reading *shi* 使 instead of *xing* 行.
^{f} Following the *Longxi jingshe* and *Xiao wanjuanlou congshu* versions of *Discourses* in inserting *zu* 卒.
^{g} Following *Qunshu zhiyao*, 46.19b, in inserting *zhi* 之.

emoluments! That being so, a ruler is also aware that if he has a horse he must first mount it before he goes far, and that if he has a physician he must first consult him before his illness is cured. What is the reason, then, that when it comes to having worthies, he does not know that he must first use them before order prevails? Is it that worthies are difficult to recognize? If so, then why seek them from faraway places? Are they easy to recognize? If so, why is the ruler unable to use them? Surely it is not because there are not enough of them available to be employed that he first wants to increase their number? Again, what utter delusion!

It is not for their strength that worthies are praised by other people. For strongmen to be of use there must be many of them, but the value of worthies does not depend on there being many of them. Thus a king has seventy thousand troops and is assisted by the six ministers.*[9] For this reason Shun had five ministers and the whole empire was ordered, and King Wu of Zhou had ten capable ministers and the whole world was brought to submission.[10] Is this not proof of the efficacy of using but a few ministers? Further, even though the rulers of the six states did not use the services of worthies, when they summoned them, they still practiced ceremonial with a full expression of feeling and did not dare to slight them.† In the case of Wang Mang [r. 9–23], however, not only was he incapable of using the services of worthies, but even when he summoned them to his court, he was still unable to speak to them. As a person, on the inside, Wang Mang was full of treachery and perversity; ostensibly, however, he admired the rightness of antiquity. He also invited famous classical scholars to serve him and sought to appoint to office gentlemen skilled in the methods of administrative control. Because his policies were

---

*That is, the minister of state, the minister of education, the minister of rites, the minister of war, the minister of justice, and the minister of works.
†The six states: the major states conquered and absorbed by the state of Qin at the end of the Warring States period: Qi 齊, Chu 楚, Yan 燕, Zhao 趙, Han 韓, and Wei 魏.

煩教虐，無以致之。於是脅之以峻刑，威之以重戮。賢者恐懼，莫敢不至。徒張設虛名以夸海內。莽亦卒以滅亡。且莽之爵人〔也〕[h]，其實囚之也。囚人者、非必著之桎梏，而置之囹圄之謂也。拘係之、愁憂之之謂也。使在朝之人欲進則不得陳其謀，欲退則不得安其身。是則以綸組為繩索，以印佩為鉗鐵也。小人雖樂之，君子則以為辱〔矣〕[i]。

故明 (王)〔主〕[j] 之得賢也，得其心也。非謂得其軀也。苟得其軀，而不論其心也，斯與籠鳥、檻獸無以異也。則賢者之於我也，亦猶怨讎也。豈為我用哉？雖曰班萬鍾之祿，將何益歟？故苟得其心，萬里猶近；苟失其心，同衾為遠。今不脩所以得賢者之心，而務循所以執賢者之身。至於社稷顛覆，宗廟廢絕。豈不哀哉？荀子曰：「人主之患，不在乎言不用賢，而在乎誠

[h] Not following the *Concordance to Zhonglun* in introducing this character into the text.
[i] Not following the *Concordance to Zhonglun* in introducing this character into the text.
[j] Following *Qunshu zhiyao*, 46.20a, in emending *wang* 王 to *zhu* 主.

in disarray and his prescriptions were tyrannical, there was no way he could bring them to office. Thereupon he threatened them with harsh punishments and severe mutilation. Being terrified, none of the worthies dared not to present himself for service. Wang Mang merely established a false reputation to boast to the world, and in the end, he, too, was destroyed. Furthermore, in bestowing titles on people, he was actually imprisoning them. This does not mean that he necessarily had them put in jail, secured in fetters and handcuffs. Rather, this means that he bound them in service and made them miserable. He created a situation where, although the people at his court wanted to make recommendations, they were unable to explain their plans, and when they wanted to retire, they were unable to do so in safety. This was to use the cord on a seal as a rope and the seal as a lock.* While petty men delighted in being thus secured in office, gentlemen found it humiliating.

Thus clear-sighted rulers gain the service of worthies by gaining their hearts. This is not called "gaining their bodies." If they gain their bodies, with no regard for their hearts, then this is no different from caging a bird or an animal. Worthies will thus regard one such as he to be like a hated enemy. Would they be likely to allow themselves to be used by one such as he? Even if he said that he would grant them an emolument of ten thousand bushels, of what use would it be? Thus if a worthy's heart is obtained, then, even if he were ten thousand miles away, it would be as if he were close by. If the ruler should lose the worthy's heart, however, then, even if he were sharing the same quilt, it would be as if he were far away. Nowadays, rulers do not cultivate ways to obtain the hearts of worthies, instead devoting their efforts to look for ways to hold on to their bodies, even to the point where the state collapses and offerings are no longer made in the ancestral temple. Is this not lamentable? Xunzi said: "Calamity for a ruler lies not in saying that he will not employ the worthy, but rather in really

---

*That is, the cord and seal of an official.

不用賢^k。言〔用〕賢者、口也，(知)〔卻〕^l賢者、行也。口、行相反，而欲賢者〔之〕進，不肖者〔之〕^m退，不亦難乎！夫照蟬者務明ⁿ其火、振其樹而已。火不明，雖振其樹無益也。人主有能明其德者，則天下其歸之、若蟬之歸火^o也。」善哉言乎^p！

昔伊尹在田畝之中，以樂堯、舜之道。聞成湯作興，而自夏如商。太公避紂之惡，居於東海之濱。聞文王作興，亦自商如周。其次則甯戚如齊，百里奚入秦，范蠡如越，樂毅遊燕。故人君苟脩其道義，昭其德音，慎其威儀，審

---

^k Standard versions of the modern *Xunzi* text give the following reading for the first part of this quotation: "*Ren zhu zhi huan bu zai hu yong xian er zai hu cheng bi yong xian* 人主之患不在乎用賢而在乎誠必用賢." Following *Discourses*, rather than the *Xunzi* text, which is, in any case, questionable. See the comments of Lu Wenchao 盧文弨 (1717–96) and Wang Niansun 王念孫 (1744–1832), cited by Wang Xianqian, *Xunzi jijie*, 9.10b (464).

^l Following the *Xunzi* text and *Qunshu zhiyao*, 46. 20b, in reading *fu yan yong xian zhe* 夫言用賢者 rather than *yan xian zhe* 言賢者, and following the *Xunzi* text in reading *que* 卻 rather than *zhi* 知.

^m Not following the *Concordance to Zhonglun* in inserting *zhi* 之 after *zhe* 者 (twice). The *Xunzi* text reads *zhi zhi* 之至 instead of *jin* 進 and includes the character *zhi* 之 between *zhe* 者 and *tui* 退. I have followed the unemended *Discourses* reading.

ⁿ The *Xunzi* text reads *yao* 耀 for *zhao* 照 and includes the character *zai* 在 after *wu* 務.

^o The *Xunzi* text reads *jin* 今 before *ren zhu* 人主 and *ming* 明 after *gui* 歸. I have followed the unemended *Discourses* reading.

^p Following *Qunshu zhiyao*, 46.20b, in emending *hu* 乎 to *ye* 也.

not employing them. To say that one will employ the worthy is a matter of speech; to decline to employ them is an action. If one's speech and actions contradict one another, yet one wants to promote the worthy to office and have the unworthy withdraw, is this not problematical? Those who seek for cicadas by torchlight need only to devote their efforts to keeping the fire bright and to shaking the tree. If their fire is not bright, then, even if the tree is shaken, it will be to no avail. If there were a ruler capable of illuminating his virtue, then the whole realm would rally to him, just as cicadas come to fire."[11] Such wonderful words!

In the past, Yi Yin worked in the fields, thereby delighting in the way of Yao and Shun. When he heard of the rise of Cheng Tang, he went from the Xia to the Shang.* Taigong escaped Zhou's wickedness and lived by the Eastern Sea. When he heard of the rise of King Wen, he also ceased his affiliation with the former dynasty and went from the Shang to the Zhou.†[12] Later there were Ning Qi, who went to Qi, Boli Xi, who entered Qin, Fan Li, who went to Yue, and Yue Yi, who sojourned in Yan.‡[13] A good ruler cultivates his way and appropriateness, and makes resplendent his reputation. He is careful to maintain his awe-inspiring demeanor and reviews

---

*According to Zhou traditions, after Yi Yin had left the court of the wicked last ruler of Xia, he subsequently became the minister of Cheng Tang, the dynastic founder of the Shang dynasty. See *Mencius*, 5A.7, *Shiji*, 3.94.
† Jiang Taigong helped both King Wen and King Wu to establish the Zhou dynasty. On Jiang Taigong, see also "Examining the Selection of High Officials," note 2, and "Titles and Emoluments," note 9, and page 127, note †.
‡Ning Qi: see "Examining the Selection of High Officials." Originally a native of Yu 虞, Boli Xi's worth was not recognized by the duke of that small state, leading to its destruction in 654. Boli Xi then moved to many states before he became prime minister of Qin under Duke Mu (r. 659–621) for seven years. See *Mencius*, 5A.9, 6B.6. Fan Li, a native of Chu and contemporary of Confucius, went to Yue and helped King Goujian (r. 496–465) to destroy Wu. On Fan Li and King Goujian, see *Shiji*, fascicle 41. On Yue Yi, see "Be Careful of the Advice One Follows."

其教令，刑無頗僻，獄無放殘，仁愛普殷，惠澤流播，百官樂職，萬民得所，則賢者仰之如天地，愛之如親戚，樂之如塤(箎)〔篪〕，歆之如蘭芳。故其歸我也，猶決壅導滯水注之大壑。何不至之有〔乎〕[q]？

苟蠹穢暴虐，馨香不登。讒邪在側，佞媚充朝。殺戮不辜，刑罰濫害。宮室崇侈，妻妾無度。撞鐘舞女，淫樂日縱。賦稅繁多，財力匱竭。百姓凍餓，死莩盈野。矜己自得，諫者被誅。內外震駭，遠近怨悲。則賢者之視我容貌也，如魑魅；臺殿也，如狴犴；采服也，如衰絰；絃歌也，如號哭；酒醴也，如潲滫[r]；肴饌也，如糞土。從事舉錯，每無一善。彼之

---

[q] Not following the *Concordance to Zhonglun* in inserting this character into the text.
[r] Following Yu, *Zhuzi pingyi bulu*, 81, in emending *xiu di* 潲滫 to *xiu sou* 潲溲.

his instructions and edicts. He is impartial in meting out punishments and has no excess of cruelty in his prisons. His loving concern for other people is extensive and abundant, and his kindness and beneficence flow freely to be disseminated widely. Under him, the one hundred officials find contentment performing the duties of their offices, and the myriad people attain their proper station. If this is the situation, then worthy men will look up to him as they do to heaven and earth, love him as they do their parents, take pleasure in him as they do in the ocarina and bamboo flute, and delight in him as they do in the orchid's fragrance. Accordingly, they will rally to one such as he just like the body of dammed-up water which is channeled through the opening of a dike when it is opened up to flow into a large reservoir—what could possibly stop them from coming?

A bad ruler is crude, debauched, violent, and cruel, and the fragrance of the sacrificial offerings does not rise to the spirits. The slanderous and wicked surround him, and the fawning and obsequious fill his court. The innocent are mutilated and killed, and punishments and penalties are used indiscriminately to harm people. The palace chambers become places devoted to extravagance, and there is no limit to the number of concubines and consorts. There are dancing girls who dance to the beating of bells, and the pursuit of decadent pleasures grows more unrestrained by the day. Levies and taxes are frequent and numerous, and the state's wealth and strength become depleted. The people freeze and starve to death, and human corpses fill the countryside. He is arrogant and conceited, and kills those who remonstrate with him. Terror reverberates inside and outside, and those near and far are resentful and mournful. If this is the situation, then worthy men will regard his countenance as if it were the countenance of a malevolent spirit; his terraces and pavilions as if they were prisons; his ceremonial costumes as if they were mourning garb; his music and songs as if they were screaming and crying; his wine as if it were urine, and his food as if it were manure. No single good will come of any of his actions. Given that their

惡我也如是，其肯至哉？

　　今不務明其義，而徒設其祿。可以獲小人，難以得君子。君子者、行不喻合，立不易方。不以天下枉道，不以樂生害仁。安可以祿誘哉？雖強搏執之而不獲已，亦杜口佯愚，苟免不暇。國之安危將何賴焉？故《詩》曰：「威儀卒迷，善人載尸。」此之謂也。

hated of him is such, would they be willing to come and serve him?

Nowadays, rulers do not devote their efforts to illuminating what is right, but merely to setting up emoluments. Although they are thus able to obtain the services of petty men, they have difficulty in obtaining the services of gentlemen. Gentlemen do not pander to others to gain short term advantage but "stand firm and do not change their direction."[14] Not even to gain the world would they "distort the way," nor would they seek to enjoy life at the expense of humaneness.[15] How could emoluments possibly entice them? Even if they were forcibly detained, that would not be the end of the matter, for they would also refuse to speak, feign madness, or hope to excuse themselves on the pretext that they are otherwise engaged. Then upon whom will the stability of the state rely? Thus the *Book of Odes* says: "With their awe-inspiring demeanors gone completely / Good men are arranged like corpses."[16] This is what is meant.

《賞罰第十九》

政之大綱有二。二者何也？賞、罰之謂也。人君明乎賞罰之
道，則治不難矣。夫賞罰者、不在乎必重，而在於必行。必
行、則雖不重而民〔肅〕^a。〔必〕不行〔也〕^b、則雖重

^a Following *Qunshu zhiyao*, 46.22b, in inserting *su* 肅.
^b Not following *Concordance to Zhonglun* in inserting *bi* 必 before *bu* 不 and
ye 也 after *xing* 行.

19. Rewards and Punishments

Consonant with a late Han resurgence of interest in the methods of reward and punishment and administrative control associated with such pre-Qin thinkers as Shang Yang, Han Fei, and Shen Buhai, Xu Gan argues for the importance of ensuring that appropriate rewards and punishments be duly meted out in a timely fashion. Like Han Fei, who had likened rewards and punishments to two handles of political control, Xu Gan regards rewards and punishments as two "fundamental tenets of government." He stresses that it is not important that rewards be large and punishments severe; rather, what is important is that they always be carried out, and that the appropriate people are rewarded or punished. If this is not done, then the people "will hold the state's laws in contempt, relying instead on that which they themselves uphold." Furthermore, punishments and rewards should be neither too numerous nor too few, neither too light nor too heavy. In the final part of the essay, Xu Gan draws an analogy between rewards and punishments as a means of governing the people and the reins and whip used to drive a team of horses.

There are two fundamental tenets of government. What are they? They are called rewards and punishments.[1] If a ruler understands the way of rewards and punishments, then it is not difficult to bring about order. The effectiveness of rewards and punishments does not depend on their necessarily being heavy but on the certainty of their being carried out. If it is certain that they will be carried out, then even though they are not heavy, the people will pay close attention to them. If they are not carried out, then even if they are in great measure,

而民怠。故先王務賞罰之必行〔也〕。《書》曰：「爾無不信，朕不食言。爾不從誓言，予則孥戮汝，罔有攸赦。」

　　天生烝民，其性一也。刻肌虧體，所同惡也；被文垂藻，所同好也。此二者常存，而民不治其身，有由然也。當賞者不賞，當罰者不罰。夫當賞者不賞，則為善者失其本望，而疑其所行。當罰者不罰，則為惡者輕其國法，而怙其所守。苟如是也，雖日用斧鉞於市，而民不去惡矣。日錫爵祿於朝，而民不興善矣。

　　是以聖人不敢以親戚之恩而廢刑罰，不敢以怨讎之忿而廢慶賞。夫何故哉？將以有救也。故《司馬法》曰：「賞罰不踰時，欲使民速見善惡之報也c。」踰時且猶不可，而況廢之者乎？

　　賞罰不可以踈，亦不可以數。數則所及者多，踈則所漏者多。

c The modern version of the *Simafa* text reads: *shang bu yu shi yu min su de wei shan zhi li ye fa bu qian li yu min su du wei bu shan zhi hai ye* 賞不踰時欲民速得為善之利也罰不遷列欲民速覩為不善之害也.

the people will pay no attention to them. Hence the former kings made sure that rewards and punishments were necessarily carried out. The *Book of Documents* says: "On no account disbelieve me; I will not renege on what I have said. If you do not obey what I have sworn in my oath, I will kill all of you, even your sons. I will spare no one."[2]

"Heaven gave birth to the multitudes of people" and their natures are one.[3] People are alike in detesting facial branding and limb mutilation, just as they are alike in being fond of decorative clothing and ornaments. If these two are always present, yet the people do not behave properly, there is a reason. The reason is that those who deserve to be rewarded are not rewarded and those who deserve to be punished are not punished. If those who deserve to be rewarded are not rewarded, then those who do good will begin to lose sight of what they had originally aspired to and so question what they are doing. When those who deserve to be punished are not punished, then those who do wrong will hold the state's laws in contempt, relying instead on that which they themselves uphold. Should things come to this, then, even if there are daily executions in the marketplace, the people would not stop doing wrong; and should titles and emoluments be bestowed daily at court, the people would not initiate doing good. This is why sages did not dare to make exceptions when meting out punishments and penalties to a family member or relative simply because of their love for them, nor did they dare to make exceptions in bestowing prizes and rewards on an enemy or rival simply because of their hatred for them.[4] Why is this? Because they wanted rewards and punishments to rectify matters. Thus *Simafa* says: "When rewarding and punishing, one must not be tardy, so that the people will see quickly how good and bad are repaid."[5] If even tardiness is unacceptable, how much more unacceptable would it be to abandon rewards and punishments completely?

Rewards and punishments cannot be too few, nor can they be too numerous. If they are too numerous, then many people will receive them, while if too few, then many people will be

賞罰不可以重，亦不可以輕。賞輕則民不勸，罰輕則民亡懼。賞重則民徼倖，罰重則民無聊。故先王明(庶)〔恕〕以(德)〔聽〕[d]之，思中以平之，而不失其節〔也〕。故《書》曰：「罔非在中，察辭於[e]差。」

夫賞罰之於萬民，猶轡策之於馴馬也。轡策〔之[f]〕不調，非徒遲速之分也，至於覆車而摧轅。賞罰之不明也，則非徒治亂之分也，至於滅國而喪身。可不慎乎！可不慎乎！故《詩》云：「執轡如組，兩驂如舞。」言善御之可以為國也。

[d] Following *Qunshu zhiyao*, 46.22b, in emending *shu* 庶 to *shu* 恕 and *de* 德 to *ting* 聽.
[e] The modern version of *Documents* reads *yu* 于 not *yu* 於.
[f] Not following *Concordance to Zhonglun* in inserting this character into the text.

left out. Nor can rewards and punishments be too light or too heavy. If rewards are too light, then the people will not be encouraged to do good; while, if punishments are too light, then the people will not be afraid. If rewards are too heavy, then people will benefit gratuitously; while if punishments are too heavy, then people will be forced into a hopeless situation. Because of this, the former kings sought to clarify the bearing of individual circumstances, and to be balanced when they weighed up their judgments, thereby always maintaining a sense of appropriate measure.* Thus the *Book of Documents* says: "[In trying criminal cases] the aim is nothing but to rely on balanced adjudication. Inconsistencies in statements must be examined."[6]

Rewards and punishments to the myriad people are like the reins and whip to a team of horses. If the reins and whip are not applied properly, this will not only affect the pace of the team, it could even lead to the carriage's overturning and its shafts' snapping. If rewards and punishments are not made clear, this will not only affect how well ordered the people are, it could even lead to the destruction of the state and the death of the ruler. Can one afford not to be careful? Thus the *Book of Odes* says: "He holds the reins as if they were silken strings/ The two outside horses move as if dancing."[7] This is saying that, if driven properly, the state can be well governed.

---

*Clarify the bearing of individual circumstances: the point seems to be that the king would attempt to put himself in the shoes of the person being judged to see whether a particular level of reward or punishment was, in fact, appropriate to that individual's circumstances. For example, in a particular case there may have been extenuating circumstances.

《民數第二十》

治平在庶功興。庶功興在事役均。事役均在民數周。民數周、為國之本也。故先王周知其萬民眾寡之數，乃分九職焉。九職既分，則劬勞者可見，怠惰者可聞也。

20. Population Figures

Xu Gan here argues that an accurate census of a country's population figures is fundamental to orderly rule. This is because the public works undertaken by the state are dependent on corvée labor, and the corvée service, in turn, is dependent on an accurate knowledge of the population figures of each part of the country for it to operate equitably and efficiently. He is explicitly critical of contemporary government for failing to "understand the importance of showing commiseration for the people" by failing to provide them with appropriate administrative infrastructures. One of a number of consequences of this failure was that the national census register became inaccurate. In turn, the growing numbers of people who were not included on the register became homeless, and so engaged in activities ranging from "robbery and theft to raiding and pillage."

The stability of orderly rule depends on a variety of public works being established. These various public works depend upon an evenly distributed selection of those undertaking corvée service. This evenly distributed selection depends upon having a complete tally of the population figures. Having a complete tally of the population figures is fundamental for a state. Hence having a complete knowledge of the size of the population of the myriad people who lived in their states, former kings made the division of nine occupations for their people.* Being divided into the nine occupations, the industrious could be seen and the lazy could be heard about, and

* See "A Rebuke of Social Connections," note 2.

然而事役不均者，未之有也。事役既均，故民盡其心而人竭
其力。然而庶功不興者，未之有也。庶功既興，故國家殷
富，大小不匱，百姓休和，下無怨疚焉。然而治不平者，未
之有也。故曰：「水有源，治有本。」道者審乎本而已矣。
《周禮》：「孟冬司寇獻民數於王[a]。王拜而受之。登於天府。
內史、司會、冢宰貳之。」其重之如是也。

今之為政者，未知恤已矣。譬由無田而欲樹藝也。雖有良
農，安所措其（疆）〔彊[b]〕力乎？是以先王制六（卿）〔鄉[c]〕、六
遂之法，所以維持其民，而為之綱目也。使其鄰比，相保相
（愛）〔受[d]〕，刑罰慶賞，相延相及。故出入存亡，臧否順逆，
可得而知矣。如是姦無所竄，罪人斯得。

迨及亂君之為政也，戶口漏於國版，夫家脫於聯伍。

[a] Standard versions of the modern *Zhouli* text read: *si kou ji meng dong si si min zhi ri xian qi shu yu wang* 司寇及孟冬祀民之日獻其數於王. I have translated the *Discourses* reading.
[b] Following the *Longzi jingshe, Guang Han Wei congshu, Zengding Han Wei congshu, Xiao wanjuanlou xongshu*, and *Bozi* versions of *Discourses* in emending *jiang* 疆 to *qiang* 彊.
[c] Following the *Longxi jingshe, Xiao wanjuanlou congshu*, and *Liangjng yibian* versions of *Discourses* in emending *qing* 卿 to *xiang* 鄉.
[d] Following the *Zhouli*, 12.13a, 10.22b, in emending *ai* 愛 to *shou* 受.

so there was never an unevenly distributed selection of those undertaking corvée service. The selection of those undertaking corvée service being evenly distributed, people were willing to exert themselves fully, in both heart and body, and a great variety of public works was always established. Because these various public works were established, the state became wealthy and prosperous, wanting in nothing, the people were happy and harmonious, those below did not harbor enmity or ill-feeling toward those above them, and so it was never the case that rule was unstable. Thus it is said, "Water has its source and order has its root." The way is nothing more than examining into this root. According to the *Rites of Zhou*: "In the first month of spring, the minister of justice presented the population figures to the king who respectfully received them. . . . They were given to the keeper of temple treasures. The royal secretary, accountant, and minister of state had copies."[1] They placed this much importance on them.

Nowadays those who govern simply do not understand the importance of showing commiseration for the people. Take the example of a farmer wanting to plant crops but having no fields in which to do so. Even if he were a good farmer, where could he apply his strength? On account of this, the former kings instituted laws pertaining to the six administrative divisions of nearby districts and the six administrative divisions of external districts, so as to support their people by providing them with an administrative infrastructure.[2] This enabled neighborhoods to watch over one another and punishments, penalties, blessings, and rewards to spread out and reach everybody.[3] Accordingly, who came and who went, who was alive and who had died, who was to be praised and who was to be criticized, and who was compliant and who was recalcitrant could all be known. In this way, wicked people had no means of escape and criminals were thereby caught.

When reckless rulers governed, however, there were households that were missed when compiling the national census register and families which were left out of the *lian* and *wu*

避役者有之，棄捐者有之，浮食者有之。於是姦心競生，偽端並作矣。小則盜竊，大則攻劫。嚴刑峻法不能救也。

故民數者，庶事之所自出也。莫不取正焉。以分田里，以令貢賦，以造(罷)〔器ᵉ〕用，以制祿食，以起田役，以作軍旅。國以之建典，家以之立度。五禮用脩、九刑用措者。其惟審民數乎？

ᵉ Following all other versions of *Discourses* in emending *ba* 罷 to *qi* 器.

mutual responsibility groups.* There were those who avoided corvée service, those who were neglected and passed over, and those who became vagrants. Thereupon hearts of deceit multiplied in profusion, and the seeds of falsehood sprouted together. These people engaged in activities ranging from robbery and theft to raiding and pillage. Harsh laws and Draconian punishments were to no avail.

Hence it is from population figures that a variety of affairs are derived. These affairs all rely on these figures for accuracy. This is the case, whether it be apportioning land for fields and dwelling areas; ordering the payment of tributes and taxes; manufacturing implements and utensils; regulating emoluments and salaries; raising numbers of people for hunting and corvée service; or conscripting troops for battalions and armies.[4] The ducal state relies on population figures to establish its canons, and the minister's domain relies on them to set up its standards. The five rituals depend on population figures for their practice, and the nine punishments depend on them in order to be applied.[5] Does the significance of population figures, then, lie just in checking the numbers of the people?

---

*The term *wu* refers to a grouping of five people, while *lian* can refer to a grouping of either ten families or eight villages (*lü* 閭). See *Zhouli*, 12.13a.

逸文輯錄《復三年喪》

天地之間，含氣而生者，莫知乎人。人情之至痛，莫過乎喪親。夫創巨者其日久，痛甚者其愈遲。故聖王制三年之服[a]，所以稱情而立文，為至痛極也。自天子至于庶人，莫不由之。帝王相傳，未有知其所從來者。

　　及孝文皇帝天姿謙讓，務崇簡易。其將棄萬國，乃顧臣子，令勿行久喪，已葬則除之，

[a] Emending *fu* 服 to *sang* 喪.

21. Reinstitute the Three-Year Mourning Period

In this essay, which survives only in abridged form, Xu Gan argues for the reintroduction of the observance of the three-year mourning period for emperors, on the grounds that the loss of one's parents was the most painful of human emotions. Tracing the origins of a reduced mourning period for emperors to the testamentary edict of Emperor Wen (r. 180–157 BCE), Xu Gan argues that Emperor Wen did not intend this reduced period to become established in perpetuity. He proceeds to praise Emperor Ming (r. 57–75 CE) for his brief reintroduction of the three-year mourning period and complains that the tradition was not continued when Emperor Ming died.

Of all the creatures in the world that live by breathing, none has greater awareness than man, and of all the most painful of human emotions, none is more painful than losing one's parents.[1] When a wound is severe, it remains for many days; when pain is acute, the recovery is slow.[2] For this reason, the sage kings instituted the three-year mourning period so as to be able to express their feelings and establish a formal ceremonial for that expression, because, at that time, the pain of grief is most intense.[3] From the son of heaven to the common people, all observed the mourning period. The emperors and kings of later ages passed on the tradition but knew not whence it came.

Coming to the time of Filial Emperor Wen [r. 180–157], he was by natural disposition modest and deferential and tried to follow simple ritual procedures. When he was about to die, he looked around at his ministers and instructed them not to conduct a long mourning but to cease as soon as he was bur-

將以省煩勞而寬群下也。觀其詔文，唯欲施乎己而已，非為
漢室創制喪禮而傳之於來世也。後人遂奉而行焉，莫之分
理。至乎顯宗，聖德欽明，深照孝文一時之制。又惟先王之
禮〔之[b]〕不可以久違。是以世祖祖崩，則斬衰三年。孝明既
沒，朝之大臣徒以己之私意忖度嗣君之必貪速除也。檢之以
太宗遺詔，不惟孝子之心。哀慕未歇，故令聖王之迹，陵遲
而莫遵。短喪之制，遂行而不除。斯誠可悼之甚者也。

　　滕文公、小國之君耳。加之生周之末世，禮教不行。猶能
改前之失，咨問於孟軻，而服

[b] Not following the *Concordance to Zhonglun* in inserting *zhi* 之.

ied, wishing to save trouble and effort and to make it easier for those below him. Looking at his testamentary edict, however, it would appear that he desired the reduced mourning period to apply only to him; he was not seeking to institute a mourning ceremony for the Han dynasty to transmit to posterity.*[4] Those who later followed the letter of his testamentary edict failed to distinguish the principle of his action. Coming to Emperor Ming [r. 57–75], his sagely virtue, reverence, and brilliance profoundly illuminated the fact that the mourning period proposed in Filial Emperor Wen's edict was merely a temporary measure.[5] He also thought that the ceremonials of the former kings should not be abandoned for long. Thereupon when Emperor Guangwu [r. 25–57] died, he wore mourning clothes for three years.[6] When Filial Emperor Ming died, however, the chief ministers considered only their personal interests, conjecturing that the new ruler would certainly be desirous of a speedy finish to the mourning ceremonies. Yet when their actions are examined in the light of Emperor Wen's testamentary edict, they were not motivated by feelings of filial piety.[7] Heartfelt expressions of grief and longing had not been fully exhausted. Hence this led to a situation in which the footsteps of the sage kings came to be worn away and were no longer observed, while the short mourning period came to be implemented and was not done away with. This is really something very much to grieve about!

Duke Wen of Teng [late fourth century BCE] was the ruler of a small state; moreover, he lived at a time when the Zhou had declined and the edifying transformation of ritual was no longer conducted. Yet despite this, he was still able to alter mistakes that had been made in the past. Having consulted Mencius about funeral arrangements, he decided to observe

---

* In Wen's testamentary edict (157 BCE) he asks that the customary three-year mourning period be reduced to thirty-six days. See *Shiji*, 10.433–34; *Hanshu*, 4.132; Dubs, *Former Han Dynasty*, 1:268–70. The reduced mourning period seems to have been generally observed during the reigns of emperors Jing (r. 157–141) through to Yuan (r. 49–33).

喪三年。豈況大漢配天之主，而廢三年之喪，豈不惜哉！且作法於仁，其弊猶薄。道隆於己，歷世則廢。況以不仁之作，宣之於海內，而望家有慈孝，民德歸厚，不亦難乎？《詩》曰：「爾之教矣，民胥〔放°〕〔效〕矣。」

　　聖主若以遊宴之間，超然遠思，覽周公之舊章，咨顯宗之故事。感蓼莪之篤行，惡素冠之所刺，發復古之德音，改太宗之權令。事行之後，永為典式，傳示萬代，不刊之道也。

 ᶜ Following the *Sibu congkan* edition of *Qunshu zhiyao* in emending *fang* 放 to *xiao* 效. The Mao version of the ode reads *xiao* 傚.

the three-year mourning period.[8] Is it not lamentable that in the case of the Great Han rulers who are the equals of heaven, the three-year mourning period has been abolished? When a ruler models himself on humaneness, the harm done in abolishing the mourning period is at first weak. Consequently, when the way flourishes in the ruler's person, it will take several generations before the way is abandoned. How much more difficult will it be to expect compassion and filial piety to be fostered in families, and the common people's virtue to rally to a sincere ruler, if the decision to abolish the three-year mourning period is motivated by nonhumane considerations, and its practice propagated throughout the country?[9] The *Book of Odes* says: "If you set an example / The people will imitate you."[10]

If, between moments of sporting and banqueting, your highness rises above his ordinary concerns to reflect on distant matters, you may then examine the old regulations of the duke of Zhou and inquire into the precedent of Emperor Ming.[11] If you are moved by the sedulous behavior of the son in the ode "The Tall Tarragon" and despise that which is being satirized in the ode "The Undyed Cap," then you will become inspired to revive ancient words of virtue and change Emperor Wen's decree.* After this has been implemented, it will serve forever as a canonical exemplar to be transmitted and shown to countless generations as an immutable way.

---

*"The Tall Tarragon": *Odes*, Mao no. 202. In this ode a son grieves the death of his parents and the fact that he has not been able to repay his debt to them. "The Undyed Cap": *Odes*, Mao no. 147. According to the preface, 7.2.5b, the ode is directed against those who cannot see out the three-year mourning period.

逸文輯錄《制役》

昔之聖王制為禮法，貴有常尊，賤有等差，君子小人，各司分職，故下無(潛)〔僭[a]〕上之忿，而人役財力，能相供足也。往昔海內富民、及工商之家，資財巨萬。役使奴婢，多者以百數，少者以十數。斯豈先王制禮之意哉！

夫國有四民，不相干黷。上者勞心，工農商者勞

[a] Following Qian, "Reading Notes," in emending *qian* 潛 to *jian* 僭.

22. Regulate the Allotment of Corvée Laborers

As with the previous essay, this essay also only survives in abridged form. Xu Gan's chief complaint is that the system of slave ownership had become corrupted. Originally, one's social rank determined whether one could keep slaves, and if so, how many. During the Han dynasty, however, rich people, "as well as the families of craftsmen and merchants," had come to own large numbers of slaves, while many of the scholar-gentry class ("men of education and social standing") had none. In particular, Xu Gan criticizes rich merchants and powerful families for owning disproportionately large numbers of slaves. The merchants were able to do this through their wealth, while the powerful families were able to do so through land annexation.

In the rites and laws instituted by the past kings, the noble held positions that were always honored, while the lowly held positions that were hierarchically ranked. The gentleman and the small man each served in a different office. Hence there was no transgression by those below encroaching upon the authority of those above them, and so corvée labor and the strength derived from wealth were able to be supplied in sufficiency. Yet in the past, rich people in this country, as well as the families of craftsmen and merchants, possessed huge assets of wealth. Those with many slaves had as many as several hundred, while those with few still had as many as ten or so. Surely this was not the intention of the former kings in instituting the rites?

A country has four divisions of people, and they are mutually independent. "Men of education and social standing work with their minds, while craftsmen, farmers and merchants work

力。「勞心之謂君子，勞力之謂小人。君子者治人，小人者治於人。治於人者食人，治人者食於人。百王之達義也。」

今夫無德而居富之民，宜治於人且食人者也。役使奴婢，不勞筋力。目喻頤指。從容垂拱。雖懷忠信之士，讀聖哲之書，端委執笏，列在朝位者，何以加之？且今之君子，尚多貧匱，家無奴婢。(既[b])〔即〕其有者，不足供事，妻子勤勞，躬自爨烹。其故何也？皆由罔利之人與之競逐。又有紆青拖紫并兼之門使之然也。

夫物有所盈，則有所縮。聖人知其如此，故裒多益寡，稱物平施，動為之防，不使過度，是以治可致也。為國而令廉讓君子不足如此，而使貪人有餘如彼，非所以

[b] Following Qian, "Reading Notes," in emending *ji* 既 to *ji* 即.

with their strength.[1] He who works with his mind is called a gentleman, while he who works with his strength is called a small man. The gentleman rules others, while the small man is ruled by others. Those who are ruled by others feed them, while those who rule others are fed by them. This is a general principle accepted by all the one hundred kings."[2]

Nowadays, those people who are without virtue, yet occupy wealthy social positions, should be the ones who are ruled by others and who feed others. There are those who have slaves and do not exert the strength of their muscles. They communicate with their eyes and indicate with their faces. They take everything at a leisurely pace, having nothing to do. Even if they were men of education and social standing who held loyalty and sincerity close to their hearts, read the books of the sages and wise men, and stood at court in their ceremonial apparel holding their *hu* tablets, why should they be allowed to have more slaves?* Moreover, nowadays there are still many gentlemen who are impoverished and have no slaves in their homes. Even if they have them, they are too few to do all the work, so the gentleman's wife must work hard, making the fire and doing the cooking herself. What is the reason? It is all the fault of those men who caste their nets for profit and vie with the gentleman, and also the fault of those high-ranking and noble families who annex other people's land—it is they who have brought about this situation.[3]

Because things wax, they also wane. The sage understands this, hence he "reduces what is in surplus and increases what is deficient. He weighs things and distributes them evenly," so that, even if they should move, he has taken precautions to ensure that such movement will not be excessive.[4] In this way, good order can be achieved. If a state is run so that honest and deferential gentlemen do not have sufficient slaves while greedy men have more than they need, then this is not the way to

---

* The *hu* tablet was used by officials to record matters during court. By extension, it became a symbol of official rank.

辨尊卑、等貴賤、賤財利、尚道德也。

今太守令長得稱君者，以慶賞刑威，咸自己出也。民畜奴婢，或至數百，慶賞刑威，亦自己出。則與郡縣長史又何以異？夫奴婢雖賤，俱含五常。本帝王良民，而使編戶小人為己役。哀窮失所，猶無告訴。豈不枉哉？今自斗食佐吏以上，至諸侯王，皆治民[c]人者也。宜畜奴婢。農工商及給趨走使令者，皆勞力躬作、治於人者也。宜不得畜。

昔孝哀皇帝即位，師丹輔政，建議令畜占宅奴婢者有限。時丁傅用事，董賢貴寵。皆不樂之。事遂廢(覆)〔罷[d]〕。夫師丹之徒，皆前朝知名大臣，

[c] Reading *min* 民 here as excrescent, on the grounds of parallelism with the next sentence, and *Mencius*, 3A.4.
[d] Following Qian, "Reading Notes," in emending *fu* 覆 to *ba* 罷.

distinguish the respectable from the inferior, to grade the noble and the lowly, to disdain wealth and profit or to esteem the way and virtue.

Nowadays, those governors of commanderies and district magistrates who come to be called rulers completely arrogate to themselves the authority to bestow rewards and mete out punishments. Similarly, those ordinary people who keep slaves—some having as many as several hundred—also arrogate to themselves the authority to bestow rewards and mete out punishments, so how are they different from the senior officials of the commanderies and prefectures? Although slaves are base in rank, they nevertheless possess the five constant virtues.* Originally they were the good people of emperors and kings, but they have been entered into the household registers of small men as their personal slaves. Forlorn and destitute, they have lost their homes and yet have no one to tell of their plight. Have they not been wronged? Nowadays, officials from the ranks of assistant and personnel paid in pecks right up to the feudal lords and kings are all those who rule people, so it is right that they should keep slaves.⁵ As for farmers, craftsmen, and merchants, as well as runners and messengers, they are all engaged in physical labor, plowing and cultivation, and are thus the ones ruled by people. It is right that they not be allowed to keep slaves.

In the past, when Filial Emperor Ai [6–1 BCE] had acceded to the throne, Shi Dan, who was assisting in government, recommended that it be decreed that a restriction be placed on who could keep slaves for farm and household labor. At that time, Empress Dowagers Ding and Fu controlled government, and Dong Xian was valued and favored. Neither was happy with Shi's recommendation and so the matter was dropped.⁶ Shi Dan's supporters were all well known and important ministers of the Former Han dynasty, and they were all gravely

---

* That is, humaneness, rightness, propriety, understanding/knowledge, and living up to one's word.

患疾并兼之家。建納忠信，為國設禁。然為邪臣所抑，卒不施行。豈況布衣之士，而欲唱議立制。不亦遠乎！

concerned about those families who had annexed other people's land. Their recommendations were loyal and sincere, and it was for the good of the state that they proposed restrictions on slave ownership. Their recommendations, however, were blocked by wicked ministers, and in the end were not implemented. How much more remote is the chance that restrictions will be set in place now that it is left to a poor scholar to follow in their footsteps and champion it as the right thing to do?

Notes

Introduction by Dang Shengyuan

1. *Siku quanshu zongmu*, 19.773.
2. See the unsigned preface to *Discourses* (translated in Makeham's introduction to this volume); Xie, "Ni Wei taizi Yezhong ji shi," *Wenxuan*, 30.30a; Li Shan's commentary to Yang, "Da Linzai hou jian," *Wenxuan*, 40.14a.
3. This is described in the unsigned preface written for *Discourses*.
4. *Sanguozhi*, 21.599.
5. Preface to *Discourses*.
6. Quotation: Cao Zhi's (192–232) letter to Yang Dezu, *Wenxuan*, 42.12b. Qingzhou was the ancient name of the region from which Xu Gan hailed.
7. Xu You was famous for declining the sage-king Yao's offer to cede the throne to him, and then becoming a hermit.
8. Cao Pi's (187–226) letter to Wu Zhi (177–230), *Wenxuan*, 42.9a–b.
9. *Sanguozhi*, 21.599.
10. Preface to *Discourses*.
11. Inner sageliness refers to personal moral cultivation; outer sageliness refers to issues of statecraft.
12. Preface to *Discourses*.
13. *Discourses*, "Reinstitute the Three-Year Mourning Period," "Wisdom and Deeds."
14. *Discourses*, "Regulate the Allotment of Corvée Laborers."
15. *Discourses*, "Regulate the Allotment of Corvée Laborers."
16. Discourses, "Distinguishing Between Premature Death and Longevity."
17. Discourses, "Cultivating the Fundamental."
18. The "teaching of names" is a complex issue. In a general sense, it can be understood to refer to the ethos responsible for fostering the cultivation of a virtuous reputation.
19. *Discourses*, "An Examination of Disputation."
20. *Sanguozhi*, 1.31; 1.49, commentary.
21. The six virtues: wisdom, humaneness, sageliness, rightness, balance, and harmony. The six types of virtuous conduct: filial respect, friendship, maintaining harmony with one's nine degrees

of relatives, being close to one's in-laws, trusting one's friends, and commiserating with the sufferings of others. The six arts: ritual, music, archery, charioteering, writing, and arithmetic. See "Ordering Learning."

Introduction by John Makeham

1. In the unsigned Preface to *Discourses*, 4a, it is recorded that Xu Gan died in the second month of spring 218, at the age of 48. This, however, is in conflict with *Sanguozhi*, 21.602, and Cao Pi's 曹丕 (187–226) letter to Wu Zhi 吳質 (177–230), written on March 17, 218 (in *Wenxuan*, 42.8b; *Sanguozhi*, 21.608), which indicate that he died in 217. Given that the pestilence which killed so many of Xu's contemporaries was in 217, that is the more likely date. If we also accept the account in the preface that he lived forty-eight years (Chinese reckoning), he would have been born in 170. The Chinese citation text accompanying my translation is based on the text in Lau and Chen, *Concordance to Zhonglun*, which, in turn, is based on the *Sibu congkan* edition of *Discourses* and collated against a number of other texts. I have noted all instances where the readings I follow differ from those adopted in the Chinese text reproduced in the *Concordance to Zhonglun*. My punctuation of the Chinese text reflects my interpretation of the text, not the punctuation adopted in the *Concordance to Zhonglun*. For a study of editions of *Discourses* and the text's history, see my *Name and Actuality*, Appendix A, and also my entry in Loewe, *Early Chinese Texts*, 88–93.

2. For Xu Gan, the bond between name and actuality was not something to be prescribed by convention or artificially determined by human culture because it is part of the cosmic order. He believed that if that bond were broken or artificially prescribed, the repercussions would affect not only the sociopolitical order but also the moral order. See *Name and Actuality*.

3. The other six Masters were Kong Rong 孔融 (153–208), Ruan Yu 阮瑀 (165–212), Ying Yang 應瑒 (d. 217), Chen Lin 陳琳 (d. 217), Liu Zhen 劉楨 (d. 217), and Wang Can 王粲 (177–217).

4. The content of the preface makes it clear that the author was a contemporary of Xu Gan. Zhang, *Zhuzi dagang*, 63b, speculates that the author was one of Cao Zhi's 曹植 (192–232) coterie. Yan, *Quan Sanguo wen*, 55.1360, suggests that the author is the

same Mr. Ren whose comments to selections from *Discourses* are recorded in *Yilin*. He further suggests the possibility that this Mr. Ren is Xu Gan's contemporary Ren Jia 任嘏.

5. According to the *Zhengyi* commentary to *Shiji*, 74.2344, Mencius's style was Ziyu 子輿, yet Zhao Qi's 趙岐 (d. 201) preface to *Mencius* (in Jiao Xun, *Mengzi zhengyi*, 4), says that he did not know what Mencius's style was.

6. Ju Prefecture was the southeast region of modern Shouguang County, in Shandong Province.

7. There are several different accounts of which particular qualities constitute the nine virtues. Many of the qualities identified in the *Yi Zhoushu*, 3.10b, account—doing one's best on behalf of others, respectfulness, living up to one's word, firmness, softness, harmoniousness, resoluteness, constancy, and compliance—are cited more frequently in *Discourses* than those identified in other accounts. Nonetheless, the probable reference is to the account of nine virtues listed at *Zuo Commentary*, Zhao 28, 52.29a. Xu Gan cites these with approval in "Attend to the Fundamentals."

8. This and the following references to Xu Gan's age are based on the Chinese reckoning.

9. The *Book of Changes*, the *Book of Documents*, the *Spring and Autumn Annals*, the *Book of Odes*, and the *Book of Rites*.

10. The canons of the state were the various institutions and standards which constituted the public face of government and through which it exercised its authority.

11. "Name": following the *Xue Chen* and *Han Wei congshu* versions of *Discourses* in emending *shou* 售 to *ming* 名.

12. The Five Classics plus the *Book of Music*, which was no longer extant in Xu Gan's time. Some scholars have argued that such a text never existed; others have argued that it was one of the texts purportedly destroyed in the famous book-burning incident of 213 BCE.

13. Allusion to *Analects*, 19.22. References to the *pian* (chapter) and *zhang* (passage) divisions in the *Analects* follow those used in the Harvard-Yenching Institute Sinological Series.

14. Allusion to *Analects*, 19.22.

15. "Take simplicity too far": allusion to *Analects*, 6.2. "The others" are those who practice various cultivation techniques, such as those who were followers of Lao-Zhuang teachings.

16. Allusion to *Analects*, 7.16.

17. Allusion to *Analects*, 8.5.

18. Allusion to *Analects*, 14.35.

19. Dong Zhuo was a famous general. On these events, see Beck, "Fall of Han," 345–48.

20. The "arts" referred to are presumably ritual, music, archery, charioteering, writing, and mathematical reckoning. See also "The Fundamental Principles of the Arts."

21. Presumably some part of his native Beihai.

22. Yan Yuan was one of Confucius's best-known disciples. He died young and did not take up an official career. The author of this preface seems to imply that Xunzi was essentially a scholar and was not interested in promoting his philosophy to the various rulers of the day. This is not historically accurate. See Knoblock, *Xunzi*, 1:1–35. In "Examining the Selection of High Officials," Xu Gan even maintains that Xunzi never held office: "In the past, Xun Qing, who lived in the Warring States period, was a man of brilliant intellectual talents. He transmitted the ancient traditions of Yao and Shun, modeled himself on and glorified the teachings of Wen and Wu, took Confucius as his teacher, and made clear the way to bringing chaos to order. Despite this, the rulers of the various states thought him impractical and out of step with the changing times; to the end of his days none of them was prepared to employ him." In "Destruction of the State," Xu contradicts this claim about Xunzi.

23. Following the *Liangjing yibian, Xue Chen,* and *Han Wei congshu* versions of *Discourses* in emending *yu* 玉 to *wang* 王. Cao Cao was a famous general and founder of the Wei kingdom.

24. It would have been during this period that he wrote "Xuzheng fu 序征賦" and "Xizheng fu 西征賦" (in Ouyang, *Yiwen leiju*, 59.1069–70). "Xuzheng fu" was written upon returning north with Cao Cao's army after its defeat at Chibi in 208. "Xizheng fu" was probably written in 211, upon returning with Cao Cao's army after its defeat of Ma Chao 馬超. See Liu, *Jianan wenxue*, 34–35, 39–40.

25. On the "floodlike vital energy," see *Mencius*, 2A.2.

26. These comments appear to relate to a passage in "Valuing Words."

27. This is an allusion to *Analects*, 10.27. Early commentators give a number of different interpretations of this passage. It would appear that the author of the preface understood it as did the later *Lunyu jijie* editors, who, in their commentary to this passage, explain that Zigong misconstrued Confucius's sigh upon seeing some birds on the ridge of a mountain to mean that

Confucius wanted to eat them. Accordingly, Zigong captured, cooked, and served them to Confucius. In fact, Confucius had been sighing at his own failure to have encountered the right times. The point that the author of the preface is making is that, like Confucius, Xu Gan has been misunderstood by those close to him.

28. See the biographical note on Xu Gan appended to the biography of Wang Can in *Sanguozhi*, 21.599. Generally, I have followed Hucker's renderings of official titles as given in his *Official Titles*.

29. See *Jinshu*, 44.1249; *Sanguozhi*, 19.557.

30. *Sanguozhi*, 21.599. This is now part of Pingding County, Shanxi.

31. See *Sanguozhi*, 1.24, commentary; 1.32; 1.44; 1.49–50 (translated in *Name and Actuality*, 142–43).

32. *Liji*, 53.14a ("Daxue").

33. *Liji*, 53.14b.

34. An allusion to the story of Bian He of Chu, who is said to have presented a precious jade to King Li of Chu (the historical records do not record a king by this name). When an appraiser told the king that it was really only a piece of ordinary stone, the king ordered Bian He's foot to be cut off. These same events were repeated when King Wu of Chu (r. 740–690) came to the throne, and only when King Wen of Chu (r. 689–677) came to the throne was it recognized that the jade was genuine and Bian He had told the truth. See *Han Feizi*, "He shi 和氏". Presumably the line "Master He himself is not without fault" is a criticism of Xu Gan for failing to have identified who his true friends were.

35. An allusion to preparing to take up office after a period of disengagement.

36. *Wenxuan*, 24.2a–b. Xu is wishing for a true friend to help him obtain office and recognition, yet his friends are all similarly powerless—they are waiting for friends to help them, just like Xu Gan.

37. For Li Shan's comment see *Wenxuan*, 42.9b.

38. See Cao Pi's letter to Wu Zhi in *Sanguozhi*, 21.602. The *Wenxuan* version of this letter, and Li Shan's commentary to the same, sometimes have "twenty chapters" and sometimes "more than twenty chapters," depending on the edition of *Wenxuan* used.

39. Cf. also "Examining Falsity," where Xu Gan laments that since the demise of Confucius "the point of balance for ordered human relationships has remained in an unsettled state."

1. Ordering Learning

1. Considering the Eastern Han arrangement of the *Analects*, it is perhaps no coincidence that *Discourses* was also edited into twenty chapters.

2. The *locus classicus* for this view is the celebrated passage at *Zuo Commentary*, Duke Xiang 24, 35.22a–24a: "In Spring of the twenty-fourth year, Mu Shu went to Jin. Fan Xuan met him and asked, 'The ancients had a saying, To die but not to perish; what does it mean?'... 'I [Mu Shu] have heard it said that the establishment of a virtuous legacy is of the utmost importance; second, is the establishment of meritorious deeds; and third, is the establishment of one's words.'" The Confucian concern for securing a favorable posthumous reputation is already found in *Analects*, 15.20: "The gentleman is distressed at the prospect of dying and leaving behind him a name that does not match his actual qualities."

3. According to *Zhouli*, 14.2a–8a, there were two officials responsible for the education of the sons of the ruler's kinsmen and high officials: the palace master (*shishi* 師氏) and the palace protector (*baoshi* 保氏). See also Hucker, *Official Titles*, nos. 4494, 5302.

4. *Zhouli*, 10.14b, describes the six virtues, the six types of virtuous conduct, and the six arts as ways for educating the people generally, not especially the ruler's kinsmen and sons of high officials. In interpreting the six types of virtuous conduct I have followed Zheng Xuan's 鄭玄 (127–200) commentary.

5. *Mencius*, 3A.4, lists five bonds of human relationships which link individuals and define their roles in a number of key relationships: "Familial love between father and son, rightness between ruler and subject, distinction between husband and wife, precedence of the old over the young, and trust between friends." The bonds, *renlun* 人倫, were variously known as *lunchang* 倫常, *wulun* 五倫, and *tianlun* 天倫. In the Han dynasty, we find similar pairs of relationships institutionalized as the three cardinal bonds (*sangang* 三綱) and the six secondary bonds (*liuji* 六紀). The cardinal relationships are those between ruler and subject, father and son, and husband and wife. See *Chunqiu fanlu*, 12.8b–10a; *Bohutong*, 7.15a; Tjan, *Po Hu T'ung*, 2:559; Fung, *Chinese Philosophy*, 2:42–43. For a study of the three cardinal bonds in the Han dynasty, see Tanaka, "Byakko tsū," 121–37. The six secondary bonds (a category of five secondary bonds

is referred to in *Chunqiu fanlu*, 10.14b, but is not elucidated) are the relations between paternal uncles, elder and younger brothers, relatives of the same surname, maternal uncles, teachers and elders, and friends. See *Bohutong*, 7.15a; Tjan, *Po Hu T'ung*, 2:559; Fung, *Chinese Philosophy*, 2:44.

6. *Documents*, 14.26b.

7. On the pitch pipe and its various uses see *Hanshu*, 3.959, 966–69; DeWoskin, *Song*, 46, 49, 59–60, 64–66, 70–71, 80–83; Major, *Heaven and Earth*, 110–18; "Celestial Cycles," 121–31, passim; von Falkenhausen, *Suspended Music*, 314–15.

8. Mao no. 218; Karlgren, *Odes*, 172.

9. Cf. *Analects*, 15.31: "Caught up in thinking, I once went all day without eating and all night without sleeping. It was, however, of no use. My efforts would have been much better spent learning." Cf. also *Xunzi*, 1.8a: "I once spent a whole day in thought but it was not as valuable as a moment spent learning."

10. The quoted passage is not found in any earlier extant work. It is possibly based on *Analects*, 2.15: "If one learns without reflecting, it will come to nothing; if one reflects without learning, one will be plagued by uncertainty."

11. Cf. *Xunzi*, 17.15b–16a, and *Han Feizi*, 19.9a.

12. See *Hanshi waizhuan*, 8.14a, for a similar sentiment.

13. Zixia was a disciple of Confucius's. This passage is not found in any earlier extant work.

14. The aphorism on commitment is a reworking of *Mencius*, 2A.2.

15. *Changes*, Qian hexagram, "Images," 1.8a.

16. For these terms, see the *Changes*, Qian hexagram, "Judgment," 1.1a. The terms are also used in other texts. This passage is not found in any earlier extant work.

17. Cf. *Changes*, "Appended Statements (A)," 7.31a: "Confucius said, 'In order to reveal fully what they meant, the sages established images'"; and *Changes*, "Appended Statements (A)," 7.18a: "The way of the gentleman can be as an official or a recluse, without words or with." These two sentences of the *Discourses* text (here rendered in English as one sentence) are also related to the following passage from the "Appended Statements (A)," 7.18a: "Confucius said: 'In the way of the gentleman, sometimes he will go forth, sometimes he will stay put. Sometimes he will be silent, sometimes he will speak out.'"

18. "He will be able to fathom the reasons underlying both the obscure and the evident": cf. *Changes*, "Appended Statements (A),"

7.9a: "Looking up we use the *Book of Changes* to observe the markings in heaven, and looking down we use it to examine the patterns on earth, and so are able to fathom the reasons underlying both the obscure and the evident."

19. Mao no. 288. In rendering this line, I have followed the Lu interpretation, not the Mao interpretation. See Wang, *Shi sanjiayi*, 2:1042–43; Huang, *Maoshi Zhengjian*, 406.

20. On Fu Xi, see *Hanshu*, 20.863. On his creation of the eight trigrams, *Changes*, "Appended Statements (B)," 8.4b, relates the following account: "In antiquity, when Bao Xi 包犧 [Fu Xi] ruled the world, looking up he observed the images in heaven and looking down he observed the models on earth. Observing the markings of birds and animals and of what was suitable on earth, from nearby he appropriated what he found on his own body, and from faraway he appropriated what he found in other things. Thereby he began to create the eight trigrams so as to comprehend the potency of divine intelligence and to categorize the myriad things according to their true condition." For a similar account, see also *Shuowen jiezi*, 15.1a.

On Sui Ren, see *Han Feizi*, 19.1a; Knoblock, *Xunzi*, 3:29: "In the most ancient times . . . the people lived on fruit, berries, mussels, and clams—things that sometimes became so rank and fetid that they hurt the bellies of the people and many were afflicted with diseases. Then a sage appeared who drilled with sticks to produce fire to transform the rank and putrid foods. The people were so delighted by this that they made him ruler of the world and called him the 'Fire-drill Man.'" See also, *Bohutong*, 1.11a; *Fengsu tongyi*, 1.2a. Apparently, different types of wood were suitable for use as drilling sticks in different seasons. See *Zhouli*, 30.9b, where the commentary cites a passage from *Zouzi* 鄹子.

On Cang Jie, see *Shuowen jiezi*, 15.1a; Lewis, *Writing and Authority*, 272–73. For the story of Emperor Xuan and the phoenix, see *Lüshi chunqiu*, 5.8a–b; *Shuoyuan*, 19.21a–b; *Hanshu*, 21A.959, *Fengsu tongyi*, 6.6b–7a; DeWoskin, *Song*, 59–61.

21. For this anecdote, see *Analects*, 5.8: "Confucius said to Zigong, 'Who is the better, you or Hui [Yan Yuan]?' He replied, 'How could I dare to compare myself to Hui? He hears one point and is able to know all about a subject. I hear one point and can infer only a second.' Confucius said, 'You are no match for him. Neither you nor I can match him.'"

22. For a biographical introduction to kings Wen and Wu, see Knoblock, *Xunzi*, 2:32–36. On Cheng Tang, see Chang, *Shang Civilization*, 9–10; Allan, *Heir and Sage*, chapter 4. On the lord of Xia, see Allan, *Heir and Sage*, chapter 3. On Yao and Shun, see Allan, *Heir and Sage*, chapter 2.

23. The Six Classics are the *Book of Odes*, the *Book of Documents*, the *Book of Changes*, the *Books of Rites*, the *Book of Music*, and the *Spring and Autumn Annals*. Obviously, Xu cannot mean that the Six Classics were passed on book by book from the time of Yao and Shun, but rather that there was an accretion of sagely teachings and principles that was gradually incorporated in the classics. Alternatively, the sentence could be translated as "The Six Classics are books that show how the sages follow one another."

24. Similar lists are found in *Hanshi waizhuan*, 5.11a and *Qianfulun*, 1.1a–b. *Lüshi chunqiu*, 4.5a, also has a similar list but does not include Confucius. In Han times some scholars maintained that the Six Classics had been meaningfully edited, arranged, and commented on by Confucius and thus were the vehicle of his particular visionary teachings. Therefore these scholars did not subscribe to the view that by the time that the classics had been passed down to Confucius they were already essentially complete. See Pi, *Jingxue lishi*, 81–90; Jiang, *Jingxue zuanyao*, 182–83.

25. On "utensils and weapons" (*qixie* 器械), see Zheng Xuan's commentary and Jia Gongyan's 賈公彥 (*fl.* 650) subcommentary to *Zhouli*, 7.2a, under Manager of Writings (*sishu* 司書); see also Hucker, *Official Titles*, no. 5769. *Xun* 訓 means "to gloss"; "to explain." *Gu* 詁, also variously written as 古 or 故 in the *xungu* compound, means "old and ancient words or terms." See Zhang Yi 張揖 (early third century CE), *Zazi* 雜字, cited in *Jingdian shiwen*, 29.1b (3:1594); and Wang, *Guantang jilin*, 5.4b–5b. Knechtges, "Correspondence," 312, n. 14, quotes Paul L-M Serruys's dissertation, "Prolegomena to the Study of the Dialects of Han Times According to *Fang-Yen*," where Serruys defines *xungu* as "explanations of old words and characters (*ku*), and readings (*hsün*).... It stressed the study of separate words in their original meaning and their relations with modern words (*yen*)."

26. Two extant examples of Han *zhangju* commentaries are Zhao Qi's 趙岐 (d. 201) *Mengzi zhangju* 孟子章句 and Wang Yi's 王逸 (c. 89–c. 158) *Chuci zhangju* 楚辭章句. For an explanation of "sections and sentences" as the term is used in the former

work, see Jiao Xun's (1763–1820) commentary to Zhao Qi's preface to *Mencius*, in Jiao's *Mengzi zhengyi*, 16. In Wang's *zhangju* commentary to *Chuci*, comments are appended after lines of verse and before each chapter, not after. Neither displays the excesses of earlier works of this genre that contemporary critics found so objectionable. It nevertheless remains unclear whether there was a definitive *zhangju* commentary form and the extent to which this form changed during the Han. Lin, "Liang Han zhangju zhi xue chongtan," 277–97, argues that a distinguishing characteristic of the *zhangju* commentary was that the form and content of the appended commentaries and glosses followed a format or pattern associated with the exegetical principles of a particular lineage of teachers (*shifa* 師法) and scholastic lineage (*jiafa* 家法), and was passed on from teacher to student. He further characterizes it as a "closed system" that was rigidly adhered to and which did not seek to introduce alternative interpretations. His account expands on a thesis originally proposed by Dai, "Jingshu de yancheng," 77–96. One of the major charges directed against the *zhangju* commentary by contemporary critics was its long-windedness. Some commentaries ran into hundreds of thousands of words, as commentators adduced extraneous writings either for the purposes of criticism or to support their own interpretation. See Makeham, *Name and Actuality*, 115. For a summary of other criticisms, see Nylan, "*Chin wen/Ku wen*," 114.

27. Female scribe: see *Zhouli*, 8.3a–b; Hucker, *Official Titles*, no. 4345. Junior eunuch: see *Zhouli*, 7.32a–24a; Hucker, *Official Titles*, no. 4254.

2. Establishing Models and Exemplars

1. Cf. *Analects*, 20.2; Lau, *Analects*, 203, mod: "The gentleman, with his robe and cap adjusted properly and dignified in his gaze, has a presence which inspires people who see him to be filled with awe."

2. According to Zheng Xuan's commentary, *Zhouli*, 21.6b, in Zhou times official rankings were also distinguished by nine different designs that were embroidered on these costumes. A *huang* 璜 is a semicircular piece of jade. It is actually half a *bi* 璧, a disc of jade with a hole in the center. Such jade ornaments were hung on strings suspended from a girdle around the waist, and would

tinkle while the wearer was walking. See Hsia, "Shang Jades," in Chang, *Shang Archaeology*, 214, 222–23, 231.

3. In early China, two types of tally were used, the *qi* 契 tally and the *fu* 符 tally. The *qi* tally was used to record the details of a transaction where goods were exchanged. Details of the contract were carved or written on wooden or bamboo slips. The slip was then split, with the creditor and debtor each retaining one portion of the tally. The *fu* tally had a variety of uses, including the authorization of military commands (such as the mobilization of troops), exit and entry at border passes, and admittance to and from the inner chambers of the palace. *Fu* tallies were made from a variety of materials, including jade, wood, silk, and metal. The tally was either written on in ink or engraved and then cut in two. See also my discussion in *Name and Actuality* thought, 76–77 and notes. Although probably inspired by a similar metaphor in "The Tally of Replete Potency" (De chong fu 德充符) in *Zhuangzi*, Xu Gan uses the metaphor to describe how inner virtue (*de*) becomes externally manifest on the gentleman's countenance. In "The Tally of Replete Potency," however, the point that Zhuangzi is making is that *de* (which has more the sense of a power or potency) will be made manifest to others regardless of the physical appearance of a real man of *de*.

4. In *Discourses* I understand *ren* 仁 (humaneness) to mean a concern for others, and *yi* 義 (rightness) to mean the sense of what is the right course of action in each circumstance.

5. Musical entertainment: literally, "the music of feathers and pipes." The *yu* 羽, a feathered fanlike object, and the *yue* 籥, a three-holed flute, were two objects that served as props or dancing accessories, and which distinguished the types of dances they were used in as being either cultural or martial. See *Zhouli*, 24.6a–b, and subcommentary.

6. Both lines are from *Analects*, 20.2.

7. Mao no. 256.

8. The *Documents*, 17.9b; Legge, *Chinese Classics*, 5:501.

9. This passage is clearly related to the notion of "vigilant solitariness" (*shendu* 慎獨). Compare, for example, the following passage from the first section (*zhang*) of *Zhongyong*: "Therefore the gentleman is alert and careful when he cannot be seen, and apprehensive when he cannot be overheard. There is nothing more manifest than the hidden, and nothing more obvious than the obscure. Thus the gentleman is vigilant when by himself." Of

the word *du*, Tu, *Centrality and Commonality*, 109, writes: "In *Chung-yung*, the term *tu*, which may mean being alone, seems to refer to the self in terms of its singularity, uniqueness and innermost core. . . . This quality of personhood can be better appreciated when one is physically alone, but the focus of the recommendation is the essential "solitariness"—the singularity, uniqueness, and innermost core—of the self." This same sense is apparent in *Discourses*. For a more general discussion of *shendu* in late Warring States and early Han philosophical literature, see Shimanori, "Shindoku no shisō," 145–58.

10. Mao no. 7.

11. See *Documents*, 11.6b; Legge, *Chinese Classics*, 3:545, 548.

12. *Zuo Commentary*, Ai 15, 59.24b; Legge, *Chinese Classics*, 5:843.

13. Mao no. 35.

14. The quotation is from *Documents*, 2.6b, and the comment on Cheng Tang is a line from *Odes*, Mao no. 305.

15. *Odes*, Mao no. 303: "Then, long ago, Di appointed the martial Tang / To regulate the boundaries throughout the four quarters. / In those four quarters he had sway over all the local rulers, / And came to possess the nine regions." The "nine regions" are also variously referred to as *jiuyou* 九囿, *jiuyou* 九有, and *jiuzhou* 九州. *Erya*, B.9b, list the nine regions as: Ji 冀, You 幽, Yan 兗, Ying 營, Xu 徐, Yang 楊, Jing 荊, Yu 豫, and Yong 雍 (雝).

16. Paraphrase of *Documents*, 14.3a.

17. *Changes*, Guan hexagram, "Judgment," 3.8b; translation based on Shaughnessy, *I Ching*, 155 and notes.

18. Xu's commentary follows the wording of *Changes*, "Judgment," 3.9a.

19. This story is related in the *Gongyang Commentary*, Zhuang 12, 7.13b–14a: "Wan of Song had previously fought and been taken captive by Duke Zhuang of Lu. Duke Zhuang had brought him back to Lu, but released him in the military compound. After a few months he let him return to Song. After returning, Wan was promoted to counselor. When playing a game of chess with Duke Min of Song, he had many women by his side. Wan said, 'The Marquis of Lu [Lord Zhuang] is very fine and handsome. Of all the various lords, only he is fit to be a ruler.' Because the women had heard this, Duke Min became very jealous. Addressing the women he said, 'This is one who was formerly a prisoner of the Marquis of Lu; of course he praises him. How could the Mar-

quis of Lu possibly be as handsome as all that?' Wan became angry, struck Duke Min and broke his neck."

20. *Zuo Commentary*, Xuan 10, 22.13a: "Duke Ling of Chen, Kong Ning and Yi Hangfu were drinking in the house of the Xia family, when the duke said to Hangfu, 'Zhengshu looks like you.' 'He also looks like your lordship,' was the reply. Zhengshu overheard these remarks and was very angry. When the duke was leaving, Zhengshu shot and killed him from the stable."

21. *Zuo Commentary*, Wen 18, 20.11b–12a: "When Yi of Qi was still son of the duke [that is, before he became marquis], he had some trouble with the father of Bing Chu about some fields in which he did not get the better. Therefore, when he became marquis, he had Bing Chu's father's grave exhumed, and the feet of the corpse cut off, while he employed Chu as his charioteer; and although he took Yan Zhi's wife, making her his own, he carried Zhi with him as the third attendant in his chariot. In summer, in the fifth month, the duke went to the pool of Shen to swim. Bing and Yan were bathing in the pool when Bing struck Yan with a whip. When Yan got angry, Bing said to him, 'Since you allowed your wife to be taken from you without being angry, how does a crack like that hurt you?' Yan replied, 'I wonder how much pain one who saw his father's feet cut off— yet was unable to express his hatred—is able to tolerate?' They then plotted to murder Duke Yi and hide his body among the bamboo."

22. *Zuo Commentary*, Xuan 4, 21.18b–19b, relates the story as follows: "The people of Chu presented a large turtle to Duke Ling of Zheng [r. 605]. Gongzi Song [Zigong] and Zijia were on their way to an audience with the duke when Zigong's index finger began to move. He showed it to Zijia, saying, 'On other occasions when my finger has done this, I have been sure to taste some extraordinary dish.' When they entered the palace, the cook was about to cut up the turtle, and they looked at each other and laughed. The duke asked the reason and Zijia told him. When, however, the duke was feasting the other senior officers on the turtle, although he invited Zigong, he did not give him any. Zigong was angry, so he dipped his finger into the cauldron, tasted the turtle, and went out. This so enraged the duke that he wanted to kill him. Zigong then plotted with Zijia about first killing the duke; but Zijia said, 'Even in the case of an old animal, one shrinks from the task of killing it, thus how much

more should you shrink from killing your ruler!' The other turned round and threatened to implicate Zijia on some other matter so that Zijia became afraid and agreed to go along with his plan. Zigong murdered Duke Ling in the summer." See also *Shuoyuan*, 6.20a–b; *Shiji*, 42.1767.

23. Mao no. 207.

24. "Harmony but not conformity": cf. *Analects*, 13.23; Leys, *Analects*, 64, slightly modified: "The Master said, 'The gentleman seeks harmony but not conformity; the small man seeks conformity, but not harmony.'" For an original and still stimulating discussion of the virtue of *shu* (putting oneself in another's shoes; reciprocity), see Fingarette, "One Thread," 373–405.

25. *Documents*, 17.3a.

26. Cf. *Liji*, 52.1a ("Zhongyong"): "The way should not be strayed from even momentarily. If it could be strayed from it would not be the way."

27. *Analects*, 14.12, is the *locus classicus* for the term "fully cultivated person" (*chengren* 成人): "Zilu asked about the fully cultivated person. The Master said, 'A man as wise as Zang Wuzhong, as free from desires as Meng Gongchuo, as brave as Zhuangzi of Bian, and who refines these qualities with ritual and music may be said to be a fully cultivated person.'"

28. This story is recorded in *Zuo Commentary*, Xi 11, 13.18a: "The Zhou king sent Duke Wu of Shao and Guo, the royal secretary, to present the marquis of Jin with a jade tally. He received the jade tally with an air of indifference. On his return to court, Guo said to the king, 'The marquis of Jin will not have any descendants to inherit his position. Your majesty presented him with the jade tablet yet he received it with an air of indifference. In thus dissipating himself first, is he likely to have anyone succeed him?'" As predicted, he did not. See *Hanshu*, 27.2A.1357.

29. *Zuo Commentary*, Xi 28, 16.23b–24a: "The king ordered ministers Yin Shi and Wang Zihu and Royal Secretary Shu Xingfu to put in writing the appointment of Marquis Wen of Jin as overlord of the feudal lords. . . . The words of the appointment were, 'The king says to his uncle, Respectfully obey the king's orders so as to give stability to the states in every quarter. Uncover and punish those evil men who would do me harm.' The marquis thrice declined to accept the charge before finally doing so, saying, 'I, Chonger, twice prostrate myself in obeisance and kow-

tow. And so do I receive and will carry out the great, glorious, and excellent charge of the son of heaven.'"

30. *Zuo Commentary*, Cheng 14, 27.18b: "The marquis feasted Chengshu of Ku [Xi Chou] and Ning Huizi directed the ceremonies. Chengshu of Ku behaved arrogantly and Ning Huizi said, 'The family of Chengshu of Ku is likely to perish. In antiquity, entertainment and feasts were occasions to observe a man's awe-inspiring demeanor and to judge whether he would meet with misfortune or fortune. . . . Now he acts with arrogance— this is the path leading to misfortune.'" His subsequent death is recorded in the *Spring and Autumn Annals*, Cheng 17.

31. *Zuo Commentary*, Xi 33; 17.17a–17b: "At the beginning, Jiu Ji was passing by Ji village and saw Que of Ji weeding in a field and his wife bringing his food to him. She was very respectful. They treated each other with the same degree of respect one would show a guest. Jiu Ji therefore took him back with him, and told Duke Wen of Jin [r. 636–628], 'Respect is the expression of accumulated virtue. He who has respect is sure to have virtue. Virtue is used to put the people in order. I entreat your lordship to employ him.'... Duke Wen made Xi Que a counselor of the third army." See also *Guoyu*, 11.1a.

32. *Zuo Commentary*, Zhao 1, 41.10b–11a: "The chief minister [Prince Wei, who later became King Ling of Chu (r. 540–29)] feasted Zhao Meng and recited the first stanza of the 'Great Brilliance' [*Book of Ode*s, Mao no. 236]. Zhao Meng recited the second stanza of the 'Little Bird' [*Book of Odes*, Mao no. 196]. When the feast was over, Zhao Meng said to Shuxiang, 'The chief minister regards himself as king. What is going to happen?' Shuxiang replied, 'The king is weak and the minister is strong. He shall probably succeed but he will not die a natural death.' 'How is that so?' 'When strength overcomes weakness and is satisfied in doing so, the strength is not appropriate. Of strength which is not appropriate, doom will come quickly.'" In choosing to sing the first stanza of the ode "Great Brilliance," Prince Wei was intimating that he was in a position to receive the mandate ordained by heaven, hence justifying his overthrow of the old, weak king.

33. "Drunk with Wine," Mao no. 247, is a song of praise and thanks to the host. This story is recorded in *Zuo Commentary*, Xiang 27, 28.17: "Wei Pi of Chu went to Jin to participate in formalizing the covenant and the marquis of Jin entertained him. As he was

about to leave the feast, he recited 'Drunk with Wine.' Shuxiang said, 'It is proper that the Wei family should continue to have descendants in Chu. Being commissioned by the ruler, he is not unmindful to show his keen intelligence. Zitang [Wei Pi] will one day rule his state.'"

34. "Bickering Quails" is Mao no. 49. In singing this ode, Liang Xiao (Boyou) revealed his intention to usurp the position of his lord, the Earl of Zheng. For an informative account of this story, see Van Zoeren, *Poetry and Personality*, 64–66. "Insects in the Grass" is Mao no. 14. Zizhan's recital took place at the same feast hosted by the Earl of Zheng. *Zuo Commentary*, Xiang 27; Legge, *Chinese Classics*, 5:533, 534, mod: "Zizhan then sang 'Insects in the Grass' and Zhao Meng said, 'Good for a lord of the people, but I am not sufficient to answer to it. …' When the entertainment was ended, Wenzi [Zhao Meng] said to Shuxiang … 'The rest of them will all continue for several generations, and the family of Zizhan will be the last to perish.'"

35. Cf. *Liji*, 30.13a; *Hanshi waizhuan*, 1.7b.

36. Cf. *Zuo Commentary*, Zhao 11, 45.20a–b; Legge, *Chinese Classics*, 5:634, mod: "Minister Shanzi met with Han Xuanzi at Qi. The minister's gaze was lowered and he spoke slowly. Han Xuanzi said, ' … The places at court audiences are fixed, and those at meetings elsewhere are marked by flags. There is the collar of the upper garment and the knot of the sash. The words spoken at meetings and audiences must be heard at the places marked out and manifestly designated, so that the order of the business may be made clear. The looks must be fixed on the space between the collar and the knot, in order that the bearing and countenance may be properly guided. The words are intended for the issuing of orders, and the bearing and countenance to illustrate them. Any error in either of these is a defect.'"

3. *Cultivating the Fundamental*

1. Cf. *Analects*, 14.24: "In antiquity, men undertook learning as a means of cultivating themselves; nowadays they undertake learning with an eye to others." *Beitang shuchao*, 83.1b, cites a now lost passage from *Xinxu* in which a similar saying is attributed to Mozi.

2. For a similar view see *Han Feizi*, 7.4b; for a contrary view, see *Laozi*, sections 22 and 24.

3. *Changes*, Fu hexagam, "Judgment," 3.18b; translation based on Shaughnessy, *I Ching*, 115.

4. *Liji*, 60.3a ("Daxue"): "Sincerely renew oneself daily, again and again renew oneself, and still yet again."

5. *Changes*, "Appended Statements (A)," 7.13a–b.

6. This passage is not found in any earlier extant source.

7. This passage is not found in any earlier extant source. For biographical details on Confucius's disciple, Zizhang, see *Shiji*, 67.2203–4.

8. As with many Han thinkers, Xu Gan here employs *xing* in the sense of inborn nature. This understanding of *xing* is not necessarily incompatible with the view that *xing* is also concerned with an entity's spontaneous tendency of development.

9. Mao no. 238; Karlgren, *Odes*, 191, mod.

10. According to *Lüshi chunqiu*, 7.1a, in the seventh month of the year, *mengqiu* 孟秋, at dusk, the Southern Dipper appears in the sky, marking the beginning of the return of the cold time of the year. According to *Liji*, 16.8a, in the sixth month of the year, *jixia* 季夏, at dusk, the Fire Stars appear in the sky, marking the beginning of the hottest part of the year.

11. *Changes*, Sheng hexagram, "Judgment," 5.9b; translation based on Shaughnessy, *I Ching*, 117.

12. Cf. *Changes*, Sheng hexagram, "Images," 5.10a: "By following his virtue, little by little the gentleman becomes noble and great."

13. This passage is not found in any earlier extant source.

14. This passage is not found in any earlier extant source.

15. Cf. the closely related concept of *shendu*, discussed in "Establishing Models and Exemplars."

16. Alternatively: "The four seasons bring things to completion in silence. They do not do anything and yet people have faith in them." Cf. *Analects*, 17.17: "The Master said, 'Does heaven say anything? The four seasons proceed and the one hundred things are born. Yet does heaven say anything?'"

17. *Changes*, Qian hexagram, "Words of the Text," 1.20a; Lynn, *Changes*, 138, mod: "The great man is someone whose virtue is consonant with heaven and earth, his brightness with the sun and moon, his consistency with the four seasons, and his prognostications of the auspicious and inauspicious with the workings of ghosts and spirits."

18. The *shi* 士 were the cultured strata who comprised the leading families at the village level. They were also part of a larger, na-

tional class of *shi* that was hierarchically divided, according to the level at which the *shi* was engaged as an official. Thus Huan Tan divides the *shi* into five grades: those at the village level, those at the prefectural level, those at the provincial and commandery level, those at the capital, and those whose actions affected the whole country (presumably the three dukes). *Yilin*, 3.10a–b; Pokora, *Hsin-lun*, 15–16. Xu Gan was himself a member of this élite class, and when he refers to *shi*, he is speaking as an insider. For discussions of the term *shi* in the Han dynasty, see Ch'ü, *Han Social Structure*, 101–7; Yu, *Zhishi jieceng*, 109–65, 205–75, passim; Ebrey, "Economic and Social History," 630–32; Connery, *Empire of the Text*, chapter 2.

19. Paraphrase of *Analects*, 8.7.

20. This view is discussed further at the end of "Titles and Emoluments." A related view can be traced to at least the time of Xunzi. Cf. the following passage from *Xunzi*, 20.6b–8a: "When Confucius was traveling south to Chu, he ran into difficulties in the region between Chen and Cai. For seven days he had nothing cooked to eat, and his soup of wild herbs had no grain to thicken it. All of his disciples had a hungry look about them. Zilu came forward and asked, 'I have heard that "heaven will repay those who do good with good fortune and those who do bad with misfortune." Now, master, for a long time you have been accumulating virtue, building up a store of rightful acts and cherishing morally excellent behavior. Why, then, do you find yourself in such distressing circumstances as this?'

"Confucius replied, '… Although the angelica and the orchid grow in the forest, they do not lose their fragrance because there are no people there to appreciate them. The gentleman engages in learning, not so that he may become successful, but rather so that, when times are hard he will not be at a loss as to what to do; so that, in times of anxiety, he will remain steadfast of purpose; and so that, by knowing that misfortune and good fortune follow one another, just as do ends and beginnings, his heart will not become confused. Being either a worthy or a good-for-nothing depends on ability; the decision to act or not to act depends on the man; meeting with success or not depends on timing; and life and death depend on circumstances beyond individual control. Now, if a person does not encounter the right times, then even if he were a worthy, would he be able to make any headway? If, however, he should encounter the right times, what difficulties what he have? Hence, the gentleman broadens his learning, deepens his strategies, cultivates his person, and makes his conduct proper while he waits for the right times.'"

21. In the Han dynasty, three different types of *ming* were distinguished: concordant *ming* (*sui ming* 遂命); predetermined *ming* (*shou ming* 壽命 ; also variously called *zheng ming* 正命 , *shou ming* 受命, and *da ming* 大命); and converse *ming* (*zao ming* 遭命). Concordant *ming* is the view that good fortune is a reward for moral actions, and misfortune a retribution for wrongdoing. This is contrasted with predetermined *ming*, which is the view that someone's *ming* is fixed and determined before birth, and converse *ming*, which is the view that good actions are repaid with misfortune and bad actions are repaid with good fortune. Xu Gan's position is that so long as concordant *ming* prevails, good actions will serve to increase a person's longevity. Nevertheless, when the proper set of circumstances does not prevail, virtuous actions are liable to be repaid by misfortune; he uses the word *ming* to describe this situation. This sense of *ming* is clearly what I have described as "converse *ming*." According to Xu Gan, converse *ming* is brought about by departures from regularities (*bian shu* 變數). Thus if discord between name and actuality is on a large enough scale, then these abnormalities will proliferate, and human affairs, including longevity, will be determined by converse *ming*. For accounts of the Three *Ming* Theory, see, for example, *Chunqiuwei yuanmingbao* 春秋緯元命包, in Ma, *Yuhan shanfang*, 4:2110; *Xiaojingwei shoushenqi* 孝經緯授神契, in *Yuhan shanfang*, 4:2148; Zhao Qi's 趙岐 commentary to *Mencius*, in Jiao, *Mengzi zhengyi*, 2:877–78; *Bohutong*, "Shouming"; *Chunqiu yan Kong tu* 春秋演孔圖, in Yasui and Nakamura, *Isho shūsei*, 4A:55–56.

22. *Changes*, Kun hexagram, "Images," 5.12b.

23. This passage is not found in earlier extant sources.

24. Mao no. 201; Karlgren, *Odes*, 152, mod.

25. Mao no. 252.

4. *The Way of Humility*

1. An allusion, perhaps, to the so-called "warning vessel" (also called the "tilting vessel") which is described in *Xunzi*, 20.1a–b; Knoblock, *Xunzi*, 3:244, mod., as follows: "When Confucius was inspecting the ancestral temple of Duke Huan of Lu [r. 711–

694], there was a vessel that inclined to one side. He asked the temple caretaker, 'What vessel is this?'

"The caretaker replied, 'It is said to be a warning vessel.'

"Confucius said, 'I have heard that when the warning vessel is empty, it inclines, when half-full it sits upright, and when full it overturns.' Turning to a disciple, he said, 'Pour some water into it.'

"The disciple ladled some water and poured it in. Half-full, it stood straight; full, it turned over; empty, it leaned at an angle. Confucius heaved a sigh and said, 'Oh! That which is full cannot fail to turn over!'

"Zilu said, 'I should like to ask if there is a method for maintaining fullness?'

"Confucius said, 'Keen intelligence and sagely wisdom should be guarded by feigning stupidity; merit that is universally renowned should be guarded by an attitude of deference; courage and strength which give support to the age should be guarded by fear; and abundant wealth should be guarded by frugality.'"
See also *Shuoyuan*, 10.3b–4a; *Kongzi jiayu*, 2.13a–b; *Huainanzi*, 12.19a–b; *Hanshi waizhuan*, 3.18b–19a.

2. Cf. *Laozi*, section 9; Lau, *Tao Te Ching*, 65, which also appears to be alluding to the warning vessel: "Rather than fill it to the brim by keeping it upright/Better to have it stopped in time."

3. *Changes*, Xian hexagram, "Images," 4.2b. In his subcommentary Kong Yingda 孔穎達 (574–648) writes: "To say, 'The gentleman receives others with emptiness,' means that he models himself on the Xian hexagram, of which the upper trigram is a marsh and the lower trigram a mountain. He is able to open his mind, be free of prejudices and be receptive of others. When people are influenced in this way, none will fail to respond."

4. Mao no. 53.

5. "Fearing that he might never attain it": allusion to *Analects*, 16.11: "Confucius said, '"Seeing what is good, I pursue it as if I might not attain it. Seeing what is not good, I stay away from it as I would if my hand were about to enter boiling water." I have seen a person such as this and I have heard claims such as this.'"

The *locus classicus* of the term "movement toward the good" (*qian shan*) is *Mencius*, 7A.13; Lau, *Mencius*, 184: "The people daily move toward the good without realizing who it is that brings this about." Xu Gan applies the concept to the gentleman.

6. *Changes*, "Appended Statements (B)," 8.14a. "That son of the

Yan family" refers to Confucius's disciple Yan Hui 顏回. See also *Analects*, 6.2: "Duke Ai asked, 'Which of your disciples is fondest of learning?' Confucius replied, 'There was one Yan Hui; he was fond of learning. He did not vent his anger on others nor make the same mistake twice.'"

7. *Changes*, Pi hexagram, "Judgment," 2.23b; Lynn, *Changes*, 211, mod.

8. Paraphrase of the "Commentary on the Images" to this hexagram, 2.24a; Lynn, *Changes*, 212, mod: "The inner [lower trigram] is *yin* and the outer [upper trigram] is *yang*; the inner is soft and the outer is hard. Inside is the small man and outside is the gentleman. The way of the small man is waxing and the way of the gentleman is waning."

9. Cf. *Analects*, 19.3: "One of Zixia's students asked Zizhang about making friends. Zizhang said, 'What does Zixia have to say about it?' The student replied, 'Zixia says: Befriend those with whom it is proper to do so but refuse to have anything to do with those whom it is not.' Zizhang said, 'That is different from what I have been told: The gentleman respects the worthy and is accommodating toward the ordinary people. He commends those who are skilled and pities those who are incompetent. If I were one of the great worthies, which people should I not accommodate? If I were one of the unworthy, others would, in any case, refuse to have anything to do with me, so what need should I have to refuse to have anything to do with them?'" Note that the Han stone classic version of *Analects, Lunyu yishu*, 10.2a, and *Jingdian shiwen*, 24.21a (3:1389), twice in this passage read *ju* 距 rather than *ju* 拒.

10. Mao no. 257.

11. This whole section about the eyes repeats a standard trope.

12. For the duties of the scribes see *Liji*, 29.7a–b. For the duties of the music master and blind musicians, cf. *Guoyu*, 17.11b–12a: "When traveling by chariot, there are the admonitions of gallant soldiers. In the palace, rules of behavior toward officials and elders must be observed. When reclining at the low table, there are admonitions inscribed on it. When seated or reclined, there are officers in close service. When presiding over ceremonies, there are the directions given by the music master and the grand historian. When resting, the music master and the blind musicians recite odes and so guide and serve the king. The historians did not fail to record and the blind musicians did not fail to

recite." See also *Zuo Commentary*, Xiang 14, 32.18b–19b. For an account listing the types of vessels and implements on which admonishments were inscribed, see *Da Dai liji*, 6.2a. The *yan* 筵 was an undermat, coarser and longer than the *xi* 席(蓆), which, being of a finer weave, was placed on top and sat upon. In the light of *Zhouli*, 20.8b–13b, and *Guoyu*, 17.11b, it would seem that the admonishments were more likely to have been inscribed on the low tables which were placed on the mats.

13. Paraphrase of *Guoyu*, 17.11b. For the ode, see Mao no. 256, Karlgren, *Odes*, 216–19. According to the Mao preface to this ode, it was, in part, an admonition directed by Duke Wu against King Li (r. 857/53–842/828). Yet Wu did not become duke of Wei until 812, at least sixteen years after the death of King Li. Thus, as Duke Wu reigned until 758, and the ode was composed at the very end of his reign, the statement given in the Mao preface should be discounted.

14. The Mao Preface also records that this ode, Mao no. 55, was written in celebration of Duke Wu but attributes its authorship to a member of the Zhou court.

15. *Changes*, Zhen hexagram, "Images," 5.24a; Lynn, *Changes*, 461, mod.

16. Mao no. 256.

5. *Valuing Proof*

1. This quotation is not found in any earlier extant source.

2. *Changes*, Heng hexagram, "Judgment," 4.4a; translation based on Shaughnessy, *I Ching*, 101.

3. Xu Gan here uses the wording of the "Judgment" to the Heng hexagram.

4. See also *Odes*, Mao no. 200, commentary, 12/3.20b. *Gu lienü-zhuan*, 6.17b, refers briefly to this incident, as does the following story cited in the commentary to *Xu Hou Hanshu*, 69B.6a: "*Hanshi* says: In Lu there was a man who lived alone. One night, at the onset of a fierce storm, a woman approached his house and asked to be let in. The man closed the door and refused her entry, saying, 'I have heard it said that if a man and a woman are not sixty years of age they may not occupy the same dwelling.' The woman replied, 'Why don't you learn from Hui of Liuxia?' The man replied, 'In future, then, I will learn from Hui of Liuxia's capabilities when regarding my own incapabilities.'" *Mencius*,

2A.9, quotes Hui of Liuxia as saying, "Although you stand naked by my side, how could you defile me?" As Powers notes, by "Han times he had become the paradigm of the man of integrity placed in a morally compromising situation." See his *Art and Political Expression*, 214–15, and fig. 93, which depicts this story.

5. *Taiping yulan*, 430.4a, also attributes this quotation to Zisi. For variations on the passage, see *Huainanzi*, 10.3b; Yang, *Huainanzi zhengwen*, 95; *Hou Hanshu*, 27.934.

6. Cf. *Analects*, 15.21: "That which one seeks from within is confirmation of the integrity of one's thoughts and actions."

7. Takeuchi, *Eki to Chūyō*, 68, proposes that this passage originally came from the last part of the "Shuo xia 説下" section of *Zhong-yong*. For discussion of the significance of this passage in Xu Gan's theory of naming, see Makeham, *Name and Actuality*, 15; for its connection with *xingming* 刑名 (accountability of word and deed) thinking, see ibid., 73–83.

8 On Zhou, see *Xunzi*, 10.8b; *Mencius*, 1B.8: "I have heard of the mutilation of that fellow Zhou." On Confucius as "uncrowned king," see Dong Zhongshu's 董仲舒 (c. 179–c. 104) memorial quoted in his *Hanshu* biography, 56.2509; see also *Shuoyuan*, 5.2a.

9. This ode is not recorded in any earlier extant source.

10. See *Documents*, 4.17a; *Mencius*, 2A.8.

11. On the deaths of King Li and Wu Qi, see *Shiji*, 4.142 and 65.2168, respectively.

12. This passage is not found in any earlier extant source.

13. Mao no. 196; Karlgren, *Odes*, 145, mod.

14. Paraphrase of *Documents*, 14.21b. For similar statements, see also *Guoyu*, 19.4a; *Qianfulun*, 1.11b; *Mozi*, 5.7a.

15. Mao no. 165; Karlgren, *Odes*, 108–9, mod.

16. The quotation is from *Analects*, 1.8.

17. This passage is not found in any earlier extant source.

18. Mao no. 192; Karlgren, *Odes*, 137, mod.

6. *Valuing Words*

1. On the term *fei qi ren* 非其人, see de Crespigny, "Political Protest," 6.

2. *Changes*, Gen hexagram, 5.28b.

3. Paraphrase of the "Commentary on the Images" to this hexagram, 5.28b: "'Restraining the cheeks,' so that one is to the point and correct."

4. Cf. *Analects*, 9.1: "The Master is good at guiding people on, step by step."

5. Literally, "are dependent upon the creative transformer (*zaohua-zhe*)."

6. *Analects*, 15.8.

7. I can find no other reference to this particular chariot. On the *huang* jade, see *Zuo Commentary*, Ding 4, 54.15a; Ai 14, 9.19a; *Huainanzi*, 7.4b.

8. *Xunzi*, 1.14a–b.

9. This quotation does not appear in any earlier extant source.

10. Cangwu Bing was also variously known as Cangwu Rao 倉梧嬈, Cangwu Rao 倉梧繞, and Cangwu 蒼梧. For versions of the story, see *Huainanzi*, 13.11a–b; *Shuoyuan*, 3.13a; *Kongzi jiayu*, 4.5a.

11. For versions of the story, see *Zhuangzi*, 998; *Zhanguoce*, 9.5a; *Huainanzi*, 13.10b–11a; *Shiji*, 69.2264–65; *Hanshu*, 65.2842, Yan commentary.

7. The Fundamental Principles of the Arts

1. As I have argued in *Name and Actuality*, 139, in the context of contemporary discussions of the relationship between inner ability (*cai* 才) and moral nature (*xing* 性), this view was compatible with the position that they were the same.

2. Paraphrase of the *Changes*, "Appended Statements (B)," 8.4b.

3. Leaves and branches pair with roots. The roots-branches analogy is an organic metaphor commonly employed in early Chinese writings to signify a relationship between the fundamental and peripheral parts of an entity or situation. Here it is employed to describe how the external (what can be seen) and the internal (what cannot be seen) are mutually supporting, just as the roots and branches of a tree are necessary to its continued growth.

4. Palace protector: see "Ordering Learning."

   According to Zheng Xuan's commentary to *Zhouli*, 14.6b, the five types of ritual were for auspicious offerings (*jili* 吉禮), for mourning (*xiongli* 凶禮), for guests (*binli* 賓禮), for armies (*junli* 軍禮), and for marriage (*jiali* 嘉禮).

   According to Zheng Xuan's commentary to *Zhouli*, 22.9a, the six types of music were associated with the sage kings of antiquity: the Yellow Emperor, Yao and Shun, Yu, Tang, and

King Wu of the Zhou. According to the same source and his commentary to *Zhouli*, 14.6b, the six types of music were called: *yunmen* 雲門, *daxian* 大咸, *dashao* 大韶, *daxia* 大夏, *dahuo* 大濩, and *dawu* 大武.

The five skills of archery, according to Zheng Xuan's commentary and Jia Gongyan's (*fl.* 650) subcommentary to *Zhouli*, 14.6b, 7a, were (1) White arrow (*bai shi* 白矢). The arrow is shot and hits the target so forcefully that it pierces the target and the glimmering tip can be seen on the reverse side of the target. (2) Three in a row (*san lian* 參連). One arrow is shot first and this is followed by three arrows, shot in rapid succession, such that they all hit the spot where the first arrow landed. (3) To shoot an erect arrow (*yan zhu* 剡注). The feathered end of the arrow shaft is held high with the tip lowered, so that the arrow moves in an erect fashion. (4) Yielding a foot (*rang chi* 讓尺). When a minister competes with a lord he should retreat one foot to give the lord an advantage. (5) Well formation (*jing yi* 井儀). Four arrows pierce the target in the shape of the word for well (*jing* 井).

According to Zheng Xuan's commentary and Jia Gongyan's subcommentary to *Zhouli*, 14.6b–7a, 7b, the five skills of driving a carriage were (1) To sound the *he* and *luan* bells (*ming he luan* 鳴和鸞). The *he* bells are secured to the crossbeam of the carriage and the *luan* bells to the yoke. There should be a harmony between the pace of the horses and the ringing of the bells. (2) Following the bends of a river (*zhu shui qu* 逐水曲). To maneuver while traveling on a twisty road beside a river without falling in. (3) To dance through a crossroad (*wu jiao qu* 舞交衢). When passing through or turning at a crossroad, the carriage is maneuvered as gracefully as if following dancing steps. (4) Driving the game leftward (*zhu qin zuo* 逐禽左). When hunting, the game is herded to the left of the carriage, thus enabling the lord to shoot it. (5) Passing the flag of the lord (*guo jun biao* 過君表). When various lords were assembled at meetings, their ranks were marked by different flags. Depending on the different ranks, different forms of driving etiquette would be displayed when passing these lords. (Following the interpretation of Sun, *Zhouli zhengyi*, 26.1013, which is based on Du Yu's 杜預 [222–84] commentary to *Zuo Commentary*, Zhao 11, 45.20b).

Following *Shuowen jiezi*, 15.1b (although the *liu shu* 六書 of the *Zhouli* may have referred to something entirely different), the six classes of script are (1) Pointing to the matter (*zhishi* 指事). Characters of this class are discernible on sight, and when investigated, their meanings are apparent. The characters *shang* 上 and *xia* 下 are examples. (2) Imitating the form (*xiangxing* 象形). Characters of this class are drawn to resemble a real object, and in minute detail. Examples are the characters *ri* 日 and *yue* 月. (3) Forming the sound (*xingsheng* 形聲). Characters of this class are formed metaphorically by combining the names of certain affairs with a new phonetic component. The characters *jiang* 江 and *he* 河 are examples. (4) Conjoining the meanings (*huiyi* 會意). Here characters from different classes are brought together and their meanings combined so as to express a new and different meaning. The characters *wu* 武 and *xin* 信 are examples. (5) Transferring notation (*zhuanzhu* 轉注). This refers to establishing a class of characters which, by virtue of a common origin, simultaneously share common meanings. The characters *kao* 考 and *lao* 老 are examples. (6) Borrowing (*jiajie* 假借). Characters belonging to this class in fact never existed in form, only in sound. Homophonic characters are relied upon to fill in as substitutes and carry the meaning. The characters *ling* 令 and *zhang* 長 are examples. On Xu Shen's *liu shu*, see Boltz, "Early Chinese Writing," 196–97; Pulleyblank, *Classical Grammar*, 7–8.

There can be no certainty about the nine ways of reckoning that are referred to in *Zhouli*. Of the nine that feature in *Jiuzhang suanshu* 九章算術, eight are identical with those listed by Zheng Xuan in his commentary to *Zhouli*. The ninth listed by Zheng, *pang yao* 旁要, is, according to Christopher Cullen, "a term of unknown significance." For a summary description of the nine listed in *Jiuzhang suanshu*, see Cullen, "Chiu chang suan *shu*," 16–17.

5. For Zheng Zhong's 鄭眾 (d. 83) and Zheng Xuan's differing descriptions of the sort of demeanor that was appropriate to each of these occasions ("august and magnificent," "reverent and solemn," and so on), see *Zhouli*, 14.7a, 8a.

6. See *Zhouli*, 23.7a. The term *she cai* 舍采 has been variously interpreted. See Sun, *Zhouli zhengyi*, 7:1815–19.

7. Mao no. 176; Karlgren, *Odes*, 119–20, mod.

8. The preface to this ode, 10.A.15a, says, "This ode expresses the joy of fostering talent. When the lord is able to develop and

foster talented men, then the whole world gladly rejoices in the occasion."

9. *Changes*, "Appended Statements (A)," 7.13a.

10. Cf. *Guoyu*, 11.1b: "One's external appearance is the flower of one's inner feelings and words are the vehicle of that external appearance." "Xing shou 形守," (*Shiliu jing* 十六經), line 136B, in *Mawangdui boshu*: "For this reason, words are the tally of the heart and one's facial expressions are the flower of one's heart."

11. Quotation from *Changes*, Kun hexagram, "Commentary on the Words of the Text," 1.27b.

12. Paraphrase of *Liji*, 54.9b.

13. Cf. *Xunzi*, 1.11a: "Where there are precious stones under a mountain, the grass and trees have an enhanced luster."

14. In *Xiaojing*, 10.12a, the order of the sentences is reversed.

15. Allusion to *Analects*, 8.4: "As for the details of ritual vessels, it is minor officials who are responsible for them."

16. Cf. similar passages at *Shiji*, 24.1204; *Liji*, 38.18a, 37.15b.

17. *Deyin* literally means the "sound of virtue." The term refers to the virtuous teachings of former worthies and sages which posterity is still able to "hear."

18. Mao no. 161; Legge, *Chinese Classics*, 2:246, and Karlgren, *Odes*, 104, mod. I have translated this passage as I think Xu Gan intended it to have been understood.

19. The first part of the sentence is an allusion to *Analects*, 8.4.

20. *Analects*, 7.6.

8. An Examination of Disputation

1. *Zhuangzi*, 1111: "Gongsun Long's disciples . . . were able to defeat people verbally, but could not persuade them in their hearts. Such was their limitation." For reticence, cf. *Analects*, 13.27: "The Master said, 'Firmness, decisiveness, simplicity, and reticence in speech are near to humaneness.'"

2. Xu Gan was not, of course, the first to have raised such criticisms; see, for example, Zou Yan's 騶衍 (305–c. 240) description of what constitutes good and bad disputation, *Shiji*, 76.2370, n. 2, translated by Graham, *Later Mohist Logic*, 20–21.

3. *Zuo Commentary*, Zhao 12, 53.20a; Cheng 14, 27.19a.

4. Cf. *Analects*, 17.24: "[Zigong] said, 'I detest those who consider a flurry of words to be wisdom, those who consider insolence to be courage.'" In translating this passage I have followed Zheng

Xuan's text of the *Analects*, which reads *jiao* 絞 rather than *jiao* 徼 of the standard text (which follows *Lunyu jijie*). See *Jingdian shiwen*, 3:1387 (24.20a).

5. For the quotation cf. *Changes*, Tong ren hexagram, "Images," 2.26b; Lynn, *Changes*, 217, mod.: "The gentleman associates with his own kind and clearly distinguishes among things by their categories."

6. Paraphrase of *Liji*, 13.9a–b.

7. Paraphrase of *Xunzi*, 20.2a. Similar passages are also found at *Shuoyuan*, 15.14b–15a; *Yin Wenzi*, 11b; and *Kongzi jiayu*, 2.5a.

8. First sentence: *Analects*, 15.26; second sentence: *Mencius*, 7B.37.

9. Wisdom and Deeds

1. On the bearing that this has to the inner ability and moral nature debate, see Makeham, *Name and Actuality*, 140–43.

2. Fu Xi and the eight trigrams: see "Ordering Learning," p. 11, note *. There are two traditional accounts of which statements King Wen is supposed to have appended to each of the sixty-four hexagrams in *Changes*. According to one view, King Wen was responsible for appending both the hexagram statements (*guaci* 卦辭) and the line statements (*yaoci* 爻辭). According to the other view, because many of the matters referred to in the line statements occurred after the time of King Wen, the hexagram statements should be attributed to King Wen while the line statements should be attributed to the duke of Zhou. See Kong's preface to *Zhouyi zhengyi*; Lynn, *Changes*, 4.

3. *Changes*, Li hexagram, "Commentary on the Images," 3.27a; Lynn, *Classic of Changes*, 324, mod.

4. *Documents*, 2.6b.

5. *Documents*, 2.19b. Huan Dou has been variously portrayed in the early literature as a country, a man, and a tribe. See *Lüshi chunqiu*, 20.2a, Gao You commentary; *Documents*, 2.19b, 4.18a, and *Mencius*, 5A.3. Judging by the subsequent reference to the border tribes, Xu Gan seems to have understood Huan Dou to have been a tribe. Xu Gan is presumably following the *Documents*, 2.19a–20a, account of Gong Gong as one of the men Yao rejected as a possible assistant. On legends and traditions associated with the name Gong Gong, see Knoblock, *Xunzi*, 2:330–33.

Gun: *Documents*, 2.19b. Again, judging by the subsequent reference to the border tribes, Xu Gan (as with the Kong commentary to *Documents*) seems to have understood the term *si yue* 四嶽 to refer to four leaders of various tribal lords. Gun was the father of Great Yu, the founder of the Xia. On early traditions associated with him, see Allan, *Heir and Sage*, 62–66; *Shape of the Turtle*, 69–70.

6. Four types of conduct: see *Analects*, 11.3: "Distinguished for their virtuous actions were Yan Yuan, Min Ziqian, Ran Boniu, and Zhonggong; for speech were Zai Wo and Zigong; for government were Ran You and Ji Lu; for culture and learning were Ziyou and Zixia." See also Chen, *Dongshu dushuji*, 2.4b–6a.

7. *Analects*, 5.9.

8. *Analects*, 11.4.

9. On humaneness cf. *Analects*, 12.22: "Fan Chi asked about being humane. Confucius said, 'Be concerned about people.'"

10. *Analects*, 14.16: "Confucius said, 'It was due to the strong support of Guan Zhong that Duke Huan was able to assemble the feudal lords on repeated occasions, without a show of force.'" *Analects*, 14.17: "Confucius said, 'Guan Zhong assisted Duke Huan become overlord of the feudal lords, uniting and rectifying the whole realm.'" On observance of rituals see *Analects*, 3.22: "[Confucius] said, 'Rulers of states erect gate screens; Mr. Guan did so as well. In order to promote friendly relations with the rulers of other states, rulers of states have cup stands; Mr. Guan had one as well. If Mr. Guan understood ritual, then who does not understand ritual?'"

11. *Analects*, 14.17.

12. See *Zuo Commentary*, Zhuang 9, 8.20a; *Shiji*, 32.1486.

13. *Xunzi*, 20.2a, records the following version of this incident: "When Confucius was acting as regent in Lu, on but his seventh day at court he executed Deputy Mao. One of his disciples approached him and said, 'Deputy Mao is a well-known figure in Lu. If, Master, one of the first things you do in running the government is to execute him, won't this undermine the support that others have in you?'

"Confucius replied, 'Be seated and I will tell you the reason. Not including robbery and theft, there are five types of wrongdoing that people can be guilty of. The first is the treachery born of a penetrating mind; the second is to persist obstinately in aberrant behavior; the third is to defend duplicitous speech with disputation; the fourth is to record what is scandalous and spread it; and the fifth is to become soiled by following what is wrong.

Even if a man were guilty of only one of these crimes, he would not avoid being executed by a gentleman. Deputy Mao, however, is guilty of all five!'"

14. See *Shiji*, 55.2036–48, and *Hanshu*, 40.2025–37. Zhang Liang's strategies led to the defeat of the famous general, Xiang Yu.

15. For detailed discussion of the historicity of the four Gray Beards and legends concerning them, see Vervoorn, *Cliffs and Caves*, 96–100; Berkowitz, "Patterns of Reclusion," 139–63.

16. *Analects*, 5.18. As Huang Hui, *Lunheng jiaoshi*, 1:406–7, demonstrates, *zhi* 知 was read as *zhi* 智 here by a number of Han scholars.

17 See *Shiji*, 4.131–32.

18. See *Shiji*, 4.132; 33.1518; 45.1565; and *Zuo Commentary*, Ding 4, 54.19b. According to these sources, only Guan was executed, while Cai was banished.

19. This passage is a paraphrase of *Documents*, 13.11a–13a.

20. See *Shiji*, 33.1549, and the *Jijie* commentary, 33.1550, which cites Ma Rong.

21. See *Shiji*, 34.1549. A chapter by this title, which records a speech made by the duke of Zhou, is included in *Documents*. According to Shaughnessy, it "purports to record an address made by the duke of Zhou to Grand Protector Shi [the duke of Shao], encouraging him—even entreating him—to continue in the governance of the state." For a critique of this conventional interpretation, and the possible connection this chapter has with the "Shao gao 召誥" (Announcement of the duke of Shao) chapter of *Documents*, see Shaughnessy, *Before Confucius*, 109–19.

22. *Analects*, 9.30. *Quan* literally has the verbal sense of "adjusting the distribution of two weights to leverage a balance."

23. *Mencius*, 7A.26. As in *Analects*, 9.30, the metaphor employed is that of a traditional hand-held balance used for weighing.

24. *Analects*, 18.1: "The viscount of Wei left King Zhou, the viscount of Ji became his slave, and Bigan remonstrated with him and was killed. Confucius said, 'In the Yin dynasty there were three men who realized humaneness.'"

25. *Changes*, Yu hexagram: "Firmer than a rock, he acted promptly." Kong, *Zhouyi zhengyi*, 2.36a; Lynn, *Changes*, 240, n. 5, mod., comments: "Aware of how fast incipience works, he acted promptly to banish what is evil and cultivate what is good, and so constantly preserved his rectitude." For the virtue of the vis-

count of Wei, see *Lüshi chunqiu*, 11.8b–9a, *Shiji*, 38,1607, and *Mencius*, 6A.6.

26. Cf. *Changes*, Ming yi hexagram, "Judgments," 4.14a; Lynn, *Changes*, 357, mod: "Although there is adversity within, one should be able to keep one's aim squarely on course, as did the viscount of Ji." This presumably refers to his ability to maintain his commitment to his goals and aspirations even when he was a slave under a tyrant. See also *Shiji*, 38.1610.

27. *Shiji*, 38.1607–10, records that after Bigan had remonstrated with Zhou, Zhou cut out his heart to see if it had the seven openings of a sage's heart.

28. Ikeda, "Chūron kōchū (2)," 93, n. 64, notes that of the twenty such instances recorded in the *Annals*, in not one case is a counselor criticized for having failed to use wisdom to escape death, either in the *Annals* itself or one of the three commentaries. Despite this, at least one example can be identified; see *Zuo Commentary*, Xuan 9, 22.10b, where Confucius criticizes Xie Ye, who was killed for remonstrating with Duke Ling of Chen. Xu Gan's claim that all such cases were recorded is clearly more a case of polemical strategy than fact.

29. See *Han Feizi*, 19.2a; Watson, *Han Fei Tzu*, 99, mod: "King Yan of Xu lived east of the Han river, in a territory of five hundred square *li*. He practiced humaneness and rightness. Thirty-six tributary states came with gifts of territory to pay him tribute, until King Wen of Jing [r. 689–671], fearing for his own safety, raised troops in a campaign against Xu and wiped it out. Hence, while King Wen practiced humaneness and rightness and came to rule the world, King Yan, who also practiced humaneness and rightness, lost his state. This is why although humaneness and rightness were used in the past, they are no longer used today."

30. See *Zuo Commentary*, Yin 11, 4.26b, 27a: "[The minister] Yufu asked for permission to kill Duke Huan [Yin's younger brother and successor], so that he might become chief minister. Duke Yin said, 'Up until now I have ruled as regent out of consideration of his youth; now I will return the throne to him. . . .' In the eleventh month, in preparing to sacrifice to the spirit Zhongwu, Duke Yin fasted in the Shepu gardens, lodging in the house of the officer Wei. On the fifteenth day of the eleventh month, Yufu sent thugs to murder the duke in the house of officer Wei."

31. See *Gongyang Commentary*, 11.21b: "The duke of Song had agreed beforehand with the viscount of Chu that they would go to the meeting by ordinary carriage. Gongzi Muyi remonstrated with the duke, saying: 'Chu is a barbarian state. They might be powerful but they have no sense of rightness. I beg you to go to the meeting in a war chariot.' The duke replied, 'That is not possible. I agreed with him to go to the meeting in an ordinary carriage. I myself proposed this condition. It would be unthinkable of me not to respect it.' In the end, he went to the meeting in an ordinary carriage, and, as had been suspected, Chu war chariots were waiting in ambush. So as to punish Song, he was taken captive."

32. Bozong was a minister in the state of Jin. On his death, *Zuo Commentary*, Cheng 15, 27.24b, relates the following account: "The three men surnamed Que falsely implicated Bozong. They then slandered him, killed him, and also implicated Luan Fuji. [Thereupon, Bozong's son,] Bo Zhouli, fled to Chu. Han Xianzi said, 'These Que will not escape a bad end. Good men are a bond between heaven and earth. In destroying them like this, one after the other, what are the Que waiting for, if not their own destruction?' Initially, whenever Bozong went to court, his wife was sure to implore him with the following words, 'Thieves despise their masters, and the common people hate their superiors. Your fondness for frank talk will bring you troubles!'"

33. *Odes*, Mao no. 260. The "Greater Court Odes" is the second of the four major sections of the received text of the *Odes*. Shusun Bao was a minister of the state of Lu. On his death, see *Zuo Commentary*, Zhao 4, 42.32a–34a; Legge, *Chinese Classics*, 5:598–99. The gist of this long passage is that Shusun Bao, ignorant of the treacherous nature of his illegitimate son, Niu, let him into his favor only to be starved to death by him.

34. *Changes*, Weiji hexagram, Commentary on the Images, 6.24a.

35. *Analects*, 1.3 (repeated at 17.7): "The Master said, 'Rare, indeed, is it to find humaneness behind clever words and a contrived appearance!'"

10. Titles and Emoluments

1. *Analects*, 8.13. I have translated this passage in the sense that I have taken Xu Gan to have understood it.

2. The first part of the sentence paraphrases *Liji*, 49.13b.

3. Intendants: the title *zongren* 宗人 is used here, but given the nature of the duties described, it is more likely being used to refer to the offices of the minister and vice minister of rites (*da zongbo* 大宗伯 and *xiao zongbo* 小宗伯).

4. *Odes*, Mao no. 295; Karlgren, *Odes*, 253, mod. The Mao preface to this ode says: "This ode relates to the great investiture with fiefs that was conducted in the ancestral temple. 'Lai 賚' means to bestow and describes what is conferred upon good men."

5. *Documents*, 4.21b.

6. Mao no. 130, Karlgren, *Odes*, mod. 84.

7. This same area is also given in *Liji*, 31.5b. In his commentary to *Guoyu*, 18.2a, Wei Zhao states that Shaohao was the son of the legendary Yellow Emperor. The site of his capital is said to have been in the modern county of Qufu, in Shandong. According to *Shiji*, 4.127, the name of the capital was Lu 魯, and it was King Wu, not King Cheng, who enfeoffed the duke of Zhou at this site. According to *Zuo Commentary*, Ding 4, 54.17a, and *Odes*, Mao no. 300, however, it was the duke of Zhou's son, Bo Qin 伯禽, the duke of Lu, whom King Cheng enfeoffed at this site.

8. "Mountains, rivers, lands and fields, dependencies": based on *Odes*, Mao no. 300. The passage from "ritual objects used by rulers" to "ritual vessels used by officials" is from *Zuo Commentary*, Ding 4, 54.16a. *Liu* 旒 were pennants (such as tails) suspended as accompaniments to the main flag. On the three major Han dynasty interpretations of the *di* 禘 (imperial progenitor) ceremony, and also Song and Qing dynasty interpretations, see Chow, *Confucian Ritualism*, 137–47. Xu Gan would seem to be conflating the *jiao* 郊 (suburban) sacrifice with the *di* ceremony. The authority to make this sacrifice was a privilege normally afforded only the Zhou Son of Heaven. *Liji*, 31.5b, has a reference to the duke of Lu performing this sacrifice.

9. For the biography of Jiang Taigong, see *Shiji*, fascicle 32. On legends associated with him, see Allan, "Taigong Wang," 57–99. According to *Shiji*, 32.1481, the site where Jiang Taigong was enfeoffed was called Qi 齊, now part of Linzi county in modern Shandong. See also *Shiji*, 4.127.

   According to *Zuo Commentary*, Zhao 17, 48.7a, the Shuangjiu (eagle) clan had been appointed to the hereditary office of minister of justice under the legendary ruler Shaohao.

10. Paraphrase of *Zuo Commentary*, Xi 4, 12.11a. In *Zuo Commentary* and also *Shiji*, 32.1480–81, this speech is attributed to the duke

of Shao, not to King Wu. On the rank grand preceptor (*taishi* 大師) see Hucker, *Dictionary*, no. 6213.

11. *Analects*, 8.13.

12. For an illuminating discussion of the concept of *shi* 勢 in early Chinese thought, see Ames, *Art of Rulership*, chapter 3. John Hay's definition of the term as "the power invested in authority" accords well with the sense in which Xu Gan employs it; see Hay, *Boundaries in China*, 19. For a more extensive study of this concept, see the book-length study by Jullien, *Propensity of Things*, passim, and part one in particular.

13. Allusion to *Analects*, 14.32; Waley, *Analects*, 188, mod: "Weisheng Mu said to Confucius, 'Qiu, why do you go around, perching yourself now here, now there? Is it not simply to show off the fact that you are a clever talker?' Confucius said, 'I'd never dare to consider myself a clever talker; rather, I am pressingly committed to the task of remaining steadfast in my aims.'" I interpret 14.32 in the light of *Analects*, 15.2: "Confucius said, 'The gentleman remains steadfast (*gu* 固) in hard times, while the petty man is swept away as if caught in a deluge.'"

14. *Changes*, Jing hexagram, 5.16a–b.

15. Cf. *Xunzi*, 1.8a: "If you climb up to a high place and wave, it is not because your arms have become longer that people far away are able to see you; and if you shout when upwind, it is not because you have shouted louder that people are able to hear you more clearly."

16. This sentence is based on the *Documents*, 3.19b. The term *wenzu* 文祖 has been variously interpreted; see, for example, the commentaries cited at *Shiji*, 1.22. For other references to this story see *Wenxuan*, 11.35a, Li Shan's 李善 (d. 689) commentary to He Yan's 何晏 (c. 190–249) *Jingfu dian fu* 景福殿賦, where Li quotes the late Warring States/early Han work, *Shiben* 世本; *Bo Kong liutie*, 7.21a, which quotes a third century work, *Diwang shiji* 帝王世紀; *Taiping yulan*, 872.12a, which quotes a Han prognosticatory work, *Ruiying tu* 瑞應圖. On the Queen Mother of the West, see Loewe, *Ways to Paradise*, chapter 4.

17. See *Taiping yulan*, 785.2a; *Hanshi waizhuan*, 5.8a–b; *Shuoyuan*, 18.13a–b; *Lunheng*, 5.4a. On the Bright Hall, see *Zhouli*, 31.1a–b, Kong commentary. Rulers other than the Zhou king also possessed Ming Tang. From "When the duke of Zhou " to "axe-head designs" is a paraphrase of *Liji*, 31a.2a; see also *Shiji*, 4.132.

18. *Changes*, Feng hexagram, "Judgment," 6.1a; Lynn, *Changes*, 487, mod.

19. *Mencius*, 7A:3.

20. On Ji, see *Documents*, 3.22a; *Mencius*, 3A.4; Chang, *Art, Myth, and Ritual*, 12–15, 42. On Xie, see *Documents*, 3.22a–b; *Mencius*, 3A.4; Chang, *Shang Civilization*, 3–6; *Shiji*, 3.91. On Bo Yi, see *Documents*, 3.25b; *Shiji*, 5.173; *Mencius*, 3A.4. On Fu Yue, see *Shiji*, 3.102. On the king he served, Wuding, the first king to whom oracle bone divination inscriptions can be attributed, see Chang, *Shang Civilization*, 12. For the biographies of Yan Yuan, Min Ziqian, Ran Geng, and Zhong Gong, see *Shiji*, fascicle 67.

21. Cf. the related discussion in "Cultivating the Fundamental."

22. Mao no. 191; Karlgren, *Odes*, 134, mod.

11. *Examining Falsity*

1. See *Name and Actuality*, chapter 6.

2. For a fuller description and analysis of Xu Gan's correlative theory of naming, see *Name and Actuality*, chapter 1.

3. On the concept of *qi* see, for example, Graham, *Disputers of the Tao*, 101–4; Kuriyama, *Expressiveness of the Body*, 102–4.

4. See *Shiji*, 105.2785–94, and commentaries; Morita, "Hen Jaku kō," 15–25. Kuriyama, *Expressiveness of the Body*, 154, comments: "Bian Que's name became synonymous with medical acumen partly because he could see, literally, what others could not see." Kuriyama, 162–63, has translated the *Shiji* account of the episode with Bian Que and Duke Huan.

5. For accounts in English of these philosophers, see Graham, *Disputers of the Tao*, 54–61, 33–53, on Yang Zhu and Mo Di; Creel, *Shen Pu-hai*, passim, on Shen Buhai; Hsiao, *Chinese Political Thought*, 368–424, passim, on Han Fei; Fung, *History of Chinese Philosophy*, 1:132–33, 153–59, on Tian Pian; and Graham, *Disputers of the Tao*, 82–94, on Gongsun Long.

6. The two words *ji gu* 疾固, "pressingly committed to the task of remaining steadfast in one's ambition," are from *Analects*, 14.34; see "Titles and Emoluments," note 13.

7. Crooked and upright: allusion to *Analects*, 2.19: "Duke Ai asked, 'How can the people be made to be obedient?' Confucius replied, 'If the upright are promoted and placed above the crooked, then the people will be obedient; if, however, the crooked are

promoted and placed above the upright, then the people will not be obedient.'"

8. The significance of "fear and concern" is revealed in the following passage from *Zuo Commentary*, Xiang 31, 40.23a: "When a ruler has the dignified demeanor of a ruler, his ministers fear him and are concerned about him, take him as their model and imitate him. For this reason, he is able to hold reign over his country and have his fame continue long through the ages. When a minister has the dignified demeanor of a minister, his subordinates will fear and show concern for him. For this reason, he is able to safeguard his office, protect his clan, and do what is proper for his family."

9. Allusion to the following line from *Analects*, 9.11: "With a sigh, Yan Yuan said, '… The Master is adept at leading one forward, step by step. He broadens me with culture and constrains my excesses with ritual. Even if I wanted to stop learning, I could not.'"

10. Cf. *Analects*, 1.10: "Ziqin asked Zigong, 'When the Master arrives in a particular state he is sure always to learn about its government. Is this because he asks or because he is told?' Zigong said, 'The Master is able to learn about the government because he is genial, amicable, courteous, self-effacing, and deferential. Is his method of inquiry not perhaps different from that of other people?'"

11. *Analects*, 1.16; 14.32. See also *Analects*, 4.14.

12. "Fathers stealing their son's names": perhaps an allusion to *Analects*, 13.18: "Speaking to Confucius, the governor of She said, 'In my community we have an upright man who turned in his father for stealing a sheep.' Confucius said, 'Those who are upright in my community are quite different: fathers cover up for their sons and sons cover up for their fathers.'"

13. See the entry for Zhao 20, Autumn. For details of the story, see *Zuo Commentary*, Zhao 20, 49.5a–6a; Legge, *Chinese Classics*, 5:681–82.

14. For Shu Qi see *Annals*, Xiang 21; *Zuo Commentary*, Xiang 21, 34.12a. For Mou Yi see *Annals*, Zhao 5; *Zuo Commentary*, Zhao 5, 43.13a–b. For Hei Gong see *Annals*, Zhao 31; *Zuo Commentary*, Zhao 31, 53.19b.

15. See *Zuo Commentary*, Zhao 20, 49.5a–b; Legge, *Chinese Classics*, 5:681–82.

16. *Zuo Commentary*, Zhao 31, 53.19b–20a.

17. *Xunzi*, 2.9b.
18. *Documents*, 2.19b.
19. *Guoyu*, 2.19a.
20. For further discussion, see Makeham, "Confucius and Reputation," 582–86.
21. For example, *Analects*, 12.20: "Zizhang asked, 'What must an officer be like such that he can be said to have arrived [*da* 達]?' The Master said, 'What do you mean by to have arrived?' Zizhang replied, 'This is when one is certain to have been heard of whether serving in a ducal state or a ministerial domain.' The Master said, 'That is not to have arrived; that is to have been heard of [*wen* 聞]. A person who has arrived is one who has an upright disposition and loves to do what is right, who is sensitive to what people say and observant of their expressions, and who is mindful of being deferential to others. Such a person is certain to have arrived, whether serving in a ducal state or a ministerial domain. A person who remains merely heard of, however, is one who appears to be humane even though his conduct belies the appearance. Such a person is certain to have been heard of, whether serving in a ducal state or a ministerial domain.'"
22. *Liji*, 53.14a.
23. *Liji*, 53.14b.

12. *A Rebuke of Social Connections*

1. Earlier essays written in a similar critical vein include Zhu Mu's 朱穆 (100–163) "Juejiao lun 絕交論" (*Hou Hanshu*, 43.1466, commentary n. 1, citing *Zhu Mu ji* 朱穆集); Liu Liang's 劉梁 (second century CE) "Poqun lun 破群論" (*Hou Hanshu*, 80B.2635); Wang Fu's 王符 (c. 90–165) "Jiaoji 交際" (*Qianfu lun*, chapter 30); and Cai Yong's 蔡邕 (133–92) "Zhengjiao lun 正交論" (*Hou Hanshu*, 43.1474). See the discussion of some of these essays, as well as an alternative translation of this *Discourses* chapter, in Connery, *Empire of the Text*, 118–26.
2. *Zhouli*, 2.8b, gives the following account of the nine occupations: "The first was called the occupation of the three types of agriculture [mountains, plains, and marshes]. They produced the nine cereals [broomcorn millet, millet, sorghum, rice, hemp, soybean, red bean, wheat, and barley]. The second was called the occupation of gardens. They cultivated the fruits of plants and trees. The third was called the occupation of the mountains.

They exploited the resources of the mountains and marshes. The fourth was called the occupation of livestock farming. They bred birds and beasts. The fifth was called the occupation of the one hundred artisans. They worked the eight materials [pearl, ivory, jade, stone, wood, metal, leather, and feather]. The sixth was called the occupation of commerce and trade. They maintained the abundant flow of goods. The seventh was called the occupation of married women. They made silk and hemp material. The eighth was called the occupation of servants. They collected the edible roots and fruits of wild plants and trees. The ninth was the occupation of casual laborers. They had no regular occupation but moved about, filling in for others."

Zhouli, 10.26a, gives the following account of the eight punishments: "The first is called punishment for being unfilial. The second is called punishment for quarreling with one's clan. The third is called punishment for quarreling with one's in-laws. The fourth is called punishment for not being respectful to one's teachers and elders. The fifth is called punishment for not trusting one's friends. The sixth is called punishment for not commiserating with those who are suffering hardships. The seventh is called punishment for starting rumors. The eighth is called punishment for inciting the people to rebellion."

3. Quotation is from *Changes*, Qian hexagram, "Words of the Text," 1.13a.

4. Paraphrase of *Documents*, 14.19a. I have departed from my usual translation of *shi* 士 as "men of social standing" to "members of the bureaucracy" because here the context clearly indicates a type of official rank.

5. In Zhou times, the three dukes were reportedly the grand preceptor (*taishi* 太師), the grand mentor (*taifu* 太傅), and the grand guardian (*taibao* 太保). The nine chamberlains were reportedly the junior preceptor (*shaoshi* 少師), the junior mentor (*shaofu* 少傅), the junior guardian (*shaobao* 少保), head of the ministry of state (*zhongzai* 冢宰), minister of war (*sima* 司馬), minister of works (*sikong* 司空), minister of education (situ 司徒), minister of justice (*sikou* 司寇), and minister of rites (*zongbo* 宗伯).

6. According to Wei Zhao's commentary, 5.10b, *shiyin* 師尹 was an official of the counselor rank.

7. Autumnal equinox: following Dong's commentary, *Guoyu zhengyi*, 5.24b.

8. *Guoyu*, 5.10b–11b.

9. *Zhouli*, 3.11a.

10. On the malevolent *chimei* and *wangliang* spirits, see Knechtges, *Wen Xuan*, 1:216, note to line 509.

11. According to Zheng Xuan's commentary, *Zhouli*, 39.2b, the five materials were metal, wood, leather, jade, and earth.

12. This passage is based on *Zhouli*, 10.22b.

13. Six arts: following Zheng Zhong's interpretation of *daoyi* 道義. See Zheng Xuan's commentary at *Zhouli*, 3.20b. This description of the duties of the counselor is based on *Zhouli*, 12.1a–b.

14. This paragraph is based on *Zhouli*, 12.2a–b.

15. *Documents*, 4.20b.

16. While sometimes *fayue* 閥閱 is used to mean "powerful and distinguished families," here the term means "achievements and experience." On this usage of the term, see Ōba, "Kandai ni okeru," 14–28.

17. Cf. the following comments by Ge Hong, in *Bao Puzi* (*waipian*), 20.la: "At the end of the Han, during the time of Emperors Ling and Xian, the critical selection of men of quality had become perverted and indiscriminate. Outstanding and refined men were frustrated, while the avaricious secured official positions. Names did not accord with actualities, nor was value based on worth. Those who were successful became the worthies, while those who were blocked became the fools."

18. On the status of retainers, see Ebrey, "Patron-Client Relations," 539–40.

19. Alternatively, this could be referring to scholars and officials trying to get to see high officials to secure patronage. Cf. also a related situation described in Ge Hong's account of Guo Tai 郭泰 (128–69), *Bao Puzi* (*waipian*), 46.3a; Sailey, *Master Who Embraces Simplicity*, 228: "Kuo permitted his reputation to become very great. . . . The lanes in front of his house were filled with the ruts of vermilion carriages [of officials]. In the halls were waiting rows of guests in red sashes [officials]. Imperial carriages for summoning officials filled the streets [around his house], and carts followed one another bearing memorials to the emperor [recommending that Kuo be given a high office]."

20. For a discussion of "household disciples," see Ebrey, "Patron-Client Relations," passim.

21. On Guo Tai, see Makeham, *Name and Actuality*, 173; see also endnote 19, chapter 12 of *Discourses*. On Xu You, see Graham,

Chuang-tzu, 45. For a similar contemporary account of Xu You and Chaofu, see Wang, *Qianfulun*, 343.

22. As his following comments make clear, Xu Gan reads *ren* 仁 not *ren* 人.

23. *Odes*, Mao no. 204.

13. Astronomical Systems

1. Sun, moon, and stars: following Ikeda, "Chūron kōchū (3)," 135, n.2, in interpreting *ji lü* 紀律 to be a reference to *san guang* 三光, the sun, moon, and the stars, on the basis of near parallelism in the opening passage of the *Lüli zhi* (*xia*) (Treatise on Harmonics and Calendrical Astronomy, Part 3), *Hou Hanshu*, (Treatises), 3055. Crossed the observer's meridian due south: following the interpretation of *zhong* 中, "centered," advanced by Cullen, *Astronomy and Mathematics*, 19, 41–42.

2. "Clepsydras to check the timing of star transits": on this interpretation, see Cullen, *Astronomy and Mathematics*, 42. On the celestial sphere, see Needham, *Science and Civilisation*, 3:339–82. On the *huntian* theory in the Han, see Cullen, *Astronomy and Mathematics*, 53–66; Nakayama, *Japanese Astronomy*, 35–39. On the gnomon see Cullen, *Astronomy and Mathematics*, 15–17; 40–43, 101–27. On the clepsydra, see Needham, *Science and Civilisation*, 3:313–29.

3. On the different periods for the taking of life and the nurturing of life, see Major, "Magic of *Hsing-te*," 281–92; *Heaven and Earth*, chapter 5, passim; Yates, "Body, Space," 56–80, passim.

4. Paraphrase of *Guoyu*, 18.2a–b. See also *Shiji*, 26.1257–58.

5. *Documents*, 2.9a; Cullen, *Astronomy and Mathematics*, slightly mod., 3. On the identification of Xi and He as descendents of Zhong and Li, see Zheng Xuan's comments as cited in Jia Gongyan's preface to *Zhouli*, 4–5.

6. See *Shiji*, 26.1257–58; *Analects*, 20.1; *Hanshu*, 21A.973.

7. *Shiji*, 2.85.

8. When Tang and Wu changed the mandate to rule and began to devise systems to clarify the seasonal demarcations, they respectfully followed the "heavenly numbers": paraphrase of *Changes*, Ge hexagram, "Judgment," 5.18a; Lynn, *Changes*, 445, mod.

9. On the Grand Scribe, see Hucker, *Dictionary*, no. 6018.

10. Paraphrase of *Zhouli*, 26.13a–b.

11. Paraphrase of *Zuo Commentary*, Xi 5, 12.18a.

12. *Zuo Commentary*, Wen 1, 18.3a.

13. See *Zuo Commentary*, Ai 12, 59.5a. Presumably the manager of the calendar failed to intercalate one month, hence the confusion.

14. See *Shiji*, 26.1260, and *Hanshu*, 21A.974. On the Zhuan Xu astronomical system and the question of its supposed adoption at the beginning of the Han, see Cullen, "Scientific Change," 188–90; *Astronomy and Mathematics*, 27–28.

15. On these events and their background, see Cullen, "Scientific Change," 190–201, passim; Sivin, "Cosmos and Computation," 10–11 and notes; Yabuuchi, *Temmon rekihō*, 21–27; *Shiji*, 12.483, 28.1402–3; *Hanshu*, 6.199, 21A.975–76.

16. Sivin, "Cosmos and Computation," 10–11, comments that "this particular moment of time was simultaneously the winter solstice, the first day of the Astronomical First Month, and the first day in the sixty-day cycle by which days were recorded." See also 15–19.

17. On this system, see Sivin, "Cosmos and Computation," 11–13; Yabuuchi, *Temmon rekihō*, 27–29; Ho, *Astronomical Chapters*, 113. Sivin comments: "There has been considerable discussion as to whether Liu Hsin's system was new or simply copied from that of Teng P'ing and his collaborators.… The position that Liu took over the Grand Inception calendrical methods and constants, but with great originality extended them into a universal system which became the pattern for his successors, is the only one which accounts for all the evidence. This interpretation can be traced back at least as far as Hsü Kan's *Chung lun*."

18. Developed by Li Fan 李梵 and Bian Xin 編訢 et al. See Sivin, "Cosmos and Computation," 19–33; Yabuuchi, *Temmon rekihō*, 30–35. Implemented in 85 CE; see *Hou Hanshu*, 3026–27.

19. On the selection of this year, see Sivin, "Cosmos and Computation," 21–22.

20. See *Jinshu*, 17.498; Ho, *Li, Qi and Shu*, 61; *Hou Hanshu*, 3043, n.1.

14. *Distinguishing Between Premature Death and Longevity*

1. For a more detailed analysis of "Distinguishing Between Premature Death and Longevity" and how it relates to Xu's views on

the name and actuality relationship, see chapter 8 of Makeham, *Name and Actuality*.

2. For the quotation from Confucius see *Analects*, 6.22. On Yan Yuan see *Analects*, 6.2: "The duke of Ai asked which of Confucius's disciples loved to learn. Confucius replied, 'Yan Hui loved to learn; he did not vent his anger on others nor make the same mistake twice. Unfortunately, his circumstances were mean and he has died already.'" In the biography of Boyi and Shuqi, *Shiji*, 61.2124–25, Sima Qian writes: "Of the seventy disciples, Zhongni especially singled out Yan Yuan as being fond of learning. Hui was often so poor that he could not even eat his fill of the left-over grain used in distilling, so he died prematurely. To what can heaven's rewarding of good men be compared?"

   "The house that heaps good upon good": the *locus classicus* of this saying is the *Changes*, Kun hexagram, "Words of the Text," 1.26a.

3. These words are from the famous *Zuo Commentary* passage translated in "Ordering Learning," page 296, note 2. Cf. Horace, *Odes* III, 30, 6–7: "Non omnis moriar; multaque par mei Vitabit Libitinam" (I shall not wholly die: some part of me / Will cheat the goddess of death). Translation by Michie, *Odes of Horace*. For the biography of Xun Shuang, see *Hou Hanshu*, fascicle 62.

4. *Po* 魄 refers to the sentient life which inheres in the body and sinks into the earth at death. By extension, it means the body. Cf. *Zuo Commentary*, Zhao 7, 44.13a.

5. Mao no. 33; Karlgren, *Odes*, 260.

6. Cf. *Mencius*, 6A.10; Lau, *Mencius*, 166 mod.: "On the one hand, although life is what I want, there is something I want more than life. That is why I do not cling to life at all costs. On the other hand, although death is what I loathe, there is something I loathe more than death. That is why there are troubles I do not avoid.… This is an attitude not confined to the man of worth, but common to all people. The man of worth simply never loses it."

7. Cf. *Analects*, 7.15: "Ran You said, 'Will the master assist the lord of Wei?' Zigong said, 'Well, I shall ask him.' He went in and said, 'What sort of men were Boyi and Shuqi?' 'They were worthy men of old.' 'Did they harbor enmity?' 'They sought humaneness and attained it, so what enmity was there?' When Zigong came out he said, 'The master will not assist him.'"

8. Nothing further is known of Sun Ao. Presumably he was a con-

temporary or near contemporary of Xu Gan. Xu Gan was also a native of Beihai Kingdom. On the three different types of *ming* that were distinguished in the Han times, see "Cultivating the Fundamental," note 21.

9. *Analects*, 8.9.
10. *Xiaojing*, 1.3a.
11. *Mencius*, 4B.26.
12. Quotation is from *Changes*, "Appended Statements (B)," 8.23b.
13. *Documents*, 12.24b. The other four blessings are wealth, a healthy body and freedom from worry, the cultivation and love of virtue, and the opportunity to live the full span of one's life.
14. *Odes*, Mao no. 173.
15. *Analects*, 6.23.
16. *Documents*, 16.10a–11b; Legge, *Chinese Classics*, 3:465–68, mod.
17. Quotations are from *Analects*, 15.9 (in which the order of the two sentences is reversed) and 12.7, respectively. "Zigong asked about government. The Master said, 'If they have enough food to eat and enough weapons to defend themselves, the people will have faith in their government.' Zigong said, 'If one of these three had to be taken away, which would be the first choice?' Confucius said, 'Take away the weapons.' Zigong said, 'If one of these two had to be taken away, which would be the first choice?' Confucius said, 'Take away the food. From antiquity, death has been the lot of all men. If, however, the people have no faith in their government, then it will not stay standing.'"
18. *Analects*, 17.2.
19. On Ku see Allan, *Shape of the Turtle*, 33–35, 39–41, 51–54, 58–62. On the *di* ancestors, see Knoblock, *Xunzi*, 2:267–68.
20. This quotation is from *Changes*, "Appended Statements (B)," 8.9b.
21. *Analects*, 14.12. Presumably Xu Gan is here referring to Bigan. *Analects* 18.1: "The viscount of Wei left him [the tyrant Zhou]. The viscount of Ji became a slave because of him, and Bigan was killed for remonstrating with him. Confucius said, 'In them the Yin had three humane men.'"
22. Presumably alluding to the viscount of Wei. See previous note and also *Shiji*, 3.108, 38.1610.
23. Presumably alluding to the viscount of Ji. See *Shiji*, 38.1609; *Hanshi waizhuan*, 6.1a.
24. Presumably alluding to Hui 惠 of Liuxia 柳下. See *Analects*, 18.2: "Hui of Liuxia was dismissed thrice from office when he was

chief judge. Someone said to him, 'Isn't it time for you to leave?' He replied, 'If, in the service of others, one is not prepared to bend the way, where can one go without being dismissed thrice from office? If, in the service of another, one is prepared to bend the way, what need is there to leave one's home country?'"

25. There are many people whom Xu Gan may have had in mind here.

26. Again, the number of possibilities suggests that Xu Gan had no one figure in mind.

15. Attend to the Fundamentals

1. Cf. *Liji*, 60.1a ("Daxue"): "Things have their fundamental and peripheral aspects, and affairs have their beginning and end. If one understands what comes first and what comes after, then one will draw near to the way."

2. On Wenjiang's assignations, see, for example, *Zuo Commentary*, Huan 18, 7.25a; *Annals*, Zhuang 2 (entry for winter, twelfth month); *Guliang Commentary*, Zhuang 2, 5.5b. There appears to be no evidence, however, to support the preface's account that Duke Xiang was, in fact, the illegitimate issue of the incestuous relationship between Wenjiang and her brother, Duke Zhuang.

3. The ode cited here is Mao no. 106; Legge, *Chinese Classics*, 4:162; Karlgren, *Odes*, 69, mod. Xu Gan would seem to be following the Mao interpretation of this ode. According to the preface, the ode is satirizing Duke Zhuang for being unable to restrain his mother from pursuing her extramarital relations, and hence he is referred to as the son of the marquis of Qi.

4. To this point, this whole paragraph is a paraphrase of *Zuo Commentary*, Zhao 5, 43.7a–b.

5. *Annals*, Zhao 25, 52.5a.

6. *Guoyu*, 17.7a–b.

7. *Odes*, Mao no. 241. The version quoted is based in part on the version cited at *Zuo Commentary*, Zhao 28, 52.28a, and part on the Mao version.

8. Cf. *Analects*, 7.2: "To commit things to memory without having to recite them aloud; to learn insatiably; and to instruct others tirelessly—these present me with no difficulty."

9. Cf. *Analects*, 7.22: "In a company of several people traveling, I am bound to find a teacher among them. Choose the good and

follow them, and change the ways of those who are not good."

10. *Zuo Commentary*, Zhao 28, 52.28b–29a.

11. This English sentence is an attempt to render the meaning of what appear to be two early Chinese idioms: "able to run fast enough to pursue a team of four horses" and "having the strength to smash a portcullis door." The first describes fleetness of foot and the second a feat of strength.

12. On Zhi Boyao's headstrong nature's leading to the defeat of his army and his own death, see *Guoyu*, 15, 8b–9a and *Zuo Commentary*, Ai 27, 60.27a–28b.

16. Examining the Selection of High Officials

1. See *Documents*, 5.17a–b.

2. On the encounter, see *Lüshi chunqiu*, 2.8b; 4.5b; 13.10b; 14.8b; *Zhanguoce*, 3.44b, 3.8b; *Huainanzi*, 11.2a; *Hanshi waizhuan*, 7.4b. Xu Gan's version of the encounter seems to have deliberately omitted mentioning that Jiang Taigong contrived to come to the attention of King Wen by fishing at the Wei River.

3. *Huainanzi*, 12.5b and *Lüshi chunqiu*, 19.20a, record that he was a traveling trader.

4. See *Huainanzi*, 12.5b, and *Lüshi chunqiu*, 19.20a, and *Xinxu*, 5.3b–4a. Again, Xu Gan seems to have deliberately omitted mentioning that Ning Qi is said to have contrived to come to the attention of Duke Huan by so positioning himself outside the city gate of the capital of Qi.

5. *Guanzi*, 8.10b, records only that he was recommended by Guan Zhong to be responsible for fields.

6. Allusion to *Analects*, 6.10: "What a worthy Hui is! Living in a mean dwelling with only a scoop of rice to eat and a ladle of water to drink is a hardship most people would find intolerable, yet Hui did not change that which gave him happiness. What a worthy Hui is!"

7. Unexpected events: this concept seems to be closely related to the "departures from regularities" of "Cultivating the Fundamental" and "Examining Falsity."

8. Ikeda, "Chūron kōchū (3)," 168, n.48, writes that the text is problematical here and should be emended. He may be correct, but coherence can be maintained by assuming that this is the interjection of an interlocutor.

9. See *Shiji*, 62.2132.

10. See *Shiji*, 62.2132.

11. See *Shiji*, 62.2131 and *Zhengyi* commentary.

12. Mao no. 84; Legge, *Chinese Classics*, 4:137 mod.

13. *Documents*, 5.17a–b.

14. This sentence is based on a sentence describing the fate of Mencius at *Shiji*, 73.2343. Xun Qing may, in fact, have served in several official positions. For a detailed biography of Xun Qing, see Knoblock, *Xunzi*, 1:3–35.

15. Cf. the comment of the author of the preface to *Discourses*: "[Xu Gan] was critical of Mencius for failing to keep a proper sense of his limited capacities, emulating as he did the sage's efforts to put the way into practice by traveling about the country as the guest of one feudal lord after another." Other "itinerant persuaders" might include Su Qin 蘇秦 (fourth century BCE), Zhang Yi 張儀 (fourth century BCE), and Zou Yan 騶衍 (305–c. 240). See *Shiji*, 69.2276; 70.2279; 74.2345.

17. Be Careful of the Advice One Follows

1. Cf. *Mencius*, 2A.8; Lau, *Mencius*, 84: "The great Shun went even further. He was ready to fall into line with others, giving up his own ways for theirs, and glad to take from others that by which he could do good."

2. This quotation does not appear in other extant sources. Cf. *Analects*, 2.16: "Knowledge is to recognize that when you know something you know it and when you do not know something that you do not know it."

3. On these incidents, see *Shiji*, 80.2429–2430; *Zhanguoce*, 9.34a–38a; Crump, *Chan-kuo Ts'e*, 543–47. For Yu Yi's biography, see *Shiji*, fascicle 80.

4. A number of sources relate this story. See, for example, *Zhuangzi*, 22–24.

5. See *Han Feizi* 14.4b–5a; *Shiji*, 34.1555–57; *Huainanzi*, 18.17b; *Zhanguoce*, 9.12a–b.

6. See *Gongyang Commentary*, 27.10b–13a passim.

7. Paraphrase of *Shiji*, 7.334.

8. There is a pun involving the word *bei* 北 here. *Bei* also means "defeat" (*bai* 敗).

9. Fan Zeng had warned Xiang Yu to lose no time in attacking and killing Liu Bang 劉邦 when at Hongmen 鴻門. Xiang Yu

ignored him, with the result that Liu Bang escaped and later defeated Xiang Yu. See *Shiji*, 7.311–13, 325.

10. See *Shiji*, 7.336.

11. See *Shiji*, 8.376. For an English translation of the relevant passage, see Watson, *Grand Historian*, 1:103.

18. *Destruction of the State*

1. See Zuo *Commentary*, Zhao 13, 46.8a; Watson, *The Tso Chuan*, 170–71. On Wu Ju's (also known as Jiao Ju 椒舉) exhortation to King Ling to pay attention to ritual in order to secure the allegiance of the feudal lords, see *Zuo Commentary*, Zhao 4, 42a.26a. For King Ling's praise of Yi Xiang for being able to read ancient books, see *Zuo Commentary*, Zhao 12, 45.36b–37a; Watson, *Tso Chuan*, 166. On Zige's giving advice to King Ling, see *Zuo Commentary*, Zhao 12, 45.35a–36b; Watson, *Tso Chuan*, 165–67.

2. On Taishu Yi, see *Zuo Commentary*, Xiang 14, 32.16b; on Prince Zhuan, see *Zuo Commentary*, Xiang, 26, 27.3a; on Qu Boyu, see *Zuo Commentary*, Xiang, 26, 27.3a, and *Analects*, 15.7; on Shi Qiu, see *Analects*, 15.7. On Duke Xian's fleeing Wei, see *Zuo Commentary*, Xiang 27, 32.15a.

3. On Zhao Xuanzi, Dong Hu, and Duke Ling's death, see *Zuo Commentary*, Xuan 2, 21.9a–12a; Watson, *Tso Chuan*, 76–79; on Fan Wuzi, see *Zuo Commentary*, Xiang 27, 38.12a.

4. On Zijia Ji, see "Attend to the Fundamentals," page 211, note †. On Shusun Ruo (also known as Shusun She 舍 and Zhaozi 昭子) see *Zuo Commentary*, Zhao 5, 43.3a–4b. On Duke Zhao's death in Qi, see *Zuo Commentary*, Zhao 32, 53.26a. See also "Attend to the Fundamentals."

5. On Yan Pingzhong, see *Zuo Commentary*, Xiang 25, 36.5b–6b. On the scribe of the south and Duke Zhuang's death, see *Zuo Commentary*, Xiang 25, 36.6b, 36.4b–5a; Watson, *Tso Chuan*, 143–47.

6. Sivin, *Medicine, Philosophy, and Religion*, 4: 19–29, has argued against the idea that an academy had ever been established at Jixia.

7. See, for example, *Shiji*, 46.1895; Sivin, *Medicine, Philosophy, and Religion*, 4: 22, mod.: "King Xuan was fond of gentleman-retainers (*shi*) who were literary scholars, or itinerant advisers such as Zou Yan 騶衍, Chunyu Kun 淳于髡, Tian Pian 田駢,

Jie Yu 接子, Shen Dao 慎到, Huan Yuan 環淵 and the like—twenty-six in all. He granted them all mansions and made them senior counselors. They did not engage in the work of government, but in argument and discourse. Because of this the scholarly gentlemen of Jixia, in Qi, flourished once again, even growing to several hundred or over a thousand." See also *Yantielun*, 2.13b.

8. See *Shiji*, 78.2395.

9. Seventy thousand troops: as prescribed in *Zhouli*, 28.2a.

10. Paraphrase of *Analects*, 8.19: According to the commentary attributed to Kong Anguo in *Qunshu zhiyao*, the five ministers were Yu 禹, Ji 稷, Xie 契, Gao Tao 皋陶, and Bo Yi 伯夷. According to the commentary attributed to Ma Rong in *Lunyu jijie*, the ten capable ministers were Dan 旦 the duke of Zhou 周, Shi 奭 the duke of Shao 召, Grand Duke Wang 太公望, Duke Bi 畢, Duke Rong 榮, Tai Dian 太顛, Hong Yao 閎夭, San Yisheng 散宜生, Nan Gongkuo 南宮适, and King Wen's mother.

11. *Xunzi*, 9.12b.

12. See *Mencius*, 4A.13, and *Shiji*, 32.1478.

13. See *Zhanguoce*, 9.16b.

14. *Changes*, Heng hexagram, "Images," 4.5b.

15. Cf. *Analects*, 15.9: "The Master said, 'Gentlemen of purpose and humane men do not seek life at the expense of humaneness. On the contrary, they may even have to sacrifice their own lives in order to consummate their humaneness.'" "Distort the way": cf. *Analects*, 18.2, cited in "Distinguishing Between Premature Death and Longevity," note 24.

16. Mao no. 254. Xu Gan uses these lines to describe the gentlemen forced to serve at the court of bad rulers.

19. *Rewards and Punishments*

1. Cf. *Han Feizi*, 2.4a: "The enlightened ruler controls his ministers solely by means of the two handles. The two handles are punishment and favor. What is meant by punishment and favor? Killing and mutilation are called punishment; prizes and rewards are favor."

2. *Documents*, 8.2b.

3. Cf. *Analects*, 17.2: "The Master said, 'By virtue of their natures, people are close to one another, but through habituation they

diverge.'" "Heaven gave birth … ": first line of *Odes*, Mao no. 260.

4. See, for example, the principle of "killing relatives on the basis of great rightness" (*da yi mie qin* 大義滅親) described in *Zuo Commentary*, Yin 4, 3.15b–17b. For several examples illustrating both types of behavior, see the "Zhi gong 至公 " chapter in *Shuoyuan*.

5. Paraphrase of *Simafa*, A.4b.

6. *Documents*, 19.31b.

7. Mao no. 78; Karlgren, *Odes*, 53.

20. *Population Figures*

1. *Documents*, 35.6b, 5b.

2. Nearby districts were those nearby the royal domain. The sixfold division, from small to large, consisted of neighborhoods (*bi* 比), villages (*lü* 閭), precincts (*zu* 族), wards (*dang* 黨), townships (*zhou* 州), and districts (*xiang* 鄉). See "A Rebuke of Social Connections." The external districts lay beyond the royal domain. The sixfold division, from small to large, consisted of neighborhoods (*lin* 鄰), villages (*li* 里), precincts (*zan* 酇), wards (*bi* 鄙), townships (*xian* 縣), and districts (*sui* 遂). See *Zhouli*, 15.14a.

3. This sentence is based on *Zhouli*, 12.13a.

4. According to Zheng Xuan's commentary to *Zhouli*, 11.2b–3a, a battalion (*lü* 旅) consisted of 500 troops, while an army (*jun* 軍) consisted of 12,500 troops. Proportioning land: *Zhouli*, 15.14a–b.

5. According to Zheng Xuan's commentary to *Zhouli*, 10.26b, and 14.6b, the five rituals were for auspicious sacrifices, mourning, receiving guests, the military, and marriage. According to *Hanshu*, 23.1091, and Wei Zhao's commentary at 23.1095, n. 15, the nine punishments were branding, cutting off the nose, cutting off the feet, castration, execution, banishment, fine, whipping, and flogging.

21. *Reinstitute the Three-Year Mourning Period*

1. "Of all the creatures …": paraphrase of *Xunzi*, 13.21a.

2. This sentence occurs in *Xunzi*, 13.20b.

3. The second half of this sentence is based on *Liji*, 58.1a.

4. Actually, the traditional mourning period extended only into the third year and did not require three full years. Some sources state that the requisite period was twenty-five months (*Liji*, 58.1b; 6.22a, Kong Yingda subcommentary; *Xunzi*, 13.20b), while some state that it was twenty-seven months (*Liji*, 6.22a, Kong subcommentary quoting Zheng Xuan; *Hanshu*, 4.134, n. 16, Yan Shigu commentary).

5. Xianzong was the temple name of Emperor Ming.

6. Shizu was temple name of Emperor Guangwu. On Emperor Ming's observance of the three-year mourning period, see Xie Cheng's 謝承, *Hou Hanshu*, "Treatise on Ceremonial (Liyi zhi 禮儀志)," in Zhou, *Bajia Hou Hanshu*, 1:4. Actually, by the time of Emperors Cheng 成 (r. 33–7 BCE) and Ai 哀 (r. 7–1 BCE), and then Wang Mang 王莽 (r. 9–23 CE), the three-year mourning period had already gradually started to be reintroduced (Yang, *Handai hunsang*, 239–41), and in the Eastern Han it became widely practiced, even though it was not officially endorsed. In addition to Emperor Ming, it seems that emperors He 和 ([r. 88–106]; *Hou Hanshu*, 25.876) and Ling 靈 ([r. 168–89]; *Hou Hanshu*, 10B.452) also observed the full mourning period. In 116 an edict was promulgated allowing officials with salaries of two thousand *shi* and Inspectors of Provinces to observe the three years of mourning (*Hou Hanshu*, 5.226). In 121, however, permission was revoked (*Hou Hanshu*, 46.1560–61; *Hou Hanji*, 17.136a–b). In 154 once again permission was given to the same ranks of senior officials to observe the three-year mourning period (*Hou Hanshu* 7.299; 7.302), but it was revoked again in 159 (*Hou Hanshu*, 7.304).

7. Taizong was the temple name of Emperor Wen.

8. See *Mencius*, 3A.2.

9. Cf. *Analects*, 1.9; Lau, 60, mod: "Zengzi said, 'If you reverently attend to your parents' funeral arrangements and continue the offerings to your ancestors, the virtue of the common people will rally to sincerity.'"

10. *Odes*, Mao no. 223; Karlgren, *Odes*, 177.

11. Your highness: this chapter is possibly part of a memorial submitted to Cao Cao, hence the term of address, *shengzhu* 聖主.

22. *Regulate the Allotment of Corvée Laborers*

1. For further discussion of this fourfold division, see "A Rebuke of Social Connections."

2. The passage in quotation marks is adapted based on a paraphrase of *Mencius*, 3A.4.

3. High-ranking and noble families: literally, "bound up in blue, trailing purple." This line is possibly from Yang Xiong's "Jie chao 解嘲 (Dissolving Ridicule)"; Knechtges, *Han Rhapsody*, 97. The word *qing* 青 refers to the color of the string on the seals of the nine counselors, while *zi* 紫 refers to the color of the strings on the seals of dukes and marquises. See the passage from *Dongguan Hanji* quoted in Li Shan's commentary to *Wenxuan*, 45.6a–b.

   On *bingjian* 并兼 or *jianbing* 兼并, see Ch'ü, *Han Social Structure*, 182, 198, 394, and note; Yang, "Great Families," 103–15. Ebrey, "Economic and Social History," 625, describes this as a process whereby "those with wealth took over the land of those without, either through the legal means of purchase or debt foreclosure, or through bullying tactics."

4. Quotation is from *Changes*, Qian hexagram (no. 15), "Images," 2.32a–b.

5. According to the commentary of Yan Shigu, *Hanshu*, 97A.3936–37, personnel paid in pecks were clerks whose yearly salary was less than one hundred *shi*.

6. On Shi Dan and his recommendation, see *Hanshu*, 3.1141–42; Loewe, *Crisis and Conflict*, 268 ff.

Bibliography

Allan, Sarah. *The Heir and the Sage*. San Francisco: Chinese Materials Center, 1981.

———. "The Identities of Taigong Wang in Zhou and Han Literature." *Monumenta Serica* 30(1972–73): 57–99.

———. *The Shape of the Turtle: Myth, Art, and Cosmos in Early China*. Albany: SUNY Press, 1991.

Ames, Roger T. *The Art of Rulership*. Honolulu: University of Hawaii Press, 1983.

Analects. (*Lunyu* 論語). Trad. attrib. to disciples and later followers of Confucius [trad. 551–479], but comp. over several centuries, reaching textual closure c. 150 BCE. *Shisanjing zhushu* ed.

Baopuzi 抱樸子 (The Master Who Embraces Simplicity). Attrib. Ge Hong 葛洪 [c. 280–c. 340]. *Sibu congkan* ed.

Beck, B. J. Mansvelt. "The Fall of the Han." In Twitchett and Loewe, *Cambridge History*.

Beitang shuchao 北堂書鈔 (Digest of Books in the Sui Imperial Library). Comp. Yu Shinan 虞世南 [558–638]. *Siku quanshu* 四庫全書 (Complete Collection of the Four Treasuries) ed. Taipei: Shangwu yinshuguan, 1983–86.

Berkowitz, Alan J. "Patterns of Reclusion in Early and Early Medieval China: A Study of the Formation of the Practice of Reclusion in China and its Portrayal." Ph.D. diss., University of Washington, 1989.

Bodde, Derk. *Essays in Early Chinese Civilization*. Ed. Charles Le Blanc and Dorothy Borei. Princeton: Princeton University Press, 1981.

Bohutong 白虎通 (Comprehensive Discussions in the White Tiger Hall). Attrib. Ban Gu 班固 [32–92], but possibly compiled early third century CE. *Sibu congkan* ed.

Bo Kong liutie 白孔六帖 (Bo Juyi and Kong Chuan's Compendia of Passages for Use in Composing Poetry). Comp. Bo Juyi 白居易 [772–846] and Kong Chuan 孔傳 [early twelfth century]. Facsimile of Ming dynasty Jiaqing reign period [1522–66] ed. Taipei: Xinxing shuju, 1969.

Boltz, William G. "Early Chinese Writing." In *The World's Writing Systems*, ed. Peter T. Daniels and William Bright. New York: Oxford University Press, 1996.

Book of Changes. See *Zhouyi*.

Book of Documents. See *Shangshu*.

Book of Odes. See *Shijing*.

Chang, K.C. *Art, Myth, and Ritual: The Path to Political Authority in Early China*. Cambridge: Harvard University Press, 1983.

———. *Shang Civilization*. New Haven: Yale University Press, 1980.

Chao Gongwu 晁公武 [d. 1175]. *Junzhai dushuzhi* 郡齋讀書志 (Prefecture Studio Reading Notes). Taipei: Guangwen shuju, 1967.

Chen, Chi-yun. *Hsün Yüeh and the Mind of Late Han China*. Princeton: Princeton University Press, 1980.

Chen Li 陳澧 [1818–82]. *Dongshu dushuji* 東塾讀書記 (Reading Notes from the Eastern Vestibule). *Sibu beiyao* ed.

Chow, Kai-wing. *The Rise of Confucian Ritualism in Late Imperial China: Ethics, Classics, and Lineage Discourse*. Stanford: Stanford University Press, 1994.

Ch'ü, T'ang-tsu, and Jack L. Dull. *Han Social Structure*. Seattle: University of Washington Press, 1972.

Chunqiu fanlu 春秋繁露 (Luxuriant Gems of the Spring and Autumn). Trad. attrib. Dong Zhongshu 董仲舒 [c. 179–c. 104]. *Chunqiu fanlu yizheng* 春秋繁露義證 (Proofs of Meanings in Luxuriant Gems of the Spring and Autumn) ed., comp. with notes by Su Yu 蘇輿. Taipei: Heluo tushu chubanshe, 1974.

Connery, Christopher Leigh. *The Empire of the Text: Writing and Authority in Early Imperial China*. Lanham: Rowman and Littlefield, 1998.

Creel, Herrlee G. *Shen Pu-hai: A Chinese Political Philosopher of the 4th Century B.C.* Chicago: University of Chicago Press, 1974.

Crump, J.I., Jr., trans. *Chan-kuo Ts'e*. Oxford: Clarendon, 1970.

Cullen, Christopher. *Astronomy and Mathematics in Ancient China: The Zhou bi suan jing*. Cambridge: Cambridge University Press, 1996.

———. "Chiu chang suan shu." In Loewe, *Early Chinese Texts*.

———. "Motivations for Scientific Change in Early China: Emperor Wu and the Grand Inception Astronomical Reforms of 104 B.C." *Journal of the History of Astronomy* 24(1993): 185–203.

Da Dai liji 大戴禮記 (Book of Rites of the Elder Dai). Attrib. Dai De 戴德 [first century BCE], but probably comp. first or second centuries CE. *Sibu congkan* ed.

Dai Junren 戴君仁. "Jingshu de yancheng 經疏的衍成" (The Development of Subcommentaries to the Classics). *Kong Meng xuebao* 19(1970): 77–96.

De Crespigny, Rafe. "Political Protest in Imperial China: The Great Proscription of 167–184." *Papers on Far Eastern History* 11(1975): 1–36.

———. "Politics and Philosophy Under Emperor Huan." *T'oung Pao* 66(1980): 41–83.

DeWoskin, Kenneth J. *A Song for One or Two: Music and the Concept of Art in Early China*. Ann Arbor: Center for Chinese Studies, University of Michigan, 1982.

Dong Zengling 董曾齡 [nineteenth century], ed. *Guoyu zhengyi* 國語正義 (Correct Meaning of the Conversations of the States). Facsimile of 1880 ed. Chengdu: Ba-Shu shushe, 1985.

Dubs, Homer H., trans. *The History of the Former Han Dynasty*. Baltimore: Waverly, 1938, 1944, 1955.

Ebrey, Patricia. "The Economic and Social History of Later Han." In Twitchett and Loewe, *Cambridge History*.

———. "Patron and Client Relations in the Later Han." *Journal of the American Oriental Society* 103.3(1983): 533–42.

Erya 爾雅 (Progress Toward Correctness). Third-century BCE comp. *Shisanjing zhushu* ed.

Fengsu tongyi 風俗通義 (Comprehensive Meaning of Popular Customs). Comp. Ying Shao 應劭 (c. 140–before 240). *Sibu congkan* ed.

Fingarette, Herbert. "Following the One Thread of the Analects." *Journal of the American Academy of Religion* 47.3S(1979): 373–405.

Fung Yu-lan (Feng Youlan). *A History of Chinese Philosophy*, trans. Derk Bodde. Princeton: Princeton University Press, 1951–52.

Gongyangzhuan 公羊傳 (Gongyang Commentary). Trad. attrib. Gongyang Gao 公羊高 [fifth century BCE]. *Shisanjing zhushu* ed.

Graham, A.C. *Disputers of the Tao: Philosophical Argument in Ancient China*. LaSalle: Open Court, 1989.

———. *Later Mohist Logic, Ethics, and Science*. Hong Kong: Chinese University Press and School of Oriental and African Studies, 1978.

Guanzi 管子. Attrib. Guan Zhong 管仲 [d. 645 BCE]. Composite work containing writings from the Spring and Autumn, Warring States, and probably Han periods. *Sibu congkan* ed.

Guliangzhuan 穀梁傳 (Guliang Commentary). Attrib. Guliang Zhi 穀梁志 [fifth century BCE]; probably a third- or second-century BCE work. *Shisanjing zhushu* ed.

Gu lienüzhuan 古列女傳 (The Old Ordered Biographies of Women). Purportedly that portion of *Lienüzhuan* originally comp. Liu Xiang 劉向 [79–8 BCE]. *Sibu congkan* ed.

Guoyu 國語 (Conversations of the States). Anon. work comp. from fifth- to third-century BCE writings. *Sibu congkan* ed.

Han Feizi 韓非子. Attrib. Han Fei 韓非 (d. 233 BCE), comp. by his followers. *Sibu congkan* ed.

Han Geping 韓格平. "Beitu cang Huang Pilie ba ben Xu Gan Zhonglun 北圖藏黃丕烈跋本徐幹中論" (Exemplar of Xu Gan's Balanced Discourses with Postface by Huang Pilie Held in the Beijing Library). *Wenxian* 4(1988): 277–80.

———. "Xu Gan Zhonglun jiaoji shize 徐幹中論校輯十則" (Ten Collated Passages from Xu Gan's Balanced Discourses). *Guji zhengli yanjiu xuekan* (1987.1), 42–44.

Hanshu 漢書 (History of the Former Han Dynasty). Comp. Ban Gu 班固 [32–92]. Beijing: Zhonghua shuju, 1983.

Han Ying 韓嬰 [fl. 157 BCE]. *Hanshi waizhuan* 韓詩外傳 (Outer Commentary to Han [Ying's] Recension of the Book of Odes). *Sibu congkan* ed.

Hay, John, ed. *Boundaries in China*. London: Reaktion, 1994.

He Yan 何晏 [c. 190–249] et al. *Lunyu jijie* 論語集解 (Collected Explanations of the Analects). Yuan reengraving (no date) of the Song dynasty *Shicaitang* 世綵堂 redaction of the Directorate recension. In *Wuqiubei zhai Lunyu jicheng* 無求備齋論語集成 (Comprehensive Collection of Analects Texts from the Studio Which Does Not Seek Completeness), comp. Yan Lingfeng 嚴靈峰. Taipei: Yiwen yinshuguan, 1966.

Ho Peng Yoke, *The Astronomical Chapters of the Chin shu*. Paris: Mouton, 1966.

———. *Li, Qi, and Shu: An Introduction to Science and Civilisation in China*. Hong Kong: Hong Kong University Press, 1985.

Hou Hanji 後漢紀 (Record of the Later Han Dynasty). Comp. Yuan Hong 袁宏 (320–76). *Sibu congkan* ed.

Hou Hanshu 後漢書 (History of the Later Han Dynasty). Comp. Fan Ye 范曄 (396–446). Beijing: Zhonghua shuju, 1982.

Hsiao, Kung-chuan. *A History of Chinese Political Thought*. Vol.1, *From the Beginnings to the Sixth Century A.D.* Trans. F.W. Mote. Princeton: Princeton University Press, 1979.

Huainanzi 淮南子. Comp. by scholars at the court of Liu An 劉安 (d. 122 BCE). *Sibu congkan* ed.

Huang Hui 黃暉. *Lunheng jiaoshi* 論衡校釋 (Critical Edition of

Critical Essays with Explanatory Notes). Taipei: Taiwan Shangwu yinshuguan, 1983.

Huang Zhuozhuan 黃焯撰. *Maoshi Zhengjian pingyi* 毛詩鄭箋平議 (Balanced Account of Zheng [Xuan's] Commentary to the Mao Recension of the Book of Odes). Shanghai: Shanghai guji chubanshe, 1985.

Hucker, Charles O. *A Dictionary of Official Titles in Imperial China*. Taipei: Southern Materials Center, 1986.

Ikeda Shūzō 池田秀三. "Jo Kan Chūron kōchū 徐幹中論校注 (1–3)" (Critical Edition of Xu Gan's Balanced Discourses with Annotations). *Kyōto daigaku bungakubu kenkyū kiyō* 23(1984): 1–62; 24(1985): 73–112; 25(1986): 117–200.

Jiang Boqian 蔣伯潛. *Jingxue zuanyao* 經學纂要 (Essential Notes on Classical Studies). Taipei: Zhengzhong shuju, 1969.

Jiao Xun 焦循 [1763–1820]. *Mengzi zhengyi* 孟子正義 (Correct Meaning of the Mencius). Beijing: Zhonghua shuju, 1991.

Jingdian shiwen 經典釋文 (Explanations of the Texts of the Classics). Comp. Lu Deming 陸德明 [556–627]. Shanghai: Shanghai guji chubanshe, 1985.

Jinshu 晉書 (History of the Jin Dynasty). Comp. Fang Xuanling 房玄齡 [579–648] et al. Beijing: Zhonghua shuju, 1974.

Jullien, François, *The Propensity of Things: Towards a History of Efficacy in China*. Trans. Janet Lloyd. New York: Zone, 1995.

Karlgren, Bernhard. *The Book of Odes*. Stockholm: Bulletin of the Museum of Far Eastern Antiquities, 1950.

———. "Grammata Serica Recensa." *Bulletin of the Museum of Far Eastern Antiquities* 29(1957): 1–332.

Knechtges, David R. *The Han Rhapsody*. Cambridge: Cambridge University Press, 1976.

———. "The Liu Hsin/Yang Hsiung Correspondence in Fang Yen." *Monumenta Serica* 33(1977–78): 309–25.

Knoblock, John. *Xunzi: A Translation and Study of the Complete Works*. Stanford: Stanford University Press, 1988–94.

Kongcongzi 孔叢子 (The Kong Family Masters' Anthology). Attrib. Kong Fu 孔鮒 [Qin dynasty], but widely considered to be a forgery by Wang Su 王肅 [195–256]. *Sibu congkan* ed.

Kongzi jiayu 孔子家語 (Sayings of the School of Confucius). Anon. collection of pre-Han and late Han writings, ed. Wang Su 王肅 [195–256]. *Sibu congkan* ed.

Kuriyama, Shigehisa. *The Expressiveness of the Body and the Divergence of Greek and Chinese Medicine*. New York: Zone, 1999.

Lau, D.C., trans. The *Analects*. Harmondsworth: Penguin, 1983.

———, trans. *Mencius*. Harmondsworth: Penguin, 1983.

———, trans. *Tao Te Ching*. Harmondsworth: Penguin, 1978.

Lau, D.C., and Chen Fong Ching, series eds. *A Concordance to the Zhonglun*. In *The Ancient Chinese Texts Concordance Series* (Philosophy works no. 17). Hong Kong: Commercial Press, 1995.

Le Blanc, Charles, and Susan Blader, eds. *Chinese Ideas About Nature and Society: Studies in Honour of Derk Bodde*. Hong Kong: Hong Kong University Press, 1987.

Legge, James. *The Chinese Classics*. Hong Kong: Hong Kong University Press, 1960.

Lewis, Mark Edward. *Writing and Authority in Early China*. Albany: SUNY Press, 1999.

Liang Rongmao 梁榮茂. *Xu Gan Zhonglun jiaoshi* 徐幹中論校釋 (Collation of Xu Gan's Balanced Discourses with Explanatory Notes). Taipei: Mutong chubanshe, 1979.

———. *Xu Gan Zhonglun jiaozheng* 徐幹中論校證 (Collation of Xu Gan's Balanced Discourses with Proofs). Taipei: Mutong chubanshe, 1980.

Liji 禮記 (Book of Rites). Composite collection comp. over the Han period. *Shisan jing zhushu* ed.

Lin Qingzhang 林慶彰. "Liang Han zhangju zhi xue chongtan 兩漢章句之學重探" (Section and Sentence Commentary Scholarship in the Western and Eastern Han Dynasties Revisited). In *Zhongguo jingxueshi lunwenji* 中國經學史論文集 (Collected Essays on the History of Chinese Classical Studies), ed. Lin Qingzhang. Taipei: Wenshizhe chubanshe, 1992.

Liu Zhijian 劉之漸. *Jianan wenxue biannianshi* 建安文學編年史 (Chronology of Events Pertaining to Literature during the Jianan [196–220] Period). Chongqing: Chongqing chubanshe, 1985.

Loewe, Michael. *Crisis and Conflict in Han China*. London: Allen and Unwin, 1974.

———. *Divination, Mythology, and Monarchy in Early China*. Cambridge: Cambridge University Press, 1994.

———. *Ways to Paradise: The Chinese Quest for Immortality*. London: Allen and Unwin, 1979.

———, ed. *Early Chinese Texts: A Bibliographical Guide*. Berkeley: Society for the Study of Early China and the Institute of East Asian Studies, University of California, 1993.

Lunyu 論語. See *Analects*.

Lüshi chunqiu 呂氏春秋 (Spring and Autumn of Mr Lu). Comp. under patronage of Lü Buwei 呂不韋 [d 235 BCE]. *Sibu congkan* ed.

Lynn, Richard John, trans. *The Classic of Changes: A New Translation of the I Ching as Interpreted by Wang Bi*. New York: Columbia University, 1994.

Major, John S. "Celestial Cycles and Mathematical Harmonics in the Huainanzi." *Extrême-Orient, Extrême-Occident* 16(1994): 121–31.

———. *Heaven and Earth in Early Han Thought*. Albany: SUNY Press, 1993.

———. "The Magic of Hsing-te." In Le Blanc and Blader, *Chinese Ideas*.

Makeham, John. "Between Chen and Cai: The Analects and Zhuangzi." In *Wandering at Ease in the Zhuangzi*, ed. Roger T. Ames. Albany: SUNY Press, 1998.

———. "Confucius and Reputation: A Note on Analects 15.18 and 15.19." *Bulletin of the School of Oriental and African Studies* 56.3(1993): 582–86.

———. *Name and Actuality in Early Chinese Thought*. Albany: SUNY Press, 1994.

———. "Zhong lun." In Loewe, *Early Chinese Texts*.

Mawangdui Han mu boshu 馬王堆漢墓帛書 (Han Tomb Silk Manuscripts from Mawangdui). Vol. 1. Beijing: Wenwu chubanshe, 1980.

Mencius. See *Mengzi*.

Mengzi 孟子 (Mencius). Probably comp. by Mencius's [fourth century BCE] disciples. *Shisanjing zhushu* ed.

Michie, James, trans. *The Odes of Horace*. Harmondsworth: Penguin, 1973.

Morita Denichirō 森田傳一郎. "Hen Jaku kō 扁鵲考" (An Investigation of Bian Que). *Nihon Chūgoku gakkaihō* 32(1980): 15–25.

Mozi 墨子. Comp. by members of the Mohist tradition over the fourth and third centuries BCE. *Sibu congkan* ed.

Needham, Joseph. *Science and Civilisation in China*. Vol. 3. Cambridge: Cambridge University Press, 1959.

Nylan, Michael. "The Chin wen/Ku wen Controversy." *T'oung Pao* 80(1994): 83–145.

Ōba Osamu 大庭脩. "Kandai ni okeru kōji ni yoru seishin ni tsuite 漢代における功次による昇進について" (On the Han Dynasty Practice of Granting Promotion on the Basis of Merit Ranking). *Tōyōshi kenkyū* 2.3(1953): 14–28.

Pi Xirui 皮錫瑞. *Jingxue lishi* 經學歷史 (A History of Classical Studies). Annotated by Zhou Yutong 周予同. Hong Kong: Zhonghua shuju, 1961.

Pokora, Timoteus, trans. *Hsin-lun (New Treatise) and Other Writings by Huan T'an (43 BC–28 AD)*. Ann Arbor: University of Michigan Press, 1975.

Powers, Martin J. *Art and Political Expression in Early China*. New Haven: Yale University Press, 1991.

Qian Peiming 錢培名 [nineteenth century]. *Zhaji* 札記 (Reading Notes to Balanced Discourses). *Xiaowanjuanlou congshu* 小萬卷樓叢書, 1854.

Qunshu zhiyao 群書治要 (Emendments to Key Passages in Various Books). Comp. and ed. Wei Zheng 魏徵 [580–643]. *Sibu congkan* ed.

Qu Wanli 屈萬里. *Shijing shiyi* 詩經釋譯 (The Book of Odes with Notes and Translation). Taipei: Zhonghua wenhua, 1953.

Sailey, Jay. *The Master Who Embraces Simplicity: A Study of the Philosopher Ko Hung, AD 283–343*. San Francisco: Chinese Materials Center, 1978.

Sanguozhi 三國志 (History of the Three Kingdoms). Comp. Chen Shou 陳壽 [233–97], commentary by Pei Songzhi 裴松之 [372–451]. Beijing: Zhonghua shuju, 1985.

Serruys, Paul L.-M. "Prolegomena to the Study of the Dialects of Han Times According to Fang-Yen." Ph.D. diss., University of California, Berkeley, 1956.

Shangshu 尚書 (Book of Documents). Anon. collection of documents dating from the early Western Zhou to the third or fourth century CE. *Shisanjing zhushu* ed.

Shaughnessy, Edward L. *Before Confucius: Studies in the Creation of the Chinese Classics*. Albany: SUNY Press, 1997.

———. *I Ching: The Classic of Changes*. New York: Ballantine, 1998.

Shiji 史記 (Records of the Grand Historian). Sima Tan 司馬談 [c. 180–110] and Sima Qian 司馬遷 [145–c. 86]. Beijing: Zhonghua shuju, 1982.

Shijing 詩經 (Book of Odes). Anon. collection of odes from the twelfth to sixth centuries BCE. *Shisanjing zhushu* ed.

Shimanori Tetsuo 島森哲男. "Shindoku no shisō 慎獨の思想" (Thought Pertaining to Vigilant Solitariness). *Bunka* 42.3/4(1979): 145–58.

Shisanjing zhushu 十三經注疏 (The Thirteen Classics with Annotations and Subcommentaries). Comp. Ruan Yuan 阮元 [1764–1849]. Taipei: Yiwen yinshuguan, 1985.

Shuoyuan 說苑 (Garden of Persuasions). Comp. Liu Xiang 劉向 [79–8]. *Sibu congkan* ed.

Sibu congkan 四部叢刊 (The Four Divisions Collection). Shanghai: Shangwu yinshuguan, 1919; supplements 1934–36.

Simafa 司馬法. Unattrib., fragmentary remains, possibly of Warring States origins. *Sibu congkan* ed.

Sivin, Nathan. "Cosmos and Computation in Early Chinese Mathematical Astronomy." *T'oung Pao* 55(1969): 1–73.

———. *Medicine, Philosophy, and Religion in Ancient China: Researches and Reflections*. Aldershot: Variorum, 1995.

Sun Yirang 孫詒讓 [1848–1908]. *Zhouli zhengyi* 周禮正義 (Correct Meaning of the Rites of Zhou). Beijing: Zhonghua shuju, 1987.

Taiping yulan 太平禦覽 (Royal Encyclopedia of the Taiping Era). Comp. Li Fang 李昉 [925–96] et al. Beijing: Zhonghua shuju, 1985.

Takeuchi Yoshio 武内義雄. *Eki to Chūyō no kenkyū* 易と中庸の研究 (Studies on the Book of Changes and the Doctrine of the Mean). Tokyo: Iwanami shoten, 1943.

Tanaka Masami 田中麻紗已. *Ryō Kan no shisō no kenkyū* 兩漢の思想の研究 (Studies on the Thought of the Two Han Dynasties). Tokyo: Kembun shuppan, 1986.

Tu Wei-ming. *Centrality and Commonality: An Essay on Chung-yung*. Monographs for the Society for Asian and Comparative Philosophy, no. 3. Honolulu: University of Hawaii Press, 1976.

Twitchett, Denis, and Michael Loewe, eds. *The Cambridge History of China*. Vol. 1, *Ch'in and Han Empires, 221 B.C. – A.D. 220*. New York: Cambridge University Press, 1986.

Van Gulik, R. H. *The Lore of the Chinese Lute*. Tokyo: Sophia University Press, 1940.

Van Zoeren, Steven. *Poetry and Personality: Reading, Exegesis, and Hermeneutics in Traditional China*. Stanford: Stanford University Press, 1991.

Vervoorn, Aat. *Men of the Cliffs and Caves: The Development of the Chinese Eremitic Tradition to the End of the Han Dynasty*. Hong Kong: Chinese University Press, 1989.

Von Falkenhausen, Lothar. *Suspended Music: Chime-Bells in the Culture of Bronze Age China*. Berkeley: University of California Press, 1993.

Waley, Arthur, trans. *The Book of Songs*. New York: Grove, 1960.

Wang Chong 王充 [27–c. 100]. *Lunheng* 論衡 (Critical Essays). *Lunheng jiaoshi* 論衡校釋 (Critical Edition of Critical Essays with Explanatory Notes) ed., comp. with notes by Huang Hui 黃暉. Taipei: Taiwan shangwu yinshuguan, 1983.

Wang Fu 王符 [c. 90–162]. *Qianfulun* 潛夫論 (Essays of a Recluse). *Sibu congkan* ed.

Wang Guowei 王國維. *Guantang jilin* 觀堂集林 (Collected Writings from the Hall of Contemplation). Taipei: Shijie shuju, 1961.

Wang Xianqian 王先謙 [1842–1918]. *Xunzi jijie* 荀子集解 (Collected Annotations on Xunzi). Taipei: Yiwen yinshuguan, 1973.

———. *Shi sanjiayi jishu* 詩三家義輯疏 (Collected Reconstructed Subcommentaries to Three Traditions of Interpretation of the Book of Odes). Beijing: Zhonghua shuju, 1987.

Watson, Burton, trans. *Han Fei Tzu*. New York: Columbia University Press, 1964.

———, trans. *Records of the Grand Historian*. New York: Columbia University Press, 1961.

———, trans. *The Tso Chuan: Selections from China's Oldest Narrative History*. New York: Columbia University Press, 1989.

Wenxuan 文選 (Selections of Refined Literature). Comp. Xiao Tong 蕭統 [501–31]. Beijing: Zhonghua shuju, 1977.

Xiao Dengfu 蕭登福, ed. and annot. *Xinbian Zhonglun* 新編中論 (New edition of Balanced Discourses). Taipei: Guoli bianyiguan, 2000.

Xiaojing 孝經 (Classic of Filial Piety). Attrib. Zeng Shen 曾參 [early fifth century BCE], probably a late Warring States writing. *Shisanjing zhushu* ed.

Xinxu 新序 (Newly Edited Sequential Arrangement of Sayings). Comp. Liu Xiang 劉向 [79–8]. *Sibu congkan* ed.

Xu Gan 徐幹 [170–217]. *Zhonglun* 中論 (Balanced Discourses). The citation text I have used is Lau and Chen, series eds., *A Concordance to the Zhonglun*. Other editions referred to are:

- *Bozi quanshu* 百子全書 (Complete Book of the One Hundred Masters). Shanghai: Saoye shanfan, 1919.
- *Guang Han Wei congshu* 廣漢魏叢書 (Expanded Collection of Han and Wei Texts). Cut by He Yunzhong 何允中 [fl. 1600], ed. Sun Yiqi 孫胤奇 [fl. 1622]. Wanli reign period [1573–1615] ed. Copy held in the National Library, Taiwan.
- *Han Wei congshu* 漢魏叢書 (Collection of Han and Wei Texts). Comp. Cheng Rong 程榮 (1447–1520). 1592 ed. Copy held in the National Library, Taiwan.
- *Jianan qizi ji* 建安七子集 (The Collected Writings of the Seven

Masters of the Jianan Period). Ed. Chen Chaofu 陳朝輔 (fl. 1768).
Taipei: Taiwan zhonghua shuju, 1971.

- *Liangjing yibian* 兩京遺編 (Lost Texts from the Two Capitals).
Comp. Hu Weixin 胡維新 (fl. 1559). 1582 ed. Copy held in the
National Library, Taiwan.
- *Longxi jingshe congshu* 龍谿精舍叢書 (The Longxi Refined Study
Collection of Texts ed.). Ed. Zheng Guoxun 鄭國勳. Yangzhou:
1917.
- *Xiao wanjuanlou congshu* 小萬卷樓叢書 (Collection of Texts from
the Small Hall of Ten Thousand Fascicles). Ed. Qian Peiming
錢培名 [nineteenth century]. 1854 ed.
- *Xue Chen* 薛晨. 1565 ed. This version of the text has been re-
produced in the *Sibu congkan* series.
- *Zengding Han Wei congshu* 增訂漢魏叢書 (Revised and Enlarged
Edition of the Collection of Han and Wei Texts). Comp. Wang
Mo 王謨 (fl. 1788). 1791 ed.

Xu Hou Hanshu 續後漢書 (Continuation of the History of the Later
Han Dynasty). Comp. Hao Jing 郝經 [1223–75]. 1841 ed.

Xunzi 荀子. Comp. by followers of Xun Qing 荀卿 [c. 335–c. 238].
Sibu congkan ed.

Xu Shen 許慎 [c. 55–c. 149]. *Shuowen jiezi* 説文解字 (Explaining the
Unit Characters and Analyzing the Compound Characters).
Beijing: Zhonghua shuju, 1979.

Yabuuchi Kiyoshi 數内清. *Chūgoku no temmon rekihō* 中國の天文曆
法 (China's Celestial Calendrical Systems). Tokyo, 1969.

Yan Kejun 嚴可均 [1762–1843], comp. and ed. *Quan shanggu Sandai
Qin Han Sanguo Liuchao wen* 全上古三代秦漢三國六朝文 (Complete Prose of Antiquity, the Three Dynasties, Qin, Han, the
Three Kingdoms, and Six Dynasties). Beijing: Zhonghua shuju,
1985.

Yang, Lien-Sheng. "Great Families of the Eastern Han." In *Chinese
Social History*, ed. E-tu Zen Sun and John De Francis. Washington: American Council of Learned Societies, 1956.

Yang Shuda 楊樹達. *Handai hunsang lisu kao* 漢代婚喪禮俗考 (Investigation of Han Dynasty Marriage and Funerary Customs).
Shanghai: Shangwu yinshuguan, 1933.

———. *Huainanzi zhengwen* 淮南子證聞 (Proofs of What I Have
Heard Concerning Passages in Huainanzi). Shanghai: Shanghai
guji chubanshe, 1985.

Yantielun 鹽鐵論 (Discussions on Salt and Iron). Comp. Huan Kuan
桓寬 [fl. 81–73]. *Sibu congkan* ed.

Yasui Kozan 安居香山 and Nakamura Shōhachi 中村璋八. *Isho shūsei* 緯書集成 (Comprehensive Collection of Apocryphal Texts). Vol. 4a. Kyoto: Kan Gi bunka kenkyūkai, 1963.

Yates, Robin D. S., "Body, Space, Time, and Bureaucracy: Boundary Creation and Control Mechanisms in Early China." In Hay, *Boundaries in China*.

Yilin 意林 (Forest of Selected Passages). Comp. Ma Zong 馬總 [d. 823]. *Sibu congkan* ed.

Yiwen leiju 藝文類聚 (Selections from Literature by Category). Comp. Ouyang Xun 歐陽詢 [557–641] et al. Shanghai: Shanghai guji chubanshe, 1982.

Yi Zhoushu 遺周書 (Bequeathed Book of Zhou). Completed first century BCE, includes much older material. *Sibu beiyao* ed.

Yuhan shanfang ji yishu 玉函山房輯佚書 (The Jade Book-Cover Studio Reconstructed Collection of Fragments from Lost Books). Comp. Ma Guohan 馬國翰 [1794–1857]. Taipei: Wenhai chubanshe, 1974.

Yu Yingshi 余英時. *Zhongguo zhishi jiecengshi lun (gudai pian)* 中國知識階層史論，古代篇 (Essays on the History of China's Intellectuals: Early Dynasties Volume). Taipei: Lianjing chuban shiye gongsi, 1984.

Yu Yue 俞樾. *Zhuzi pingyi bulu* 諸子平議補錄 (Supplement to a Balanced Account of the Philosophers). Taipei: Shijie shuju, 1973.

Zhanguoce 戰國策 (Intrigues of the Warring States). Anon. collection comp. Liu Xiang 劉向 [77–6]. *Sibu congkan* ed.

Zhang Tao 張濤 and Fu Genqing 傅根清, trans. and annot. *Shenjian Zhonglun xuanzhu* 申鑑中論選注 (Selected Translations from Extended Reflections and Balanced Discourses). Chengdu: Ba-Shu shushe, 1991.

Zhang Yongni 張詠霓. *Zhuzi dagang* 諸子大綱 (Outline of the Philosophers). N.p., 1947. Photocopy held in Fu Sinian Memorial Library, Academia Sinica, Taipei.

Zhenguan zhengyao 貞觀政要 (Record of Key Political Documents from the Zhenguan [627–50] Period). Comp. Wu Jing 吳兢 [670–794]. Shanghai: Shanghai guji chubanshe, 1978.

Zhouli 周禮 (Rituals of Zhou). Han compilation based on earlier material. *Shisanjing zhushu* ed.

Zhou Tianyou 周天游. *Bajia Hou Hanshu jizhu* 八家後漢書集注 (Collected Annotations of the Eight Commentators to the History of the Later Han Dynasty). Shanghai: Shanghai guji chubanshe, 1986.

Zhouyi 周易 (Book of Changes). Early Zhou text with Warring States and Han appendixes. *Shisanjing zhushu* ed.

Zhuangzi 莊子. Comp. by Zhuangzi's (late fourth-/early third-century BCE) followers, and includes writings from fourth?–second centuries BCE. *Zhuangzi jishi* 莊子集釋 ed., comp. Guo Qingfan 郭慶藩. Taipei: Muduo chubanshe, 1982.

Zuozhuan 左傳 (Zuo Commentary). Attrib. Zuo Qiuming 左丘明 (fifth century BCE), but possibly not compiled in its present form until c. 300 BCE. *Shisanjing zhushu* ed.

Index

References are to the text and notes of the English translation only. Thus where reference is made to a sequence of pages (e.g., 19–21), the intended reference is to the pages of translation only (i.e., 19 and 21), not to the Chinese text.

household disciples, 171
Huan, Duke: Guan Zhong and, 111, 237
humility, 51–59

Ji, viscount of, 117, 320n.24, 321n.26
Ji Lu (disciple of Confucius), 21, 23
Jian'an period (196–220), Seven Masters of the, xi, xii, 292n.3; *Discourses* and, xiii; Xu Gan and, xxix
Jiang Taigong, 127, 259, 323
Jiaxia: worthy men at, 253, 337–38n.7
Jing hexagram: quotation of, 129

Kun hexagram: quotation of "Images," 47

Lao-Zhuang thought: influence on Xu Gan, xix
learning, xxvii–xxviii, 3–15; gentleman and, 7; sages and, 5; worthies and 13; Xu Gan's attitude to, xxvii
Li hexagram: quotation of "Image," 107
Li Lou, 213
Li Shou, 213
Ling, Emperor (r. 168–89), xxxi, 153
longevity: derived from reputation, 197; humaneness and, 187, 191; three types of, 193–95

matters, rhrce, 155
memorials, xxxvi, 205
Mencius: on adjusting priorities *(quan)*, 117; followers of, 253; role of education in thought of, 3; as sage, xxx; Xu Gan's attitude to, xxxiii
Min Ziqian (disciple of Confucius), 133
ming (circumstances beyond human control), 33; status and position and, 131; three different types of, 309n.21; variability of, 47, 187
Ming, Emperor (r. 57–75): the three-year mourning period and, 277, 281
Mohists: disputation and, 97
mourning period, three-year, 277–81
music, six types of, 155, 314–15n.4

name and actuality *(ming shi)*, xxi–xxv; accord between, 121; of achievement, 141; addressed in *Discourses,* xxii; of argument, 159; breakdown between, 135, 233; correlative bond between, xxix, 135; of disputation, 97, 99; the gentleman and, 161; relationship between, 149–51; of social intercourse, 153. *See also* actuality; reputation
names: principles and, 101, 101n. *See also* reputation

sages: and the arts, 87; attainments of, 105; after Confucius, 137; emotional responses of, 109; fame seekers and, 139; importance of as models, 3; intelligence, wisdom, and, 105, 107; learning and, 5, 13; position and, 129; prohibition of fame seekers and, 143; qualities of, 117; teachings of, 199; virtue of, 9

script, six classes of, 316n.4

Shao, duke of, 115, 116, 199

shen du. See vigilant solitariness

Sheng hexagram: quotation of "Images" and "Judgment," 41

shi (man of social standing), 253, 285, 307–8n.18, 324n.12; abilities of, 119; commitment of, 49; slave ownership and, 283

Shi Dan, 287

Shi Kuang, 213

Shun (=Youyu), 13, 59n, 105; books of, 249; as commoner, 131; cultivation of virtue and, 45, 67; five ministers of, 255, 338n.10; followers of Yao and, 141; the four wicked ones and, 237, 237–38n; in his local community, 59; laws of, 137; mandate to rule and, 179; status and position and, 131; way of, 259; Yao and, 221, 225, 233; Zeng Shen (=Zengzi) and, 107

Simafa, 267

Six Classics, xii, xxix, 299n.23; Confucius's editing of, 299n.24; transmission by sages of, 13

slave ownership, xv, 283–89

small man *(xiao ren):* blaming others and, 21; disputation and, 103; the gentleman and, 285; the peripheral and, 33; reputation and, 41; way of, 151

speech/words: being contrary to the way, 239; carelessness in, 21; clever, 119; countenance and, 91; deeds/action and, 31, 39, 61, 73, 81, 259; the gentleman and, 23, 61, 75; as models, 29; virtue and, 69

Spring and Autumn Annals: Confucius and, 35, 185; quoted in text, 181, 211

strategies, 205; emperors Gaozu, Guangwu and, 243; long-term, 209; selecting and rejecting, 235, 245

structure-application *(tiyong),* 85

Sui Ren: fire making and, 11, 298n.20

Sun Ao (late second century), 191, 197, 203, 332–34n.8

talent, men of, xxv–xxvii; assessing, xxvi; Cao Cao and, xxvii; valued by sages, 113

tally: two types of, 301n.3; as metaphor for the gentleman, 17

Jiang Taigong and, 223; virtue of, 211–13

wen zhi (cultural refinement and unadorned simplicity), xii, 89; as a balanced whole, 85

worthies, 11; of antiquity, 161, 162; in the employ of a ruler, 251; intelligence, wisdom, and, 107; at Jiaxia, 253; learning and, 13; securing service of, 221, 225, 249, 257; social intercourse and, 159; status, position and, 131; ways pursued by, 203

Wu, King (r. 1049/45–1043), 13, 199; death of, 115; enterprise of, 115; Jiang Taigong and, 127; mandate to rule and, 179; ten ministers of, 255, 338n.10

Wu Zixu, xx, 189n., 243; Bigan and, 189, 191, 193, 201

Xia, Lord of. *See* Yu

Xian hexagram: quotation of "Images," 51

Xiang Yu, 235, 245–47

Xie, 131

xing (nature; innate tendencies), 7; as inborn quality, 31, 307n.8; loss of original, 137; ordering of, 5, 17, 19; of other people, 75; of plants, 151; refining, 39–41

Xu Gan (170–217): in Chinese intellectual history, xxix;

dates of, 292n.1; model curriculum of, 3; official career of, xxxvi; reputation of, xxxii

Xuan Yuan (=Emperor Xuan). *See* Yellow Emperor

Xun Shuang (128–90 CE), 189, 201

Xunzi (=Xun Qing [c. 335–c. 238]), 79, 147, 233, 257, 294n.22; admired by Xu Gan, xxxiv; disputation and, 97; as sage, xxx; Xu Gan's thought and, 3; *Xunzi*, 3

Yan Yuan (=Hui; disciple of Confucius), 13, 105, 131, 294n.22; admired by Xu Gan, xxxiv; behavior of, 53; Confucius and, 12–13, 109; contemporaries of, 189; departures from regularities and, xx; premature death of, 187, 201

Yang Xiong (53 BCE–18 CE): *Fanyan* and, 3

Yao, 13, 199; books of, 249; followers of Shun and, 141; Huan Dou, Gong Gong, Four Leaders, Gun, and, 107, 225; laws of, 137; Shun and, 221, 225, 233; way of, 259; Xu You and, 241

Yellow Emperor, 11

Yi Yi, 213

Yi Yin, 63, 131, 259

Library of Congress Cataloging-in-Publication Data

Xu, Gan, 171–218.
　[Zhong lun. English]
　Balanced discourses / Xu Gan ; English translation by John Makeham ;
introductions by Dang Shengyuan and John Makeham. — Bilingual ed.
　　cm. — (The classical library of Chinese literature and thought)

Includes bibliographical references and index.
　ISBN 0-300-09201-6 (alk. paper)
　1. Philosophy, Chinese. I. Makeham, John, 1955–. II. Title. III. Series.
　B126 .X813　2002
　181'.112—dc21　　　　　　　　　　　　　　　　　2002002885